Integrative Wildlife Nutrition

Food flows through a web of wildlife. In the arctic, willows provide nutrients for herbivores such as migrating caribou and habitat for songbirds such as snow buntings. Grizzly bears prey on salmon while herring gulls scavenge the salmon carcasses. Animal tissues decompose and return their nutrients to soils and plants as the cycle of nutrition continues. We observe changes in populations that reflect changes in food and habitat for wildlife such as geese and snowshoe hares. Our ability to effectively conserve and manage wildlife species depends on understanding their needs in changing habitats with changing food supplies. (Illustrator: T. Sayre)

Perry S. Barboza • Katherine L. Parker
Ian D. Hume

Integrative Wildlife Nutrition

Springer

Dr. Perry S. Barboza
Department of Biology and Wildlife
Institute of Arctic Biology
University of Alaska
PO Box 75000
Fairbanks, Alaska 99775-7000;
USA
e-mail: ffpsb@uaf.edu

Dr. Katherine L. Parker
Natural Resources and Environmental Studies
University of Northern British Columbia
Prince George, British Columbia V2N 4Z9
Canada
e-mail: parker@unbc.ca

Dr. Ian D. Hume
School of Biological Sciences A08
University of Sydney
New South Wales 2006, Australia
e-mail: ianhume@bio.usyd.edu.au

ISBN: 978-3-540-87884-1 e-ISBN: 978-3-540-87885-8
DOI: 10.1007/978-3-540-87885-8

Library of Congress Control Number: 2008938180

Cover design: WMX Design GmbH, Heidelberg, Germany

Cover illustration by Tamara Sayre

Printed on acid-free paper

9 8 7 6 5 4 3 2 1

springer.com

Contributors

Perry Barboza teaches undergraduate and graduate courses in physiology and nutrition of wildlife. He has also taught courses in human nutrition, biochemistry, histology and introductory biology. The principal focus of his research is the consequences of life history and environmental change on nutrition. He and his students study ungulates (reindeer, caribou and muskoxen) and waterfowl (ducks and geese) as well as non-game species in both wild and captive populations.

Katherine Parker teaches courses in wildlife ecology, animal physiology and wildlife nutrition. Her current interests incorporate nutritional ecology within extensive field studies of northern ungulates, bears and wolves. She and her students research the energetic strategies of wildlife for survival and reproduction; plant–herbivore and predator–prey interactions; and the role of individual animal requirements within large-scale ecosystems.

Ian Hume has taught courses in wildlife nutrition, eco-physiology and biochemistry in the USA and Australia. He has written over 150 research publications on the digestive physiology and nutrition of mammals and birds in international journals, and is the author of *Marsupial Nutrition* (1999) and co-author with C.E. Stevens of *Comparative Physiology of the Vertebrate Digestive System* (1995).

Edible. Good to eat and wholesome to digest, as a worm to a toad, a toad to a snake, a snake to a pig, a pig to a man, and a man to a worm.

Ambrose Bierce (1910) The Devil's Dictionary.

Preface

Nutrition spans a wide range of mechanisms from acquisition of food to digestion, absorption and retention of energy substrates, water and other nutrients. Nutritional principles have been applied to improving individual health, athletic performance and longevity of humans and of their companion animals, and to maximizing agricultural efficiency by manipulating reproduction or growth of tissues such as muscle, hair or milk in livestock. Comparative nutrition borrows from these traditional approaches by applying similar techniques to studies of ecology and physiology of wildlife. Comparative approaches to nutrition integrate several levels of organization because the acquisition and flow of energy and nutrients connect individuals to populations, populations to communities, and communities to ecosystems. *Integrative Wildlife Nutrition* connects behavioral, morphological and biochemical traits of animals to the life history of species and thus the dynamics of populations. An integrated approach to nutrition provides a practical framework for understanding the interactions between food resources and wildlife populations and for managing the harvest of abundant species and the conservation of threatened populations.

This book is for students and professionals in animal physiology and ecology, conservation biology and wildlife management. It is based on our lectures, demonstrations and practical classes taught in the USA, Canada and Australia over the last three decades. Instructors can use *Integrative Wildlife Nutrition* as a text in wildlife and conservation biology programs, and as a reference source for related courses in wildlife ecology. Each chapter introduces basic concepts such as units of measurement and allometric scaling, but we assume that the reader has completed introductory courses in biology, chemistry, ecology and animal physiology in the first or second year of undergraduate programs in the biological sciences. We describe a series of basic mechanisms in behavioral ecology, morphology and biochemistry involved in foraging, processing of foods and retention of energy and nutrients in animals. We focus on mammals, birds, reptiles and fish because most efforts in conservation and management concern those taxa. However, we also use examples from invertebrates to illustrate diversity of form and function such as chitin in the exoskeleton of insects and digestion of cellulose in crabs and termites. The broad diversity of species covered in the book may be considered in the groups listed in Table 1.1 and in the Index.

Integrative Wildlife Nutrition emphasizes the general features of vertebrates that pertain to the ecology and adaptations of wildlife. This approach builds on comprehensive descriptions of the digestive physiology (Van Soest 1994; Stevens and Hume 1995; Hume 1999) and nutritional biochemistry (Linder 1991; Groff et al. 1995; Brody 1999) of animals; including humans. It integrates cellular responses with function of the animal body and with the ecological requirements of wild animals. Our approach extends previous texts in wildlife nutrition (Robbins 1993; Klasing 1998; Hume 1999) by discussing the ecology of foraging and the evolution of structures and mechanisms. Our intent is not to overwhelm the reader with numerous details or specialization in any of the chapters, but rather to present aspects of a nutritional hierarchy in the context of ecological implications. This integrative approach emphasizes the importance of wildlife nutrition, from molecules and cells to ecosystems. Key concepts are illustrated with flow diagrams of biochemical pathways and with practical examples for monitoring the nutritional status of wild animals. We begin with a discussion of food resources for populations (Part I) and end with responses to variable environments (Part III). Part II describes the nutrients and their functions in the metabolism and structure of animals.

We use only common names of plants and animals in the text, figures and tables, but the 'List of Common and Scientific Names of Animals and Plants' at the end of the book contains a complete list of the common and scientific names of all species mentioned in the book.

October 2008 Perry Barboza, Katherine Parker and Ian Hume

Acknowledgments

This book attempts to answer some of the many questions posed by our students and our colleagues. We are most grateful for those questions and for the discussions that have prompted us to search the literature and our own work to prepare this text. We have enjoyed stimulating discussions with R.G. White, R.T. Bowyer, F.S. Chapin, D.G. Jorde, K. Klasing, M.P. Gillingham and others while preparing this perspective. The libraries and librarians at the Universities of Alaska, Northern British Columbia and Sydney provided the many documents we required. K.A. Kenyon advised and assisted in assembling the bibliography and index of the book. T. Sayre enthusiastically designed a cover illustration that encapsulated the scope of the book. We thank D. Czeschlik and A. Schlitzberger at Springer for their support during the preparation of the book. We are grateful to our families, friends and coworkers for tolerating our distraction during this project and for accepting the term 'almost done' without a precise definition. Most importantly we thank the animals we have observed because without them we would know very little about wildlife nutrition.

Contents

Abbreviations

AA	arachadonic acid
ADF	acid-detergent fiber
ADH	anti-diuretic hormone
ALT	alanine amino transferase
ANH	atrial naturetic hormone
AP	alkaline phosphate
AVP	arginine vasopressin
AVT	arginine vasotocin
BIA	bioelectrical impedance analysis
BMR	basal metabolic rate
BSA	bovine serum albumen
CCK	cholecystokinin
CLA	conjugated linoleic acid
CNS	central nervous system
CSTR	continuous-flow stirred-tank reactor
D_2	ergocalciferol
D_3	cholecalciferol
DAPA	2,6-diaminopimelic acid
DBP	vitamin D binding protein
DE	digestible energy
DEXA	dual emission x-ray absorptiometry
DHA	docosohexanoic acid
DIT	diet-induced thermogenesis
DM	dry matter or dry mass
ENSO	El Nino Southern Oscillation
EPA	eicosopentanoic acid
EUN	endogenous urinary nitrogen
FSH	follicle stimulating hormone

GC	gas chromatography
GCORT	glucocorticoid hormone
GE	gross energy
GFR	glomerular filtration rate
GLO	glucono-lactone oxidase
HDL	high density lipoprotein
LDL	low density lipoprotein
LH	luteinizing hormone
ME	metabolizable energy
MFN	metabolic fecal N
MPFR	modified plug-flow reactor
MRT	mean retention time
NAG	N-acetyl glucosamine
NAO	North Atlantic Oscillation
NDF	neutral-detergent fiber
NE	net energy
OM	organic matter or organic mass
PFR	plug-flow reactor
PSM	plant secondary metabolite
PTH	parathyroid hormone
PUFA	polyunsaturated fatty acid
QFASA	quantitative fatty acid signature analysis
RMR	resting metabolic rate
RMT	relative medullary thickness
RQ	respiratory quotient
SCFA	short-chain fatty acid
SMR	standard metabolic rate
T_3	tri-iodo-thyronine
T_4	thyroxine
TCA	tricarboxcylic acid
TDS	total dissolved solids
TOBEC	total body electrical conductivity
TT	turnover time
UV	ultraviolet
VLDL	very low density lipoprotein

Chapter 1
Introduction: Common Themes Across Diverse Taxa

Concerns about animal populations by various audiences, from elected officials and policy boards to the general public, often result in two disarmingly simple questions for wildlife biologists:

- What does a population need?
- Will that population grow or decline, and why?

The ability to answer these questions rests on a synthesis of nutritional ecology and physiology.

Nutritional ecology and physiology track the dynamic supply and demand of energy and nutrients in wildlife and their habitats. Our integrative approach to wildlife nutrition attempts to answer two general questions:

- How do animals contend with variations in the supply of resources and the environmental challenges in their habitat?
- What structures, metabolic pathways and life history parameters constrain or limit animal responses?

This first chapter introduces the topics of nutrient composition and nutrient requirements of wildlife. Table 1.1 lists the general groups of animals that are managed for control or conservation of populations. Functional relationships between food resources and animals are discussed in Part I from the scale of the population down to the individual digestive system (Chapters 2 to 5). Chemical components that provide energy substrates or tissue constituents are discussed in Part II (Chapters 6–9). Part III (Chapters 10 and 11) discusses energy flow and the adaptations of animals to changing environments and supplies of food.

1.1 Resource Supply and Organismal Demand

Supply and demand are features of both the environment and wildlife. The environment supplies food but also exerts demands on the animal. For example, low ambient temperatures increase the demand for energy to heat the body whereas

Table 1.1 General groups of animals

Taxa	Group	Energy demand	Trophic level	Typical application[1]
Fish	Marine: reef fish	Ectotherm	Carnivore	Conservation
Fish	Marine: salmon, bream, tuna, shark, herring, halibut, pollock	Ectotherm	Carnivore	Control
Fish	Freshwater: catfish, cichlids, trout, barramundi	Ectotherm	Omnivore	Control
Amphibian	Frogs, salamanders	Ectotherm	Carnivore	Conservation
Reptile	Snakes, lizards, crocodiles	Ectotherm	Carnivore	Conservation
Reptile	Iguanine lizards, chelonians	Ectotherm	Herbivore	Conservation
Bird	Passerines: songbirds	Endotherm	Omnivore	Conservation
Bird	Ratites: emus, ostrich, rhea	Endotherm	Omnivore	Control
Bird	Upland game birds: grouse, ptarmigan, pheasants	Endotherm	Herbivore	Control
Bird	Waterfowl: geese, ducks	Endotherm	Herbivore	Control
Bird	Seabirds: waders, albatross, gulls	Endotherm	Carnivore	Conservation
Bird	Cranes, raptors	Endotherm	Carnivore	Conservation
Mammal	Marine: seals, whales	Endotherm	Carnivore	Conservation
Mammal	Marsupials: grazing kangaroos	Endotherm	Herbivore	Control
Mammal	Marsupials: wallabies, wombats, possums	Endotherm	Herbivore	Conservation
Mammal	Marsupials: bandicoots, quolls	Endotherm	Carnivore	Conservation
Mammal	Rodents	Endotherm	Omnivore	Control
Mammal	Hares, rabbits	Endotherm	Herbivore	Control
Mammal	Ruminants: deer, sheep, bison, giraffes	Endotherm	Herbivore	Control
Mammal	Horses, rhinos, elephants	Endotherm	Herbivore	Control
Mammal	Primates, lemurs	Endotherm	Omnivore	Conservation
Mammal	Cats: lions, lynx	Endotherm	Carnivore	Conservation
Mammal	Bears, wolves, hyena	Endotherm	Carnivore	Control

[1]Control = monitored and often manipulated to control a population for maximum sustainable harvest or minimal adverse effects of overabundance. Conservation = monitored and often manipulated to conserve minimal viable population size.

high ambient temperatures increase the need for water to cool the animal. The patterns of energy and nutrient availability in an ecosystem provide the context for environmental supply and demand for wildlife. These patterns may be defined by the average abundance of the resource (high to low), the range of variation (broad to narrow), spatial distribution (uniform to patchy) and timing of resource availability (frequent to infrequent; constant to erratic). Wildlife diversity and abundance are high in rainforest ecosystems that are characterized by moderate temperature and high availability of water and nutrients, conditions that promote continuous plant production. Conversely, hot deserts typically support smaller populations of fewer species of wildlife because temperature and precipitation are highly variable and less conducive to plant production (Fig. 1.1)

Fig. 1.1 Primary production of plants varies widely with patterns of temperature, water availability and soils. **a** Mild temperatures and rich soils support diverse communities of plants and animals in wetlands when weather patterns are relatively stable. **b** Extreme temperatures of cold or heat combined with low and infrequent rains limit plant and animal communities in montane and desert habitats

The demands of the animal are ultimately met with food from the environment. These demands include the maintenance of body tissues that follows the genetic program of the species throughout the life of each individual. Life-history patterns reflect the allocation of energy and nutrients to the tissues and the activities and time required for survival, growth and reproduction. For example, among northern elephant seals, adult males awaiting the arrival of females at the breeding beach appear to expend little energy or nutrients but are nonetheless maintaining their muscles and organs even as they lie motionless. Female seals incur additional costs of energy and nutrients for production of milk soon after they arrive at the beach and deliver their pups (Boness et al. 2002). The costs of growth in seal pups are likewise a programmed productive demand that will continue until they in turn begin reproduction at adulthood. The ability to support maintenance or production of tissues when environmental supplies are inadequate depends on the supply of energy and nutrients from internal stores. Female seals use body fat and protein to produce milk when fasting or eating very little.

The ability of wildlife to contend with environmental variations depends on behavioral and physiological flexibility. Estuarine species of fish contend with daily tidal flows by tolerating a wide range of salinity (Spicer and Gaston 1999). Desert reptiles contend with infrequent rainfall by tolerating a wide range of internal fluid concentrations and by accumulating water stores in the urinary bladder after each rain (Bradshaw 2003). Environmental conditions that alter the abundance and timing of food and water select for an operational range that varies both among and within species. Variation within species is often associated with phenotypic plasticity, such as the amount of body fat in songbirds at the start of winter (Rogers et al. 1994). Persistent environmental change over multiple generations may favor one phenotype over another, resulting in population drift. For example, a population of birds may increase body fat or begin migration earlier as winters become colder and the period of snow cover lengthens for each generation. The combination of demography and genetics may ultimately result in speciation. In the austere and erratic environment of the Galapagos islands, beak size within a population of finches varies with the availability and form of their diet of seeds (Grant 1999). Similarly, energy expended at rest in wild mice can vary with primary plant production in their environment (Mueller and Diamond 2001). Phenotypic and genotypic differences among animals therefore alter their demands for energy and nutrients as well as their vulnerability to environmental changes.

1.2 Principal Components of Animals and Plants

The transfer of nutrients and energy from the environment to wildlife is reflected in the chemical composition of materials from soils through plants to herbivores, omnivores and predators. The elements of food are ultimately returned to the environment in excreta and tissues lost by animals throughout their lives. Tissues synthesized at one level of this trophic hierarchy are the food for the next level. The chemical composition of ingested animals and plants reflects the costs of depositing tissues as well as their value to a consumer. This section introduces the components of water, nutrients and energy in plants and animals.

Water is the principal interface between animals and their environment because organisms absorb and excrete matter across a wet interface. Removal of water (moisture) from living tissue preserves the nutrients contained in the remaining dry material (dry matter or dry mass; DM). In human agriculture, desiccation of plant seeds and stems in late summer and autumn produces grain and hay for winter storage. Similarly, the natural desiccation of seeds allows many birds and rodents to cache a stable source of food for winter. Moisture may be the largest fraction of plant and animal tissues, and is typically measured in grams per hundred grams ($g \cdot 100\,g^{-1}$) whole mass or percent (%). Water content is greatest for aquatic plants such as algae (Fig. 1.2), and for some parts of terrestrial plants such as nectar, flowers, fruits and budding leaves. Dietary sources of water may be important for hydration of animals, but excess moisture in food can dilute the nutrients and energy in a diet. Nectarivores such as hummingbirds must therefore consume large volumes of

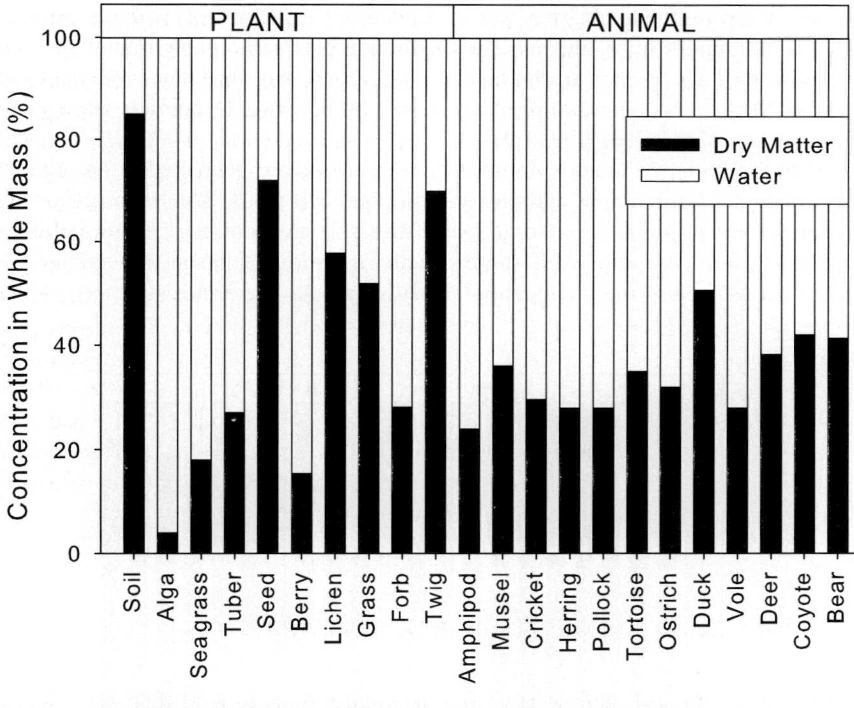

Fig. 1.2 Content of water (moisture) and dry matter in selected plants and animals in comparison with rich terrestrial soils (Chapin et al. 2002). Plants: blue-green alga (National Research Council 1983); leaves of sea grass (Mason et al. 2006); tubers of sweet potato (National Research Council 2003); seeds of corn (National Research Council 1996); blueberries (National Research Council 2003); terricolous lichen (Barboza, unpublished); aerial parts of *Schismus* grass and the leaves of the forb globemallow in spring (Barboza 1996); twigs of Barclay willow in winter (Spaeth et al. 2002). Animals: aquatic and terrestrial invertebrates (an amphipod, blue mussel and the mormon cricket) (Jorde and Owen 1990; National Research Council 2003); marine fish (herring and pollock) (Trumble et al. 2003); a reptile (the desert tortoise) (Barboza, unpublished); birds (ostrich and black duck) (Swart et al. 1993b; Barboza and Jorde 2002); and mammals (northern red-backed vole, white-tailed deer, coyote, black and brown bears) (Robbins et al. 1974; Farley and Robbins 1994; Huot et al. 1995; Zuercher et al. 1999)

water to extract sufficient nutrients from their dilute diet (Karasov et al. 1986). The water content of animal tissues is less variable than that of plants because hydration is tightly controlled in relation to exchanges of electrolytes such as sodium (Na) and potassium (K) (Chapter 9).

The dry matter of plants and animals may be divided between organic matter and minerals that are usually measured in grams per hundred grams ($g \cdot 100\,g^{-1}$) dry mass or percent (%). Organic matter (OM) includes all the combustible material that is primarily based on carbon (C), nitrogen (N), hydrogen (H) and oxygen (O). Biochemical oxidation is the process of combining substances with oxygen to yield energy and chemical byproducts. The oxidation of C in organic matter is the principal source of energy for animals. Organic matter therefore reflects the potential energy in the tissue. The non-combustible residue or ash includes most of the minerals. Minerals comprise the majority of the dry matter in soils (Fig. 1.3) whereas organic matter predominates in organisms. The distinction between organic matter and ash in organisms allows for the calculation of a crude measure of the content of structural minerals. Mineral concentrations are increased by the shell in mussels and by the bony carapace of tortoises (Fig. 1.3).

Unlike the minerals in soils, minerals in organisms are deposited in an organic matrix of protein in bone, or carbohydrate in plant cell walls. That is, much of the mineral is inert in soils, but in organisms, minerals are continuously turned over with the organic components. The most active fraction of tissue minerals is the trace minerals in enzymes of metabolic pathways. As the name suggests, 'trace

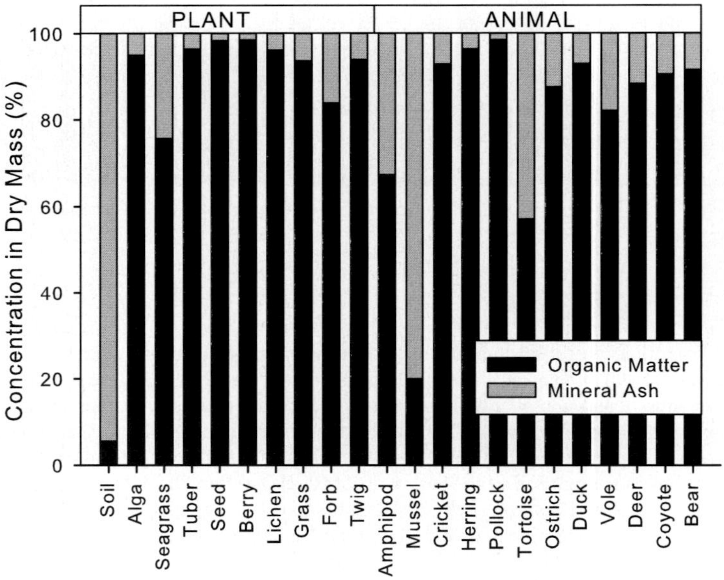

Fig. 1.3 Organic matter and mineral (ash) content of selected plants and animals. See Fig. 1.2 for details

minerals' and other nutrients such as vitamins are only a tiny fraction of the body mass and are typically measured in milligrams per kilogram ($mg \cdot kg^{-1}$) or micrograms per kilogram ($\mu g \cdot kg^{-1}$) of whole or dry mass of tissue. These units are also referred to as parts per million ($ppm = mg \cdot kg^{-1}$) or parts per billion ($ppb = \mu g \cdot kg^{-1}$). The activities of trace minerals and vitamins are discussed in Chapter 9.

The organic matrix can be divided into protein, carbohydrate and fat in a similar fashion to the labels on foods for humans and domestic animals. These three components define the relative value of foods consumed by wild animals. The units for these components are usually expressed as grams per hundred grams ($g \cdot 100\,g^{-1}$) whole mass or dry mass, but may be presented as the percentage of the organic mass (Fig. 1.4). We use the term 'crude' to indicate that precise chemical definitions are relaxed for this division: crude protein includes most compounds containing N, such as protein and nucleic acids; total carbohydrate includes sugars as well as fiber; and crude fat includes triglycerides as well as the components of cell membranes. The bulk of organic matter in plants is comprised of carbohydrate, but in animals crude protein is the principal organic component. Herbivorous animals consume foods that are quite different from their own tissues in contrast to carnivores, which digest materials that are similar to their own bodies. The consequences of this simple difference in food chemistry are discussed in relation to the functional anatomy of the digestive system in Chapter 5. The compositional difference between plants and animals partly reflects the structural roles of carbohydrates in plants and protein in animals, but components of both fractions are also involved in the intermediary

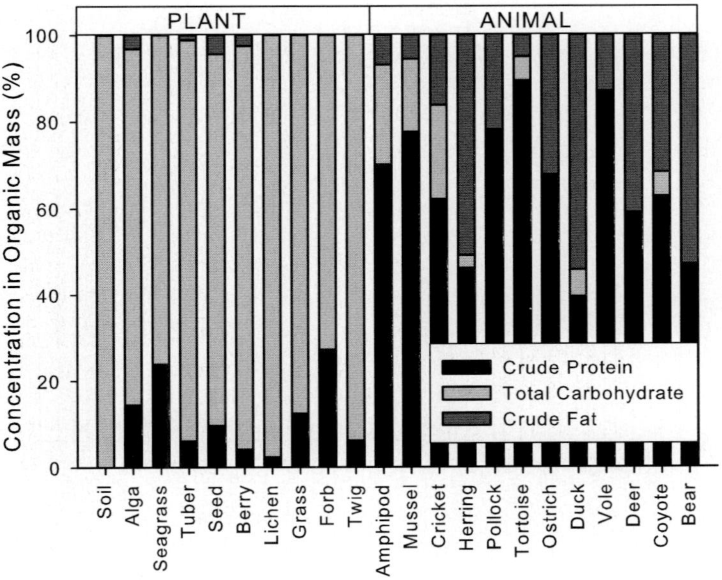

Fig. 1.4 Crude protein, total carbohydrate and crude fat in the organic mass of selected plants and animals. See Fig. 1.2 for details

metabolism of all organisms. The importance of carbohydrates and organic N in the structure and function of animals and their foods is discussed in Chapters 6 and 8.

Dietary organic matter can be oxidized as fuel or deposited in tissue for later use. Crude fat is the most energy-dense component of tissue because fat contains more C than either protein or carbohydrate, and because the C in fat is the most reduced and therefore yields the highest amount of oxidizable energy. Fat content varies widely among fish, reptiles, birds and mammals depending on the role of fat as an energy depot (Fig. 1.4). The energy available in crude fat, crude protein and total carbohydrate of tissue can be approximated from Atwater's physiological fuel values: 9 kcal·g^{-1} fat, 4 kcal·g^{-1} protein, 4 kcal·g^{-1} carbohydrate (Atwater and Bryant 1900) (Fig. 1.5). The kilocalorie (kcal) is the principal unit used in public health and agriculture in the United States of America, but the kilojoule (kJ) is more commonly used throughout the world and in the scientific literature on wildlife. We use the units of the Système Internationale throughout this book. Energy contents are presented in kilojoules by using the conversion of 4.184 kJ for each kilocalorie.

Energy contents can be calculated on the basis of dry or organic mass of food since water and ash have no 'fuel value' to animals. Large fractions of ash reduce the overall energy density on the basis of dry tissue. For example, the energy density of soil, mussels and the tortoise in Fig. 1.5 are low due to the high ash content of the dry matter. Although fuel values are used to estimate the energy content of foods for humans and companion animals such as cats and dogs, the values are best used only as a general guide for the maximum energetic content of foods for wildlife.

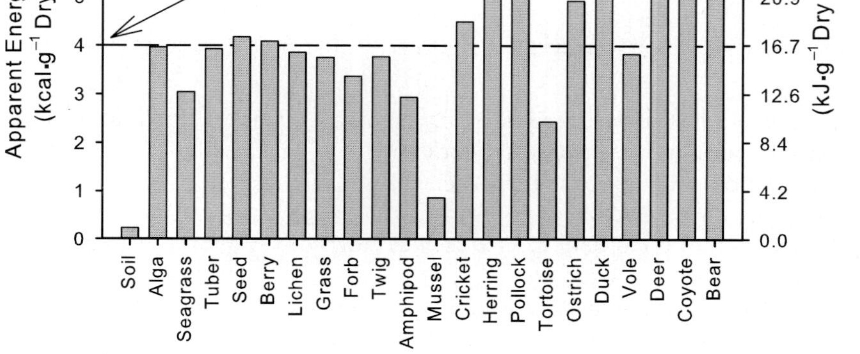

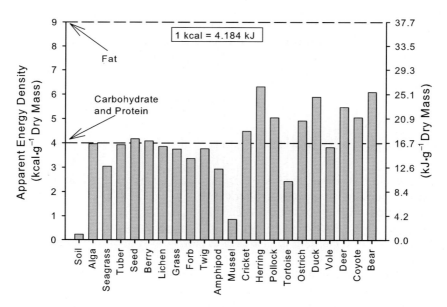

Fig. 1.5 Estimated energy content in dry mass based on Atwater's physiological fuel values (*broken lines*) (Atwater and Bryant 1900) for carbohydrate, crude protein and crude fat. See Fig. 1.2 for details

Table 1.2 The potential yield of energy is calculated with Atwater's fuel values for the corresponding amounts of fat, protein and carbohydrate in a diet. In this example we use the composition of blueberries (*Vaccinium* sp.)

Parameter	Calculation	Result
Whole mass (g)	A	2000
Dry matter (g)	B	315.8
Fat (g)	C	7.7
Protein (g)	D	13.6
Carbohydrate[1] (g)	E	241
Energy[2] from fat (kJ)	$F = C \times 9 \times 4.184$	290
Energy[2] from protein (kJ)	$G = D \times 4 \times 4.184$	227
Energy[2] from carbohydrate (kJ)	$H = E \times 4 \times 4.184$	4,041
Total 'fuel' value (kJ)	$I = F + G + H$	4,558
Energy density ($kJ \cdot g^{-1}$ DM)	$J = I \div B$	14.8
Energy density ($kJ \cdot g^{-1}$ whole mass)	$K = I \div A$	2.28

[1]Carbohydrate content is based on the non-structural components such as starch for calculations of available energy. We discuss the energy yields from fiber, starch and sugar in Chapters 6 and 10.
[2]Energy equivalents from Atwater's fuel values of $9\,kcal \cdot g^{-1}$ fat and $4\,kcal \cdot g^{-1}$ protein and carbohydrate. Kilocalories are converted to kilojoules with the factor 4.184 ($kcal \cdot kJ^{-1}$).

For example, the energy available to a bear (Family Ursidae) consuming 2,000 g of berries is calculated as 4,558 kJ in Table 1.2. The bear can therefore extract up to 1.92 kJ of energy from each gram of fresh berries. Such fuel values are then used to assess the energy yield from a food item in relation to the cost of finding and processing the food (Chapter 3). The assumptions of fuel values and the conversion of dietary substrates to energy for herbivores and carnivores are discussed in Chapters 6 through 10.

1.3 Scaling Body Size and Demands for Energy and Nutrients

The relationship between body size and resource demands is critical to wildlife science, especially when applying nutritional requirements across wide ranges of body size both among and within species. Body mass varies by over 50% among populations of Canada geese (Bellrose 1976) and white-tailed deer, as well as between sexes of dimorphic species such as caribou (Wilson and Ruff 1999). The nutrient demand of a large animal is not always a simple multiplication of the demand for a small animal because the requirement of each unit of body mass (grams or kilojoules per day per kilogram; $g \cdot d^{-1} \cdot kg^{-1}$ or $kJ \cdot d^{-1} \cdot kg^{-1}$) can vary with body size. In Canada geese the energy requirements of the smallest races in Alaska are probably higher per unit body mass than the largest races in the Mississippi flyway.

Relationships between body mass or size (M) and a dependent variable (Y) such as home range area, daily energy expenditure or capacity of the digestive tract are called allometric relationships. These relationships are of the form:

$$Y = a \cdot M^b \tag{1.1}$$

where 'a' is a constant and 'b' is a scalar for body mass. If the scalar differs from 1.00, then the relationship is curvilinear, that is, Y changes disproportionately to body mass (Fig. 1.6). Curvilinear relationships may be plotted on logarithmic axes to provide straight lines with different slopes (b) that intersect when M = 1 ($\log_{10}1 = 0$) at Y = $\log_{10}a$ (Fig. 1.6).

Scalars greater than one predict that Y increases more quickly than body mass whereas scalars less than one predict slower changes in Y as body size is gained. For example, scalars for ungulates are greater than one for home range area, equal to one for digestive tract capacity and less than one for energy demand (Reiss 1989). Large ungulates may therefore require much greater areas but use less food than the same total mass of smaller animals. The interactions between scalars for dependent variables such as food quality, energy demand and digestive tract capacity are discussed in Chapters 2 to 5.

The utility of allometric relationships can be illustrated with an example of the use of a hypothetical nutrient by Canada geese (Table 1.3). In this example, A = $100\,\text{mg}\cdot\text{d}^{-1}$ and B = 0.75. Increasing the body mass of geese by 50% from 1 kg to 1.5 kg only increases the daily demand by 36% because the scalar is less than 1. A total biomass of 1000 geese at 1000 kg will therefore use 100,000 mg each day. The same resource of 100,000 mg of the nutrient will support only 738 birds, each weighing 1.5 kg body mass (Table 1.3) because the scalar is 0.75 rather than 1.0. Only 667 birds would be supported by this resource if B was 1.0. Increasing the scalar to 1.25 further reduces the nutritional carrying capacity to 602 birds. Failure to account for body size when applying an estimate determined with one size class

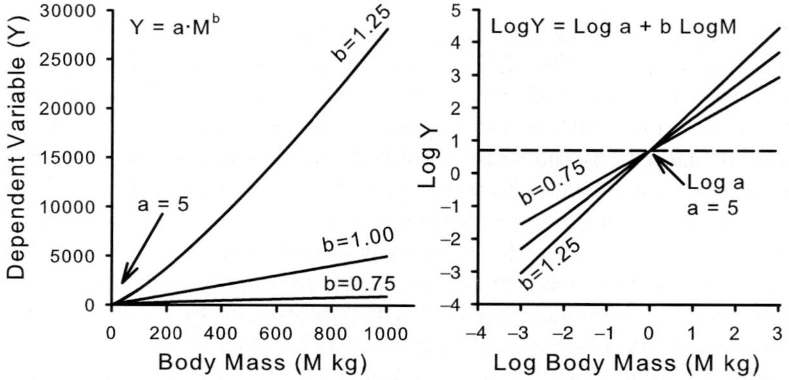

Fig. 1.6 Representative allometric relationships between a dependent variable (Y) and body mass (kg)

Table 1.3 Estimates of the number of geese that can meet their demands from a common resource when body size is increased with different scalars for the rate of nutrient use

Parameter	Calculation	Small geese	Large geese		
Rate of use ($mg \cdot d^{-1}$)	A	100	100	100	100
Scalar	B	0.75	0.75	1	1.25
Body mass (kg)	C	1	1.5	1.5	1.5
Individual demand ($mg \cdot d^{-1}$)	$D = A \times C^B$	100[1]	136	150	166
Resource available (mg)	E	100,000	100,000	100,000	100,000
Carrying capacity (number of animals)	$F = E \div D$	1,000	738	667	602

[1]D = A when C = 1 because 1 to any scalar is still 1.

to animals of a different body size may underestimate or overestimate the resources needed for a population and contribute to starvation or overabundance. We further discuss allometry and resource use by populations in Chapters 2 and 3.

1.4 Dietary Requirements and Nutritional Niche

We use the term 'requirement' to refer to the amount of water, nutrients or energy needed by an animal from its diet. Animals can temporarily use their body tissues to maintain body function during a fast, but those demands are usually referred to as minimal or fasting rates of metabolism or expenditure (Chapter 10). Food intakes ultimately support the return of body tissues after a fast, maintenance of body tissue during regular feeding, and the synthesis of new tissue during growth or reproduction. The amount of a nutrient required by an animal, therefore, increases with metabolic demands from maintenance of body mass (zero gain or loss of nutrient) to net gains of nutrients during seasonal mass gain, growth or reproduction. (Fig. 1.7).

All animals require a source of oxidizable C for energy. Intermediary metabolic pathways allow animals to derive that energy from dietary carbohydrate, protein and fat (Chapters 6 to 10). The source of nutrients for maintaining and growing body tissues may vary with intermediary metabolism, but a preformed dietary source can be essential for some species. Dietary vitamin C, for example, is required by some birds and some rodents, humans, other primates and bony fish. Essentiality of a nutrient varies among species because intermediary pathways of metabolism vary with genotype. There are two levels of essentiality: complete and conditional. Complete essentiality indicates that the animal cannot synthesize a nutrient from common precursors due to the absence of an enzyme or the production of an inactive enzyme in a pathway (Chapters 8 and 9). Vitamin C is essential for species that lack one enzyme for conversion of glucose to ascorbic acid. Conditional essentiality means that the nutrient is required in the diet when demands are elevated by growth or reproduction. The enzyme pathways are present, but they are not sufficiently active to meet the demands of a particular condition. The amino acid

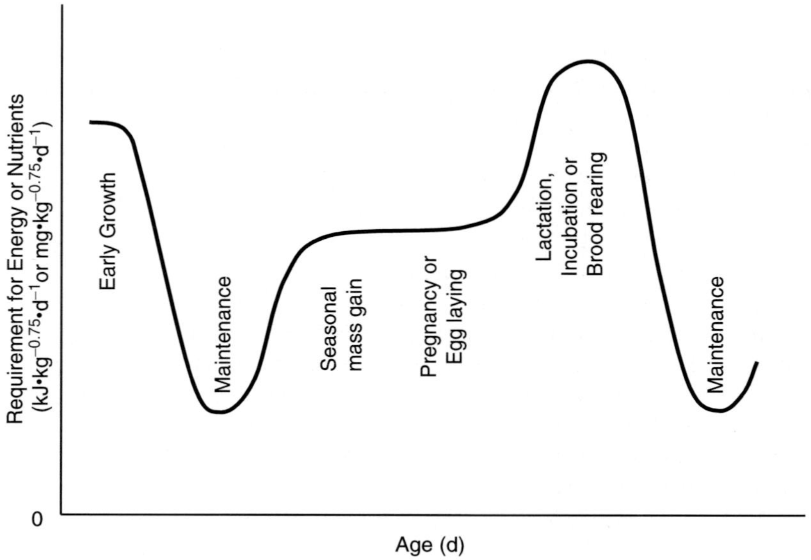

Fig. 1.7 Mass-specific requirements for energy (kJ·kg$^{-0.75}$·d^{-1}) and nutrients (mg·kg$^{-0.75}$·d^{-1}) change with metabolic demands during the life of an animal. Requirements are lowest for maintenance of body mass when animals are neither growing nor reproducing. Deposition of tissues increases requirements above maintenance during growth, seasonal mass gain or maturation, and during each phase of reproduction. This general pattern varies among species because absolute requirements and the duration of each stage vary with life history. (Adapted from Hume 1999; see Chapters 2 and 10.)

histidine is not required for rodents and humans for adult maintenance, but is required in the diet during growth. Requirements for completely and conditionally essential nutrients may be increased by interactions between nutrients and by disease. Dietary requirements for copper (Cu), for example, can be increased in young animals when rates of bacterial infection are high (Linder 1991).

Intakes of essential nutrients range from deficiency through sufficiency to toxicity. Intakes below a requirement are deficient and are accompanied by specific symptoms that directly or indirectly reflect the metabolic role of the nutrient. Night-blindness in mammals is symptomatic of vitamin A deficiency because vitamin A metabolites are required for light transmission at the retina of the eye (Chapter 9). High intakes of essential nutrients that are stored in the body may be potentially toxic. Overconsumption (toxicity) of vitamin A can lead to bone remodeling and skin loss because vitamin A metabolites also signal cellular division which is disrupted during toxicity.

Nutritional pathologies can be complex and difficult to resolve because more than one toxicity and/or deficiency may be involved. The presence of a single anomaly is usually referred to as a primary toxicity or deficiency, whereas interactions between nutrients are secondary toxicities or deficiencies. Excess intakes of zinc (Zn) in primates produce a primary toxicity with secondary effects on Cu that are symptomatic of a Cu deficiency. Normal health is only restored by treating the primary problem, in this case, the source of excess Zn consumption; the addition of Cu to the diet would only remedy some but not all pathologies. We further discuss nutritional deficiencies and toxicities for wildlife in Chapter 9.

Although the absence of toxic or deficient symptoms suggests adequate intakes, a quantitative measure of adequacy is most useful in determining the range of dietary solutions available to wildlife populations. Adequacy of diets is usually measured in relation to the requirements of an animal. Requirements may be measured through two different approaches: minimal and optimal.

A minimal requirement is commonly estimated from the relationship between nutrient balance and nutrient intake (Fig. 1.8). In this relationship, the Y axis is a measure of the nutrient retained in the body so that values above zero indicate net gains and values below zero indicate net losses. Nutrient intake (amount per time) or nutrient density (amount per mass of food) is plotted on the X axis. Minimal requirement is estimated as the nutrient intake or dietary content at zero balance. This approach is often used to define the requirement for maintenance of the body. Total requirements for activity, thermoregulation, growth or reproduction can be subsequently estimated by adding the costs of energy or nutrients for these demands to the maintenance costs (Table 1.4). We further discuss this factorial approach to estimating N and energy requirements in Chapters 8 and 10.

Requirements for growth and production are also measured with an optimal approach (Fig. 1.9) that is an extension of the minimal approach (Fig. 1.8). A productive response by the animal, such as mass gain or number of eggs, on the Y axis is plotted against the broad range of possible nutrient intakes or dietary content on the X axis (Fig. 1.9). The optimal intake or dietary content of a specific nutrient is

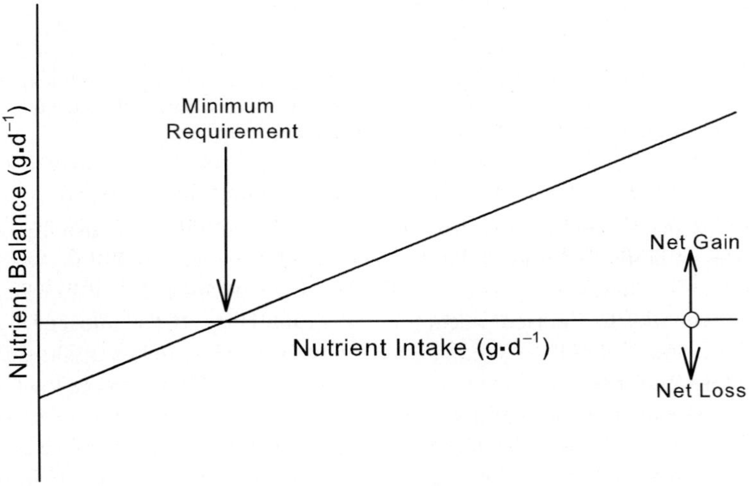

Fig. 1.8 The minimal requirement for a nutrient is estimated as the nutrient intake at zero balance (no net gain or loss)

Table 1.4 The factorial approach to estimating the total requirement for energy (kJ) in female cervids at two body sizes. Mass-specific rates for maintenance are derived from a relationship between energy intake and balance similar to that shown in Fig. 1.8

Parameter	Calculation	Small size	Large size
Body mass (kg)	A	100	150
Scalar	B	0.75	0.75
Mass-specific rate for maintenance ($kJ \cdot kg^{-0.75} \cdot d^{-1}$)	C	293	293
Maintenance cost ($kJ \cdot d^{-1}$)	$D = C \times A^B$	9,265	12,558
Activity (× maintenance)	E	0.5	0.5
Reproduction (× maintenance)	F	1.0	1.0
Additional costs ($kJ \cdot d^{-1}$)	$G = D \times (E + F)$	13,898	18,838
Requirement ($kJ \cdot d^{-1}$)	$H = D + G$	23,164	31,396

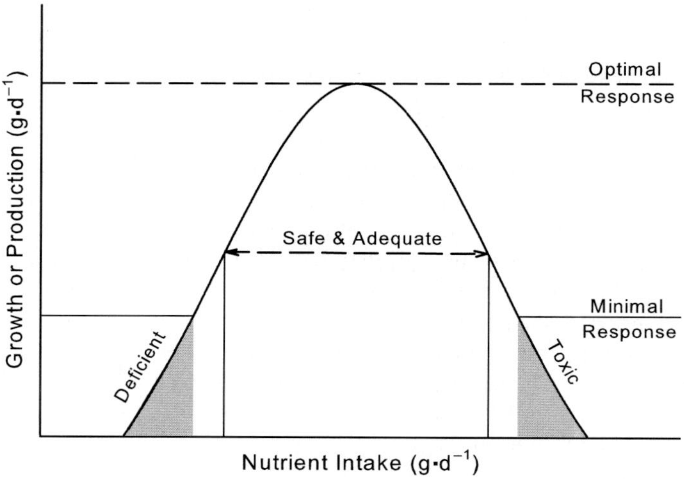

Fig. 1.9 The optimal nutrient requirement is estimated as the nutrient intake at maximum response for growth or reproduction. Intakes in the *shaded regions* are deficient or toxic because they fail to support minimum growth or production for the species. Safe and adequate intakes of the nutrient support growth or production above the minimum response

estimated at the maximum response. A range of optimal intakes is usually defined for species of domestic animals depending on environmental conditions (e.g., temperature) and genetics (e.g., race, breed). The shape of this relationship may vary with the potential for nutrient accumulation in the body. If the nutrient does not accumulate and is rapidly excreted, then the relationship reaches a plateau at the optimal response. For example, animals do not retain increasing amounts of dietary N once they reach the maximum rate of depositing N in proteins, but rather excrete the excess N to avoid toxic accumulations of ammonia (Chapter 8). Nutrients that accumulate in the body may be toxic if production is impaired when high intakes deposit large loads in the body (Fig. 1.9) (Moriarty 1999). In desert ungulates, overconsumption of milk vetch may cause toxic accumulation of selenium (Se), which impairs movement and cell function.

Food standards for humans and domestic animals typically adopt a range of safe and adequate intakes that avoid toxicity (safe) and deficiency (adequate; Fig. 1.9). The safe and adequate range is based on the intakes for minimal response with a margin for error. Compositional surveys of foods for wildlife often adopt a similar approach by relating the concentration of nutrients in each food item to the minimal or optimal requirements of domestic and wild species. Dietary requirements have been determined mostly by studies using repeated measures of domestic animals and wildlife in captivity under controlled conditions. These estimates are adjusted for application to wild populations by correcting for variables such as temperature and movement in the wild. It is also possible to measure expenditures and estimate requirements for energy and nutrients in free-living animals by combining measures of body composition (retained nutrients), food intake and expenditure using markers (see Chapters 4 and 10).

The energy and nutrients required by all life stages of a wild population may be described as a multidimensional space with an axis for each nutrient (Perrin 1994). Deficiency and toxicity delimit each axis and define the nutritional niche of the population. A nutritional niche is represented as a square for two nutrients in Fig. 1.10. The true shape of the space is probably not as linear as this representation because each response and interaction may be curvilinear for some sections of the range for each nutrient. Nonetheless, populations may be perceived as operating within the physiological range of their genotype. The niche realized by a population is equal to or smaller than the limits of the fundamental nutritional niche. The realized niche of a population describes resources that allow birth rates to meet or exceed death rates (Chase and Leibold 2003). Realized niches are therefore dependent on the abundance and distribution of food (e.g., season, population density), the ability to handle foods (e.g., defensive compounds of plants, prey avoidance), sources of mortality (e.g., predation, disease, weather) and interactions among species (e.g., direct or indirect competition, mutualism). Changes in the attributes of the population, that is, the number of phenotypes, may gradually alter attributes involved in the ability to handle foods (e.g., learning, adaptation of digestive systems) and thus shift the limits of the realized niche. Progressive changes in the attributes of the population may ultimately shift the fundamental niche as the genotype is altered. The evolution of microbial fermentation systems in the digestive tract was a fundamental shift of niche that allowed herbivores to biosynthesize essential nutrients and catabolize many potentially toxic compounds (Fig. 1.10). Evaluation of foods and nutritional resources for wildlife populations are further discussed in Chapters 2–4 and again in Chapter 11.

1.5 Summary: Introduction

The demands of wildlife for energy, nutrients and water vary in relation to supplies from the environment and the life history of each species. Nutrients deposited in plants and animals are contained in the dry fraction, which includes minerals in ash and carbohydrate, protein and fat in the organic matter. The amount of energy

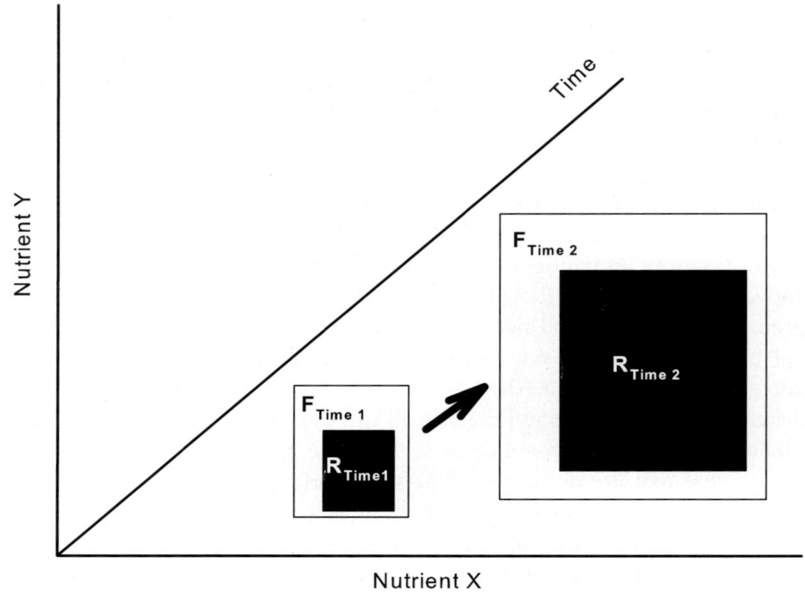

Fig. 1.10 Representation of a two-dimensional niche space with two hypothetical nutrients (*X* and *Y*) during two consecutive periods (*Time 1* and *Time 2*). The realized niche (*R*) is always equal to or less than the fundamental niche (*F*). Realized niches can change independently of the fundamental niche because the realized niche is affected by interactions both within and between species. Expansion of the fundamental niche over time permits the expansion of the realized niche. The evolution of microbial fermentation is an example of an attribute that allowed herbivores to expand both fundamental and realized niches. Similarly, contraction of the fundamental niche by the loss of an attribute may contract the realized niche. Loss of an enzyme for vitamin C synthesis may contract both fundamental and realized niches for a mammal that mainly eats seeds with little vitamin C, but not for a mammal that eats fruits rich in this vitamin

derived from organic matter can be roughly estimated from fuel values. Variation in the demands for energy and nutrients with body size can be described by allometric relationships. Nutrients are considered essential when a dietary source is required. Dietary requirements estimated by minimal or optimal approaches can be used to describe the operational range of food abundance and composition that can be used by wildlife. Changes in the requirements of animals alter their nutritional niche.

Part I
Functional Relationships

Chapter 2
Food and Populations

Wildlife biologists are often responsible for monitoring populations of free-living animals within the boundaries of parks, refuges and game management units. Captive propagation programs focus on smaller confined populations but are nonetheless ultimately concerned with understanding the relationships between wild populations and their habitats.

In evaluating population numbers with repeated inventories, two questions are usually asked: 'How many animals are in the area?' and 'How many animals do we want in the area?' Conservation programs are evaluated by gains in a population of threatened or endangered species or by declines in populations of pests and invasive species. Game management programs often aim to sustain or increase the accessibility and harvest of a population with minimal adverse effects on other resources such as plant and animal communities. Wildlife monitoring typically entails the collection of data at three levels of detail with a wide variety of techniques, from aerial surveys, capture and release of individuals, to tissue sampling of harvested animals:

- numbers and locations of individuals;
- demography (e.g., proportions of age, sex and reproductive classes);
- individual condition (e.g., body mass, growth rate, health).

Wildlife nutrition links the details of demography and individual condition to the response of a population. In this chapter we discuss the response of a population to changes in food supply and environmental demands on individuals.

2.1 Population Growth and Animal Density

The number of animals in an area changes with time and can be described as a mathematical function of time (t). Time scales for populations can be measured in minutes or hours for microbes that replicate rapidly (Chapter 6), but are typically recorded in months or years for fish and wildlife species. The size of a population at the end of a year (N_{t+1}) is related to the number of animals at the start of the annual cycle (N_t) and the activities of animals over the year: production of young (P_t), deaths (D_t), immigration (I_t) into and emigration (E_t) out of the area (White 2000):

$$N_{t+1} = N_t + P_t - D_t + I_t - E_t \quad \text{(Eq. 2.1)}$$

Changes in population size result from differences between rates of production (P_t) and death (D_t) if animal movement is accounted for by including the annual range or by excluding the possibility of movement by barriers such as oceans, mountains and deserts. Net population gain ($P_t > D_t$) is the outcome of the net supply of resources for growth and reproduction. Net loss or death ($P_t < D_t$) is a function of net demands on individuals, usually from adverse weather and disease after accounting for predation and harvest.

Incremental changes in a population may be either positive or negative and can be quantified as a proportion (R):

$$N_{t+1} = N_t \times (1+R) \quad \text{(Eq. 2.2)}$$

The maximum value of 'R' is the intrinsic growth rate of the population (R_{max}), that is, its maximal rate of net production. Intrinsic rates of growth reflect life history traits of species, such as age at first reproduction, number of offspring each year, frequency of reproduction and age of senescence or last reproduction. The more rapidly a species can replace itself during its lifetime, the higher is $R_{max.}$ Populations of species with high reproductive rates can increase quickly when offspring survive to maturity. Captive propagation programs for fecund species such as green sea turtles can produce large numbers of offspring if causes of mortality in eggs such as predation, adverse weather (temperature, water submersion) and disease are removed or minimized (Miller 1997). Conversely, captive propagation of species such as whooping cranes and California condors are much slower because only two to three eggs are produced in each clutch (Johnsgard 1983; Snyder and Snyder 2000). Organisms achieve R_{max} when supplies for production greatly exceed any demands that result in death. Introductions or expansions of individuals into new ranges can permit an exponential increase in population size (N_{t+1}). Introductions of European rabbits into Australia, brown rats, pigs and goats into Polynesia, and common brushtail possums into New Zealand all resulted in rapid increases in population sizes as founding individuals were able to sustain high reproductive rates that greatly exceeded death rates (Thomson et al. 1987; Hoddle 2004).

Populations of introduced species cannot and do not continue to grow exponentially because food and space are finite. Exponential population growth that is independent of animal density can only be sustained for short periods of time (Fig. 2.1). Increasing animal density (N_{t+1} ÷ area; individuals per hectare, individuals•ha^{-1}) ultimately decreases the food supply for each individual (food ÷ N_{t+1}; kilojoules per animal, kJ•$animal^{-1}$) and may also increase demands on individuals by increasing their exposure to diseases (pathogens ÷ $N_{t+1;}$ parasites per animal, parasites•$animal^{-1}$) and adverse weather. Net production therefore diminishes as animal density approaches a resource limit or carrying capacity (K; Fig. 2.1A). Growth rate of a density-dependent population can be expressed in relation to the maximum or intrinsic rate of growth (R_{max}) by the ratio of N_t to K:

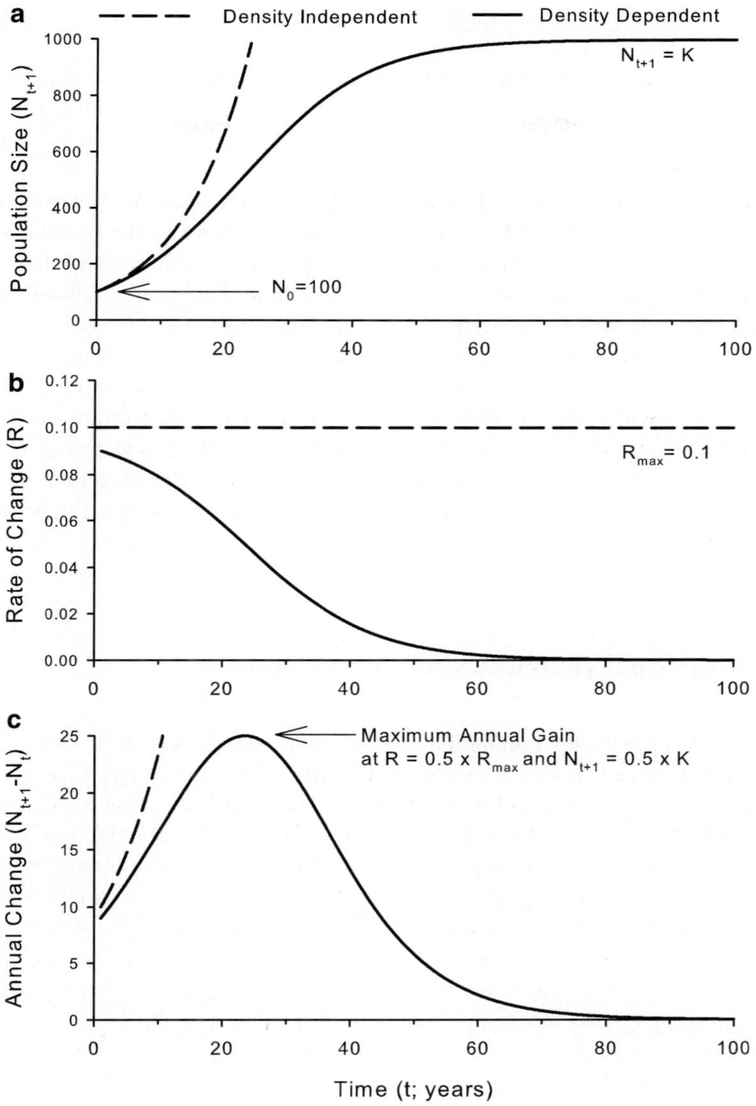

Fig. 2.1 Patterns of population growth in wildlife. **a** Changes in population number (N_{t+1}) from an initial size (N_0) of 100 individuals. Patterns are independent of animal density (*dashed line*) or dependent on a resource that limits the total number of individuals (*K*; *solid line*). **b** Annual rate of change in the population (*R*) varies for density-dependent populations. Density-independent populations grow at the maximum intrinsic rate (R_{max}). **c** Absolute annual change in population size over time. Density-independent populations increase exponentially whereas density-dependent populations reach a maximum annual gain when *R* is 50% of R_{max}. The maximum sustainable harvest of the population by humans (maximum sustainable yield) theoretically can be achieved at the maximum annual gain when the population is at 50% of K

$$R = R_{max} \times (1 - (N_t \div K)) \quad \text{(Eq. 2.3)}$$

The size of density-dependent populations is therefore expressed as:

$$N_{t+1} = N_t \times (1 + R_{max} \times (1 - (N_t \div K))) \quad \text{(Eq. 2.4)}$$

The intrinsic growth rate (R_{max}) of a population determines the time required to reach a particular number of animals and thus the rate at which the population uses food and space. Maximum annual gain in the population is achieved when R is 50% of the intrinsic growth rate ($R = 0.5 \times R_{max}$) and the population is 50% of carrying capacity ($N_{t+1} = 0.5 \times K$; Fig. 2.1B, C). The size of a sustainable harvest by humans is therefore dependent on the replacement rate of the species in relation to its consumption of food and other resources. Fast-growing species such as chinook salmon can provide large annual harvests from nutrient-rich habitats such as the Gulf of Alaska (Mundy 2005; Rodger 2006). Slower growth and lower fecundity make species such as Saiga antelope more vulnerable to overharvesting in semi-arid steppes that are not very productive habitats (Baskin and Danell 2003; Milner-Gulland et al. 2003).

2.2 Individual Demands and Food Limits

The relationship between population growth and resource use can be demonstrated with a simple model of food consumption by deer (Family Cervidae) (Fig. 2.2). If the average energy consumption is 20.3 $MJ \cdot d^{-1} \cdot animal^{-1}$, the deer population will consume 4,058 $MJ \cdot d^{-1}$ early in the growth trajectory (200 animals) and 2.5 times more when the population is at the maximum annual gain (500 animals, 50% of K; Fig. 2.2). If food availability for the area is 19,000 $MJ \cdot d^{-1}$, approximately 50% of the annual food production would be used by a population maintained at 500 animals. The unused food could be returned to the environment by decomposition or consumed by other species. This unused reserve of annual food production may also serve as a safety margin against increasing demands on the deer population.

Resource limitation is one aspect of the realized nutritional niche (Chapter 1) for a population. Energy consumed by 1,000 deer at the rate of 20.3 $MJ \cdot d^{-1} \cdot animal^{-1}$ would exceed the upper limit of food availability set at 19,000 $MJ \cdot d^{-1}$. The herd could only attain a size of 1,000 animals if the average energy consumption decreased when the population approached this food limit. In fact, the average energy consumption is a weighted average of the rates for different classes of animals within the population. In our example, the model population of deer is comprised of males (115 kg), non-breeding females (65 kg), breeding females (80 kg) and sub-adults (35 kg) that expend 293 $kJ \cdot d^{-1} \cdot kg^{-0.75}$ body mass at maintenance (no net gain or loss of body mass). If males and non-breeding females require 2.1 × maintenance to support normal activity as well as body maintenance (615 $kJ \cdot d^{-1} \cdot kg^{-0.75}$), growing

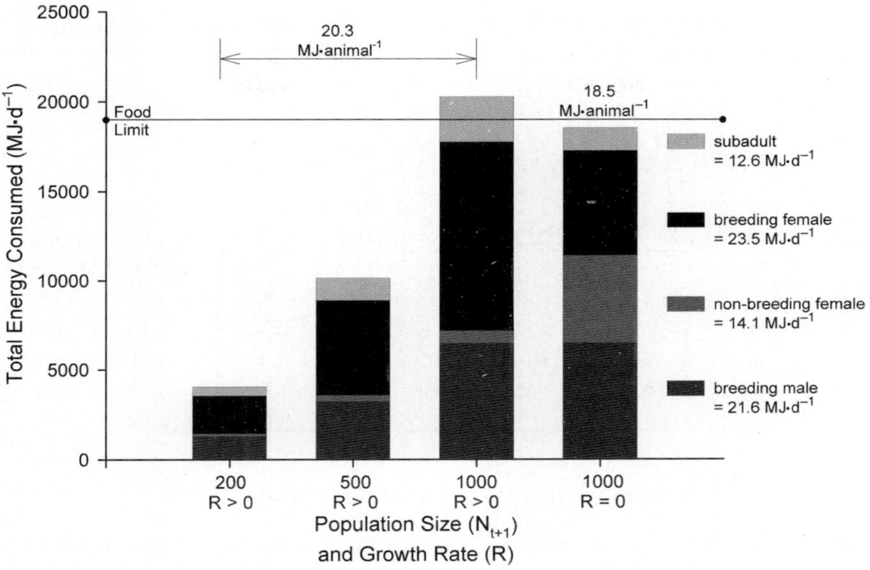

Fig. 2.2 The effect of food limitation on the demography (age class structure) of a model population of deer. The model shows total energy consumption of the population as it grows ($R > 0$) from 200 to 500 to 1,000 individuals. A population of 1,000 animals can be supported ($R = 0$) below the food limit (*solid line*) if the average consumption declines from 20.3 to 18.5 MJ • animal^{-1}. The proportion of breeding females in the population is usually reduced by food limitation because those individuals have the highest demands for energy. In this model of declining food consumption, demography changes as follows for a population of 1,000 animals: sub-adults 200 to 100; breeding females 450 to 250; non-breeding females 50 to 350; and males 300 to 300

sub-adults and breeding females would require 3 × maintenance (879 kJ • d^{-1} • kg$^{-0.75}$). In terms of daily energy requirements, growing sub-adults are projected to use almost as much as the larger non-breeding females. Breeding females use energy at the highest rates in this population even though they are smaller than males (Fig. 2.2). Such high demands for growth and reproduction are difficult to support as food becomes limited; populations near their food limits cannot support the same proportions of individuals with high productive demands. In our hypothetical population, the percentage of sub-adults declines from 20% to 10% and that of the breeding females drops from 45% to 25% as the population approaches K and population growth (R) declines to zero (Fig. 2.2).

Individual responses to resource availability can be used to predict responses at the population level (Chapter 10). Population declines (R < 0) resulting from food limitation are associated with declines in body condition of individuals and reduced deposition of energy and nutrients in fat and lean mass. Poor body condition can reduce reproductive rates if body stores of fat or protein fall below the level required for breeding (Fig. 2.3). In caribou, for example, a threshold of 6–8 % body fat in mid-winter may separate reproductive from non-reproductive females

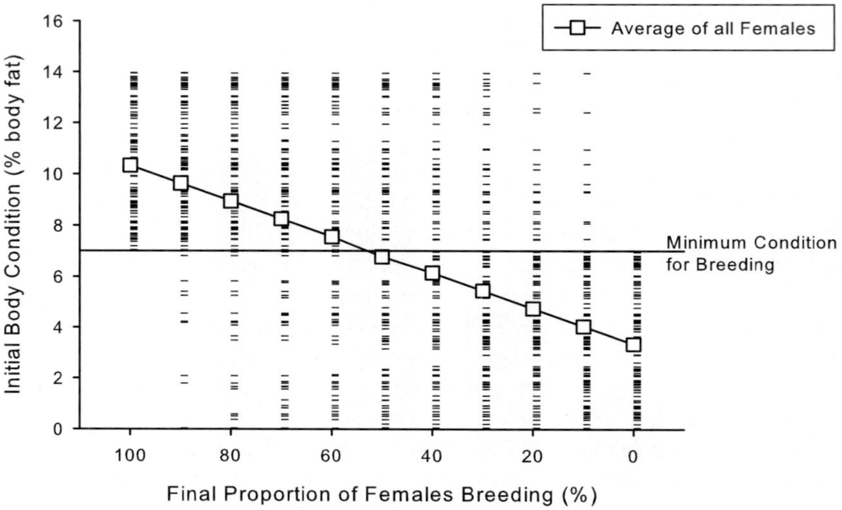

Fig. 2.3 Body condition of individual breeding animals can indicate the potential production of a population during the year. The model shows initial body fat content of 100 female deer in relation to the proportion of females breeding in the population. The simulation uses a threshold of 7% body fat for successful reproduction. Average body fat content (*open squares*) of all females declines linearly as breeding numbers and potential production decline

(Crête et al. 1993; Ouellet et al. 1997). Caribou also rely on stored body protein for fetal development (Allaye-Chan 1991; Parker et al. 2005). Similarly, snow geese rely on both body fat and protein for egg laying and incubation (Ankney and MacInnes 1978). Declines in body condition of reproductive females may therefore precede declines in population size, especially in species that use seasonal body stores to meet the high demands of pregnancy or egg production, lactation or incubation (Chapter 10). Food limitation also reduces juvenile survival, increases the age of first reproduction by constraining growth and development of young individuals and therefore decreases recruitment into the breeding cohort of the population (Eberhardt 2002). Poor body condition may increase the risk of mortality in all age classes but especially in the young and the old by increasing their susceptibility to inclement weather and disease. These differences in mortality risks among age classes may then alter the demography and the food required for the entire population.

2.3 Trophic Relationships

The flow of energy and nutrients occurs in a trophic hierarchy from primary producers such as plants to herbivores and carnivores. This transfer of energy between trophic levels depends on the efficiency of production, which is determined by the deposition of energy or nutrients in producers, and the subsequent assimilation of those products by consumers at the next trophic level. Energy that is expended by individuals for daily maintenance reduces the proportion of energy invested in replacing the population with new individuals. Consequently, ectotherms such as fish with low maintenance requirements for energy have higher production efficiencies than endotherms such as mammals and birds (Humphreys 1979). The capture of energy by consumers depends on their ability to extract energy and nutrients from food. Primary production in terrestrial plants includes energy deposited in structural carbohydrates, which are more difficult to digest by consumers than the simple structures of aquatic algae (Chapter 6). In terrestrial systems, the large investment in plant structures that have less available energy, such as wood, results in most biomass being distributed at the base of the food web. Conversely, in aquatic food webs, there is high flux of energy through the base, resulting in relatively little biomass of primary producers and therefore an inverted pyramid of biomass (Chapin et al. 2002).

The relationship between the numbers of prey and predators depends on the amount of energy in the prey and the efficiency of transfer between trophic levels. A predator such as a coyote (Family Canidae) would be able to satisfy its energy demand for 2.23 MJ•d^{-1} with fewer grouse (Family Phasianidae) at 10.8 MJ•prey $item^{-1}$ (0.2 kills•d^{-1}) than voles (Subfamily Microtinae) at 0.13 MJ•prey $item^{-1}$ (16.6 kills•d^{-1}). The size of the prey population required by a population of predators is further increased by the low efficiency of production between trophic levels. If the coyote captures only 1% of the energy in the available prey, the coyote would need to hunt from a prey base of 223 MJ•d^{-1}, which is equivalent to 21 grouse or 1663 voles each day.

Figure 2.4 models the energetic equivalents of a predator–prey relationship between a group of 50 coyotes and 1,000 grouse. Prey populations vary with annual changes in weather and food abundance that alter their rates of production (P_t) and

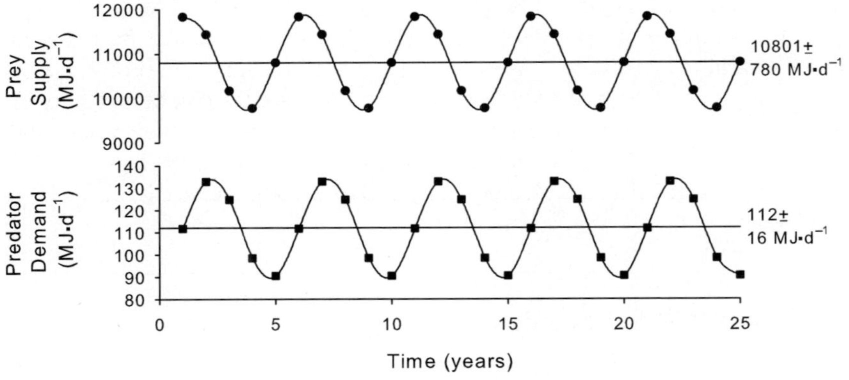

Fig. 2.4 Model of fluctuating prey supply and predator demand for energy. Average energy supply is based on 1,000 grouse predated by 50 coyotes. Predator demands are modeled with a lag of 1 year from the prey supply. Lag times vary with life history parameters such as fecundity, age of maturation and survival of the predator

death (D_t) (Bayliss and Choquenot 2003; Hudson et al. 2003). In our model the grouse population oscillates by 10% every 5 years (Fig. 2.4). Predator populations are obliged to follow the fluctuations in the supply of their food, as observed in the predator–prey cycles for lynx and snowshoe hares (Sinclair and Krebs 2003) and for barn owls and field voles (Taylor 1994). Recruitment of young into a population of lynx declines dramatically 1 year after the peak in snowshoe hare abundance (Mowatt et al. 1996). Our model of energy demands for coyotes and their population size lags behind the changes in prey supply by 1 year (Fig. 2.4).

The relationship between the numbers of predators and prey also depends on the life history parameters of each species (R_{max}), and the size and age class structure of each population. Fecund consumers that mature quickly, such as rodents and some songbirds, may respond more rapidly than ungulates and large carnivores to fluctuations in the environment and the food base (Fryxell and Sinclair 2000). The persistence of a population of prey is also dependent on its food supply and its rates of birth and death. If the food supply can support birth rates, then deaths from predation may maintain the population of prey at a level well below the food limit, as in management strategies for sustainable harvest by humans (Fig. 2.1) (Sinclair and Krebs 2003). High predation rates in small areas constrain populations of forest-dwelling caribou well below their food limits (Jenkins and Barten 2005; Wittmer et al. 2005). However, the relationship between a predator and one species of prey tends to disappear as alternative food sources become available. Figure 2.5 extends our model of 50 coyotes to include 20,000 voles as a secondary prey when the availability of the primary prey is reduced from 1,000 to 700 grouse. Multiple prey items with different patterns of abundance allow generalist predators such as coyotes to substitute one food for another as each choice declines or to meet the increasing costs of reproduction from a broader food base (Bothma and Coertze 2004).

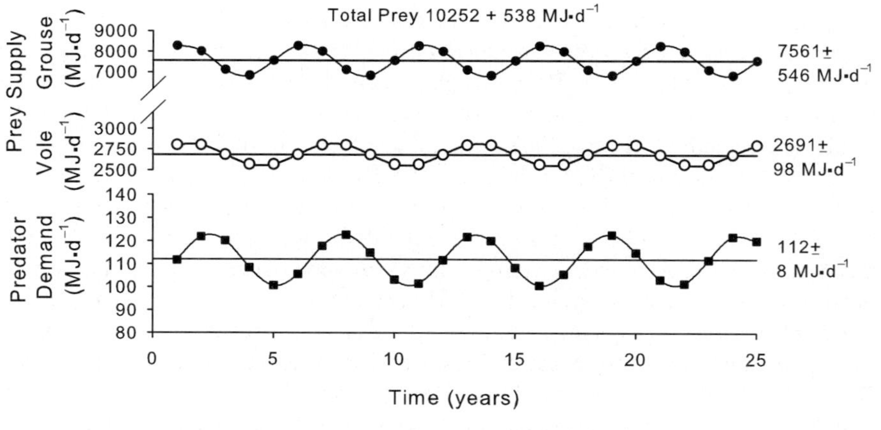

Fig. 2.5 Multiple prey items can be used to meet predator demands. The model populations for coyotes and grouse (Fig. 2.4) are extended to include 20,000 voles when grouse numbers are reduced from 1,000 to 700. The amplitude of the oscillation in predator demand is smaller than in Fig. 2.4, but the total energy demand is similar to a single prey system

An increase in the number of species of consumers and foods in a trophic system rapidly increases the complexity of trophic relationships. Trophic complexity may reinforce the structure of animal and plant communities by providing multiple feedbacks and tolerances. The number of links in a food web is not always indicative of the ability of the system to resist or recover from perturbations. Some keystone organisms exert much more influence on the organization of an ecosystem than would be predicted by their proportion of the biomass (Primack 2004). Removal of keystone predators can allow their prey to consume more plants and thus alter the plant community. Extinctions of gray wolves and grizzly bears have been associated with declines in habitat for migrant songbirds because larger moose populations have changed the structure of shrubs and trees (Berger et al. 2001a). Similarly, the reintroduction of wolves into areas from which they had previously been removed results in antipredator behavior by elk and changes in habitat selection that affect species at progressively lower trophic levels (Creel et al. 2005; Mao et al. 2005).

The absence of predators emphasizes the importance of feedbacks between the herbivore population and the plant community. Overexploitation of a food base occurs when the production of the consumer is not tightly linked to food production. The herbivore population may increase and crash repeatedly because food abundance and consumer demands are often out of phase (Fryxell and Sinclair 2000). High fecundity and rapid maturation predispose species to overutilizing habitats with low or erratic food production. Island populations of domestic sheep that produce twins tend to overconsume pastures, whereas numbers of native red deer with single births are more closely related to the cycles of plant abundance (Clutton-Brock and Coulson 2003). Slow-growing forages such as lichens may be particularly vulnerable to overconsumption. Populations of reindeer and caribou introduced to predator-free islands can increase rapidly, but eventually fall to extinction when the food base of lichens is exhausted (Klein 1987).

Large-scale movements of animals help to disconnect animal production from plant production in areas within an annual range. The food requirements of migratory birds such as Canada geese and snow geese can exceed forage production on the spring breeding grounds if more birds survive winter by feeding on agricultural fields in the south (Ankney 1996) (Fig. 2.6). Seasonal or regional depletion of the food base may be sustainable if the plants have high leaf replacement potential and if the animals show a wide range in spatial and temporal use of the area. Large grazing herds of wildebeest and zebra can exploit highly productive grasslands of the Serengeti because the consumers move to other sites as plant availability declines (Bell 1971; McNaughton 1985). Brant and barnacle geese select the growing tips of plants which delays the maturation of the plant and results in a higher-quality diet for both the migratory geese and resident European hares during summer (Van Der Wal et al. 2000; Stahl et al. 2006). Herbivores such as hippos and hairy-nosed wombats also may enhance production of some plants such as grasses, resulting in a 'grazing lawn' (Jeffries 1999), by removing competing plant species and by recycling nutrients from excreta (Van Der Wal and Brooker 2004). The optimum density of herbivores for plant production varies with the composition of the plant community

Fig. 2.6 Populations of migratory herbivores such as Canada geese are affected by food availability over very large geographic areas. Abundant food in wintering areas may reduce mortality and increase the number of birds that return to consume plants along the migration route and at spring breeding grounds. High densities of herbivores can damage plant communities by removing leaves and seeds, trampling stems and damaging roots at rates that exceed the rate of replacement by the plants. Low to moderate densities of herbivores may facilitate plant production by increasing the availability of nutrients such as N for plants

(proportion of forbs, grasses, shrubs) and the availability of nutrients and water to plants (Person et al. 2003; Stewart et al. 2006) (Fig. 2.6).

2.4 Environmental Variation

Variation in the environment changes populations of species throughout the trophic chain, from primary producers to apex predators, over different scales of time. The environmental conditions of temperature, light, water and nutrients in both soil and water drive the primary production of microbes, algae and plants in both aquatic and terrestrial habitats (Chapin et al. 2002; Diana 2004). Changes in environmental drivers within a year produce weather patterns and seasons, whereas changes in annual cycles produce climatic patterns. Climatic patterns include interactions between atmospheric pressure and ocean temperature that affect both summer and winter weather in the northern (North Atlantic Oscillation) and southern (El Nino Southern Oscillation) hemispheres.

Weather patterns have two basic effects on animal populations. They change:

1. supplies for primary plant production and the trophic chain; and
2. demands on individuals and the risk of death.

These effects are modified by the timing, duration and intensity of environmental changes in relation to the life history demands of each species in the animal community.

The duration and timing of weather conditions favorable for primary plant production vary with geographical region. Consequently the opportunity for animal production also varies with geography and thus climate. Short seasons are tightly linked to photoperiod at high latitudes whereas seasons are less distinct or differentiated towards the equator. At high latitudes, seasonal breeding is often synchronized or cued by the light cycle. Changes in day length accompany primary plant production, which coincides with lactation or incubation, and maximizing growth rates of young in a short but regular season in the Arctic. In arid environments, the window for plant growth may be limited by less regular events such as rainfall. Timing of reproduction and the ensuing production of plants and prey in deserts are therefore cued by rainfall rather than light (Wingfield et al. 1992). Further details on the adaptations of animals to the timing of food supplies and the implications of climate change are presented in Chapter 11. Mule deer that extend across a large North American geographical gradient reproduce approximately 1 month earlier in northern populations than in southern desert populations (Bowyer 1991). In temperate zones, changes in the timing of seasonal rainfall and plant growth can result in shifts in the timing of breeding by birds (Nussey et al. 2005). Increases in the length of the European summer provide longer periods for growth by red deer that enhance body size of males and the maturation of females (Post et al. 1999). Therefore, weather effects on primary production interact with both demography and population size, and are thus components of density-dependent growth.

Severe weather events such as storms and ice expose animals to greater demands for thermoregulation and body maintenance. Animals living at high density close to the food limit of their population are likely to be more vulnerable to the unpredictable risks of adverse weather than well-fed animals. Individuals that are stressed by exposure and low food availability also may be the most vulnerable to disease. Diseases that rapidly debilitate individuals are more likely to affect populations at high than low density and produce sudden declines in the population. Conversely, chronic conditions (e.g., intestinal parasites) or non-lethal infections that reduce fecundity or lifespan may only reduce annual production (Albon et al. 2002; Joly and Messier 2004b). Disease may render some animals more vulnerable to predation. The loss of these potentially infective individuals to predation rather than disease may reduce the overall death rate (D_t) by reducing the rate of transmission of the disease (Hudson et al. 2003). The combined effects of adverse weather, disease and predation can be additive, resulting in large mortality events. Winter die-offs of caribou are reported for herds confined to small foraging areas by heavy snows and extreme cold (Tyler 1986).

Direct and indirect effects of environmental variation change population patterns from the smooth curves in Fig. 2.1 to the more erratic patterns in the model in Fig. 2.7. This model demonstrates that population size is the result of a dynamic balance between multiple factors that produce a net supply or net demand that respectively decreases or increases the population (Bayliss and Choquenot 2003).

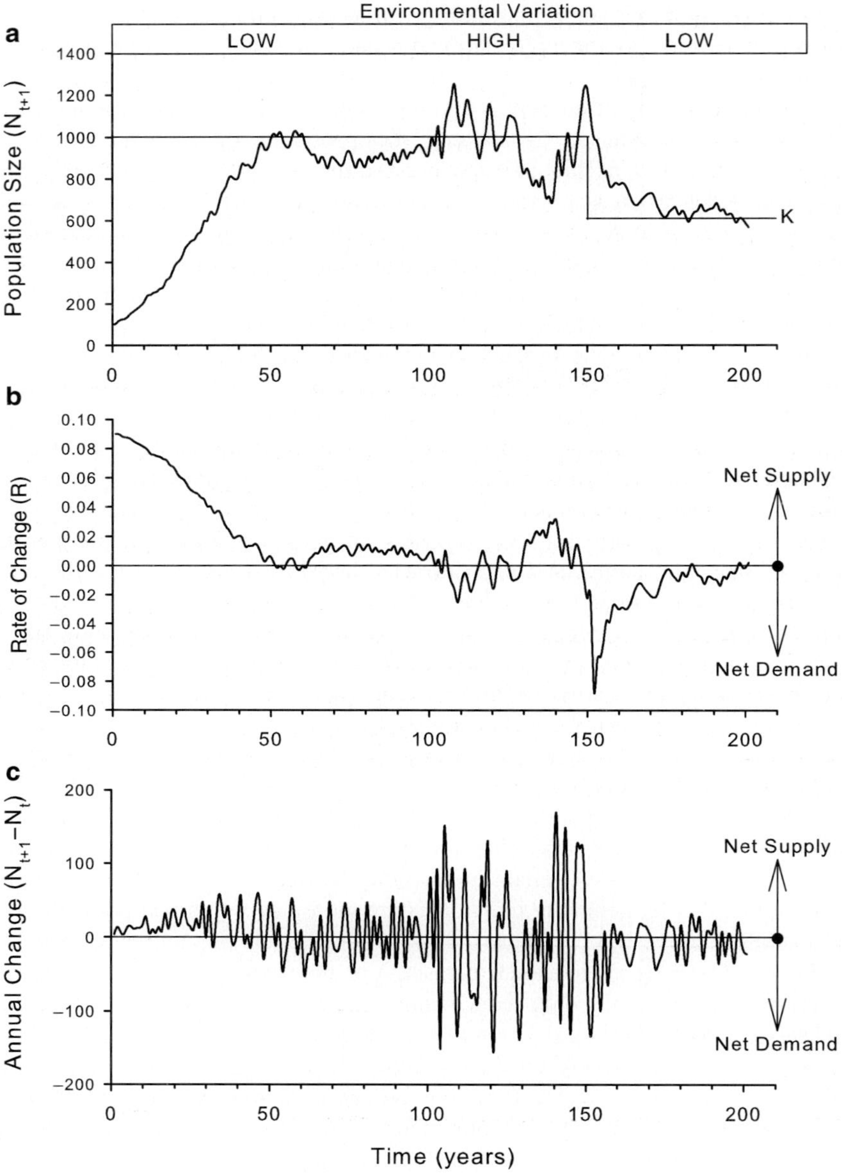

Fig. 2.7 Environmental variation affects population trends and resource limits. The density-dependent model of population growth in deer (Fig. 2.1) is extended to include periods of random variation in growth rate (*R*). Population size (**a**) is stable after 50 years during periods of low environmental variability (5% variation in *R*) similar to a benign climate. Increased variability in growth rate (15% of *R*; **b**) of a large population results in dramatic changes in the annual growth of the population (**c**). The limit for the population (*K*) is shifted from 1,000 to 600 animals at 150 years to simulate a sudden change in the habitat such as fire (**a**). Growth rate (*R*) is negative (**b**) as the population declines to the new resource limit within 50 years (**a**)

Although populations can grow in conditions of both low and high environmental variability, greatest annual changes in numbers occur when populations are large and near their food limit. Large herds are susceptible to small changes in net supply. Conversely, small populations are vulnerable to the net demands of mortality events such as adverse weather or acute diseases that reduce the number of breeding individuals. The long-term effects of a population on its plant and animal communities interact with environmental drivers to alter the food available to the population (Fig. 2.7). The nutritional niche realized by individuals is therefore an outcome of the characteristics of both the population and the environment.

2.5 Summary: Populations

Animal populations increase to limits of food and space availability in a density-dependent pattern. Animal populations are groups of individuals with different body sizes and food requirements. The adequacy of food supplies for any group of animals in the wild or in captivity depends on the number of animals and their demands for energy and nutrients in the prevailing environmental conditions. A small fraction of the energy available in a population of plants or prey is consumed by the next trophic level because energy is expended for non-productive processes at each level and because the consumer may only capture a portion of the production. Weather drives changes in primary plant production and secondarily affects risks of debility and death for animals. Environmental extremes can increase the vulnerability of small populations to greater demands (e.g., predation) and the risk of large populations to starvation.

Chapter 3
Feeding Dynamics: Functional and Behavioral Responses

Population dynamics are closely related to food supply and environmental demands on individuals. Food distribution reflects the spatial and temporal arrangement of habitat factors for each organism: light, temperature and soil nutrients for plants; food, shelter and movement of prey for animals. Consequently, the biophysical attributes of an area used by a species (e.g., altitude or depth, light conditions, wind or current, cover and substrate) are shared by the species on which it feeds. Resource selection and the area used by an animal change with season and life history demands. Foraging areas may be most important in winter and early spring for adults to maintain or gain body mass for reproduction, whereas refugia from inclement weather and predators may be most important for eggs and neonates. For waterfowl, migratory staging grounds primarily serve as feeding sites, whereas winter and spring sites provide both food and refuge.

Feeding responses are studied at multiple scales of time and space to follow changes in an animal, its habitat and food supply (Owen-Smith 2002; Parker 2003; Bowyer and Kie 2006). Feeding behaviors such as food selection or intake (in grams) are commonly quantified at the scale of 1 day (grams per day, $g \cdot d^{-1}$) and measured during foraging bouts within a day (grams per minute, $g \cdot min^{-1}$). Increases in intake over very small scales of time (e.g., rate of energy gain from a selected food; in kilojoules per minute, $kJ \cdot min^{-1}$) are amplified as duration of the feeding behavior increases to the scale of daily energy supply (kilojoules per day, $kJ \cdot d^{-1}$) for the animal; these are called 'multiplier effects' (White 1983). Animals cannot use the entire day for feeding, however, because food must be located, captured and processed before it is swallowed. Food availability may change throughout the day in response to light, water levels, wind and current. In addition, animals must rest, engage in social behaviors, tend their offspring and evade disturbances and predators (Caro 2005). Activity outside refugia may expose animals to less favorable environmental conditions (e.g., low temperatures and strong currents). Time therefore constrains feeding activity and thus the range of energy and nutrient gains that an animal can attain to meet the demands of its life history in a particular habitat.

Food abundance is usually measured at spatial scales corresponding to one or more feeding cycles: a single foraging bout for a small ungulate may be measured at the scale of square meters, whereas the area used for foraging over several days may be better considered at the scale of hectares. Consequently, available food is

P.S. Barboza et al., *Integrative Wildlife Nutrition*,

measured as biomass density (grams per square meter or kilograms per hectare; $g \cdot m^{-2}$ or $kg \cdot ha^{-1}$) or animal density (prey items per hectare; prey items$\cdot ha^{-1}$) for herbivores and carnivores, respectively. The distribution of food items within an area affects the time and effort required to locate the food. Frugivorous bats, for example, spend more energy flying between individual trees in a large forest than they would in a commercial orchard. The net gain or profit from locating food is further offset by its quality. Animals must expend additional time and effort to capture and process food items when struggling with prey or when cropping and chewing plants. The cost of handling a difficult food item can be measured as the energy expended between capture and ingestion or as the time lost in processing. A duck (Family Anatidae) may spend more time and energy consuming shelled mollusks than consuming soft-bodied worms. Energy spent breaking shells also uses time that could otherwise be spent on ingesting worms (Richman and Lovvorn 2004). Food quality is therefore reduced when the concentrations of nutrients decline in a food, as well as when the difficulty of extracting those nutrients increases.

Behavioral and physiological characteristics that maximize net gains of energy and nutrients for survival and reproduction are favored by natural selection and are thus rapidly expressed in a population. In this chapter, food abundance is linked with mechanisms that allow individuals to select and process foods to meet requirements for maintenance and production.

3.1 Functional Response and Diet Breadth

The rate at which animals can gain energy and nutrients in a habitat is constrained by the functional response that is the relationship between food intake and food abundance. The likelihood of encountering food items increases with food abundance. Functional responses are therefore characterized by curvilinear relationships that increase from zero to a range of maximal rates of food intake (Fig. 3.1). Functional responses are affected by both the requirements of the consumer and the quality of the food.

Food intakes at low abundance are highly variable and only tolerable to animals for short periods of time. Animals with high demands (kilojoules per minute, $kJ \cdot min^{-1}$) may be unable to afford long searches for food at low abundance, especially if the quality of the food (kilojoules per gram, $kJ \cdot g^{-1}$) is low and the likely gain is small. Rodents such as tree squirrels stop searching for seeds at a 'giving up density', when the rate of encountering food reaches a threshold abundance below which the costs of time and energy exceed the benefits of the harvest (Brown 1999). Elk searching for food under snow may concentrate their efforts by digging in areas where the food is usually abundant and ignore areas that are usually sparsely vegetated (Fortin et al. 2005).

Maximum rates of intake of energy and nutrients reflect the demands of the animal when food quality and abundance are high (Fig. 3.2). Food intakes follow

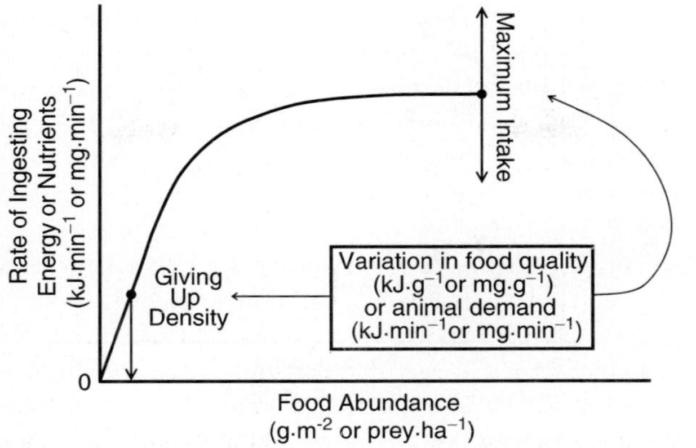

Fig. 3.1 Functional response between rate of intake of energy or nutrients in food (kJ·min^{-1} or mg·min^{-1}) and the abundance of food (g·m^{-2} or prey·ha^{-1}). The 'giving up density' of food is the threshold abundance below which animals stop feeding. Both the lower and upper limits of food intake rate are affected by changes in the quality of food (kJ·g^{-1} or mg·g^{-1}) and the requirement of the animal (kJ·min^{-1} or mg·min^{-1})

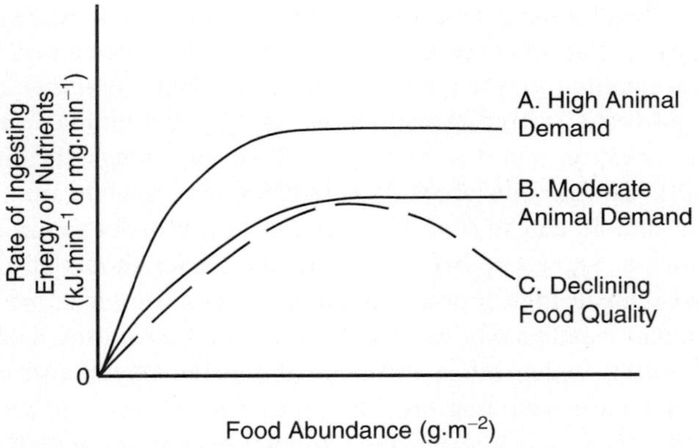

Fig. 3.2 Three contrasting functional responses of rate of food intake to changing food abundance. *A* Intake is highest when animals have the greatest demand and food quality is high. *B* Moderate demands decrease maximal intake. *C* Intakes decline from a maximum if food quality decreases with increasing abundance, as it does when forages senesce

the seasonal pattern of reproduction and maintenance (Owen-Smith 2002); for northern ungulates such as elk, intakes are lowest in winter and highest in summer (Wickstrom et al. 1984). Changes in food quality may alter the functional response curve, especially if quality changes with abundance. The growth of plants through a season increases the abundance of food, but also changes the quality of food from new leaves of high quality to senescent stems and twigs of lower quality. Functional responses may therefore fall below maximum rates of food intake if food quality declines as abundance increases (Wilmshurst et al. 1995).

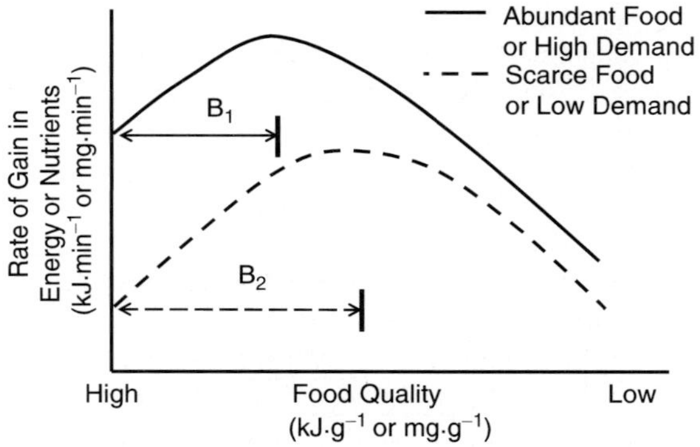

Fig. 3.3 The relationship between rate of gain of energy or nutrients and the quality of foods selected by an animal. If foods are ordered in decreasing quality, diet breadth (*B*) is the range of foods that can support maximal rates of gain of energy or nutrients. High animal demands and abundant food result in high rates of gain of energy and nutrients, which reduce diet breadth ($B_1 < B_2$). Diet breadth increases as food abundance and food quality decline because the animal must use a wider variety of foods to meet its requirements. (Adapted from Belovsky et al. 1999.)

The net gain from foraging varies with food quality and abundance. The range of foods selected by animals (diet breadth) commonly includes the highest-quality items as well as lower-quality foods that may be more abundant. Foraging areas in which abundance of all foods is high provide more choices, but animals can meet their requirements by selecting a narrow breadth of diet from amongst the highest-quality foods (Fig. 3.3). Lower food abundance reduces the absolute number of high-quality items available and thus shifts selection to a broader range that includes lower-quality items. Snowshoe hares consume the tender shoots of growing forbs in summer, but include high proportions of fibrous grass stems and willow bark when food abundance and quality are low in late winter. Low nutritional requirements allow tolerance of the widest range of food qualities. For example, ectotherms such as desert tortoises have wide diet breadth from fruits to cacti and sedges because many foods can satisfy their low demands for energy and nutrients (Barboza 1996; Van Devender et al. 2002)

3.2 Predicting Foraging Behavior with Models

Models of feeding behavior can be used to assess the management implications to a species when conditions of predation, harvest, weather, food abundance or food quality are altered. Tree harvest affects the abundance and distribution of forage plants, as well as predation, for black-tailed deer (Kie et al. 2003; Farmer et al. 2006). Consequently, changes in the amount and proportion of areas for cover and feeding may change animal distribution and therefore require changes in management. Animals may become more sensitive to disturbance as cover is reduced, but gain more energy and nutrients from each foraging bout because of an increase in forage species. Models can be used to predict the food abundance at which costs of vigilance and risk of predation decline sufficiently to allow deer to return to disturbed areas (Belovsky et al. 1999). Alternatively, modeling approaches can be used to assess the levels of disturbance and food needed to manage overabundant wildlife such as white-tailed deer in urban areas (McShea et al. 1997).

Models compare costs and benefits of foraging activities. These comparisons usually assume that energy or nutrient gain limits fitness through survival or reproduction, and food intakes range between net loss and net gain (Fig. 1.8) (Caro 2005). The value of these comparisons depends on the scale and the sensitivity of the response to a specific parameter (Kie 1999). Measures of food abundance and time spent feeding might be less useful as indicators of reproductive success than metrics of space and diversity of terrain if nesting or rearing sites are limiting.

Optimal foraging and risk-sensitive foraging are two prominent approaches used in the large number of existing foraging models. Optimal foraging models predict the lowest investment of time or effort for the greatest benefit from a single behavior or method of foraging (Belovsky et al. 1999). Risk-sensitive foraging solves for the foraging behavior with greatest outcome for fitness from a suite of responses with different probabilities of cost and benefit (Houston and McNamara 1999).

3.2.1 Optimal Foraging Time

Optimality models are used to predict the maximum rate of profit for the animal in terms of energy or nutrients for the investment of time. The simplest model is based on the marginal value theorem (Charnov 1976) that predicts the optimal time invested in feeding by an animal in a patch (Fig. 3.4). Time spent in a patch provides gains to the animal, but these diminish if feeding is prolonged and the animal begins to exhaust the food supply, that is, the animal starts to use some of its profit to continue feeding from a diminishing patch. The model also predicts that the animal should move before achieving the maximum total gain from the patch. Because the time spent traveling between patches without feeding is a cost (equivalent to negative time), the time of departure from the patch depends on the distance between patches. The graphical solution to the model is based on the greatest rate of gain (grams or kilojoules per minute, $g \cdot min^{-1}$ or $kJ \cdot min^{-1}$) after deducting the cost of travel between patches (net gain; Fig. 3.4). Net rates of gain decline as travel time (distance between patches) increases and as the optimal time in the patch increases. As patches become closer together, travel costs decline to zero, and the net gain increases to a maximum instantaneous rate (the slope of the response curve at zero). In a continuous patch, the animal can walk and eat at the maximum rate of gain.

The quality and abundance of food in a patch affect the shape of the gain curve by altering both the slope at zero and the maximum gain for the animal. Declines in food quality increase the optimal time in the patch because more food must be harvested to offset the cost of traveling to the next location (Fig. 3.5). Animals feeding in low-quality patches must therefore feed longer in each patch and visit more patches to achieve the same daily food intake as animals feeding in high-quality patches.

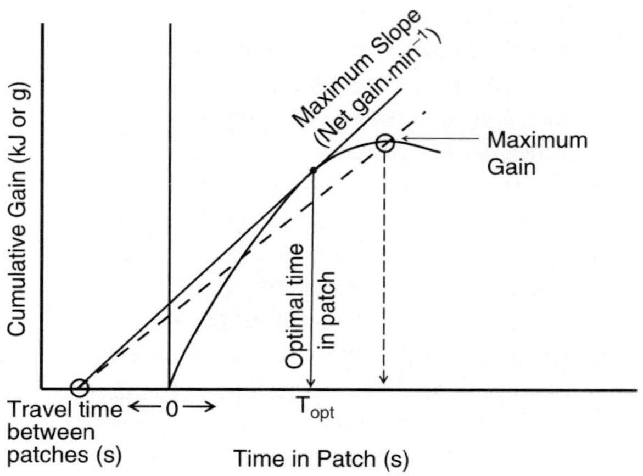

Fig. 3.4 Optimal foraging model of the cumulative gain in energy or nutrients by an animal relative to time spent feeding in a patch. Time spent traveling between patches is a cost (negative time) because nutrient gain is zero. The maximum rate of net gain is the slope of the tangent from the travel time on the *X axis* to the response curve. The optimal time in the patch (T_{opt}) is the time at that tangent to the response curve. *Broken lines* indicate the solution for maximum cumulative gain by the animal, but at a slower rate of net gain (slope) and a longer time in the patch. When travel time declines to zero, the net rate of gain increases to the slope of the response curve at zero. (Adapted from Belovsky et al. 1999.)

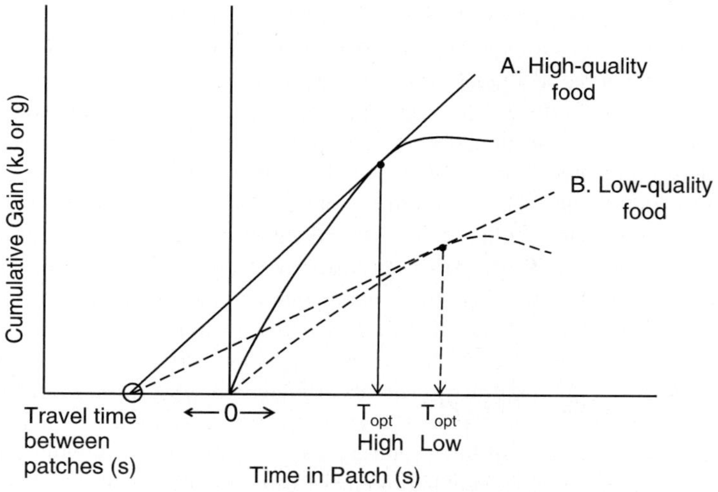

Fig. 3.5 Optimal models for feeding time in two scenarios that demonstrate the effect of cumulative gain (benefit) on optimal feeding time. *A* High food abundance or high food quality. *B* Low food abundance or low food quality. Scenario *A* provides faster gains (steeper slopes) and shorter optimal times (T_{opt}) than scenario *B*

Table 3.1 A foraging model for geese consuming two contrasting qualities of food. Travel time and daily requirements are constant between diets. Optima for time in a patch and the corresponding gain are changed by 33% from the high-quality food to the low-quality food

Parameter	Calculation	High-quality food	Low-quality food
Travel time (min)	A	0.25	0.25
Optimal patch time (min)	B	3.00	4.00
Optimal gain ($kJ \cdot patch^{-1}$)	C	17.0	11.4
Rate of net gain ($kJ \cdot min^{-1}$)	C ÷ (A+B)	5.2	2.7
Daily requirement ($kJ \cdot d^{-1}$)	D	2,000	2,000
Patches required (d^{-1})1	E = D ÷ C	118	176
Required feeding time ($min \cdot d^{-1}$)	F = E × B	353	702
Required travel time ($min \cdot d^{-1}$)	T = E × A	29	44
Foraging time ($h \cdot d^{-1}$)	G = (T + F) ÷ 60	6.4	12.4

[1]Assumes that animal travels to reach first patch.

A foraging model for geese grazing on emergent new growth (high-quality) or senescent (low-quality) sedge is outlined in Table 3.1. The model sets a constant distance between patches, but reduces both the optimal time in a patch and total energy gain by 33% when food changes from high to low quality. The resulting net gains from each patch are reduced by 50% when food quality declines. If energy demands remain at $2{,}000\,kJ \cdot d^{-1}$, geese are predicted to feed for twice as long in 50% more patches when sedges decline in quality. If animals continue to use the same number of patches when food quality declines, they would need to decrease energy demands by 33% from 2,000 to $1{,}340\,kJ \cdot d^{-1}$, but still increase feeding time by 33% from 6.4 to $8.3\,h \cdot d^{-1}$ to meet those demands.

Foraging models that optimize time or energy gain can be expanded beyond this simple example to include variable distances between patches and variable food qualities with changing animal demand (Belovsky et al. 1999; Raubenheimer and Simpson 2007). The strength of predictions from any model, however, depends on the choice of parameters and the quality of data used for each application. Optimal foraging approaches have been used to describe food selections of a wide variety of animals from moose to marine fish. Methods for measuring food intake are discussed in Chapter 4.

3.2.2 Risk-Sensitive Foraging

Foraging behavior must be considered in relation to outcomes that increase the fitness of the individual by improving survival or reproductive success. The time scale for fitness parameters is considerably longer than the instantaneous gain from food selection. Optimal solutions for moving between patches are based on the average return from each cycle of travel and feeding (Fig. 3.5), not just the maximum instantaneous rate (Fig. 3.4). Similarly, diet breadth is the outcome of both quality and abundance, not just quality alone (Fig. 3.3). The integration of feeding behavior with longer-term foraging goals is based on the state of the animal in relation to critical thresholds, such as the amount of fat required for reproduction. Foraging goals or set points provide an internal context for feeding behavior. Exposure to inclement weather and predation are external risks that also affect fitness. The trade-off between internal and external contexts is modeled as risk-sensitive foraging (Bateson and Kacelnik 1998).

Animals may use more than one foraging behavior to meet their demands. Snowshoe hares, for example, may use two foraging strategies: forage on abundant foods in the middle of a clearing with high exposure to predatory birds such as great horned owls, or forage on sparse foods under cover with low risk of predation. Foraging behaviors that provide the same level of risk with increasing gains are risk-neutral because fitness increases linearly with net gain (Fig. 3.6). Coyotes hunting spruce grouse or northern red-backed voles in the same terrain within the same foraging window incur the same fitness with either behavior, depending on the net gain. Coyotes that predate on elk calves or scavenge the kills of gray wolves may incur much greater risks than more omnivorous coyotes that subsist on some fruits as well as small prey in mesic corridors with abundant cover. Omnivory for coyotes may be a risk-averse behavior for which the probability of death is low at low rates of gain, but low food quality and abundance may limit growth or reproduction (Fig. 3.6). Predation on big game and carrion is a risk-prone behavior for coyotes; big kills

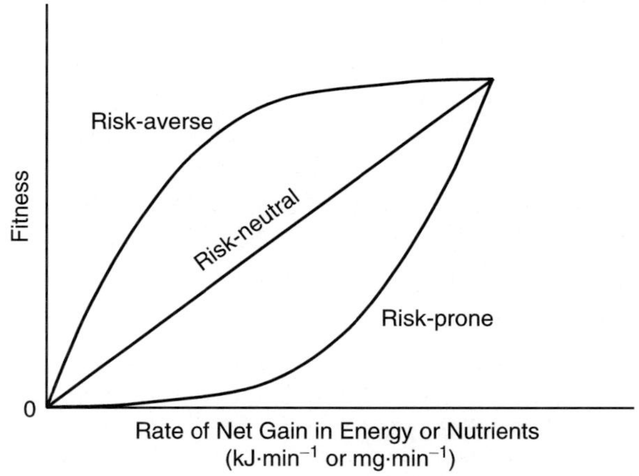

Fig. 3.6 Foraging behavior is ultimately related to fitness through the probability of survival and reproduction. Foraging and thus rate of net gain of energy and nutrients exposes the consumer to greater risks of predation or inclement weather. Behaviors are neutral with respect to risk if net gains are linearly related to fitness. Behaviors that disproportionately increase the risks of mortality or reproductive failure are risk-prone. Conversely, risk-averse behaviors are likely to provide food at disproportionately less risk to survival or reproduction. Increasing search times for larger rewards in more variable patches is risk-prone if the behavior greatly increases the chance of predation to parents or offspring with little gain. (Adapted from Bateson and Kacelnik 1998.)

represent food high in both quality and quantity that may support the gains required for reproduction, but could also incur high risk of death at lower gains when calves or kills are heavily defended by maternal elk or wolves (Fig. 3.6).

The balance between risk-prone and risk-averse behaviors varies over time, depending on the internal context or foraging goal. Figure 3.7 is a risk-sensitive solution for a songbird foraging during a limited window of time each day. The foraging window is limited by light and ambient temperatures between dawn (sunrise) and dusk (sunset). Birds in the model employ one of two contrasting foraging methods: search for high-quality foods such as insects in open terrain with high risk of predation and long search times (risk-prone), or find seeds and fruits in trees close to the nest (risk-averse). Body stores of fat or protein provide the internal context of fitness: a minimal reserve for surviving through the cold night and a greater body store for beginning reproduction. Risk-averse behavior is favored when animals have enough time to feed early in the day or when animal condition clearly exceeds the threshold for reproduction. Animals should incur more risk early in the day only to achieve this highest level of fitness by gaining body stores,

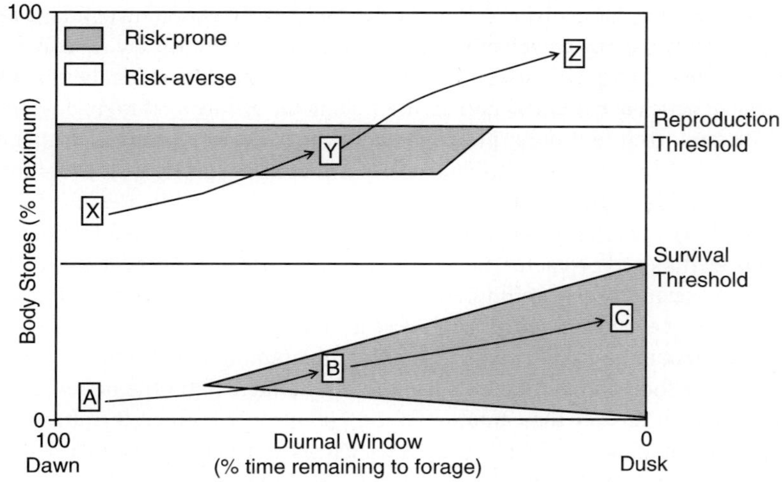

Fig. 3.7 Foraging behaviors of small birds can be classified as either risk-averse (*unshaded*) or risk-prone (*shaded*). Birds can maximize their fitness by varying their behavior through the day to satisfy the demands for survival and reproduction. Foraging is limited to a diurnal window between dawn and dusk. Net gains of foraging result in an increase in body stores such as fat and protein. Birds must rely on those body stores to survive the average night of cold (survival threshold). Additionally, those birds that meet or exceed the reproduction threshold can further increase fitness by beginning to develop reproductive tissues. Risk-sensitive foraging suggests that most birds should be cautious or risk-averse at the start of the day because they are faced with a long period of foraging (*A*, *X*). Birds with body stores below the survival threshold should become increasingly risk-prone as the day progresses and as time for foraging decreases (*B*, *C*). Birds with body stores that are close to the threshold for reproduction should use risk-prone behaviors to deposit those stores early in the day (*Y*) but become risk-averse at the end of the foraging window (*Z*). (Adapted from Bateson and Kacelnik 1998.)

but should otherwise avoid risk. Risk-prone behavior is favored late in the day for animals that are still below the survival threshold. A small subset of animals with body stores close to the survival threshold, however, may continue to avoid risk and simply conserve their meager reserves. This highly conservative option may allow some birds to survive a mild night with small fat stores. Variation in the duration and frequency of cold temperatures may alter the risk of mortality overnight and favor a shift in the internal set point for minimal reserves.

Risk-sensitive models may be extended to more complex approaches that use variable set points for survival thresholds and reproductive investments (Houston and McNamara 1999). This approach can help predict the vulnerability of small populations to acute disturbances such as hurricanes and ice storms or accidents such as oil spills.

3.3 Mechanics of Foraging

The cost of foraging includes the time and energy to locate and capture or harvest each food item. The large diversity of structural adaptations used by animals during foraging reflects the strong selective pressure to minimize costs by increasing the efficiency of finding and acquiring food. Animals detect food by using the absorption of electromagnetic energy. Most animals use light in the visible to ultraviolet spectrum to find flowers, fruit and prey; platypus and electric eel use changes in magnetic fields for prey detection; viperous snakes use infra-red absorption for detecting the warm bodies of prey. Bats, fish, marine mammals, snakes and lizards use pressure sensors to monitor vibrations of sound and motion in air, water and land for detecting prey, conspecifics and obstacles. Seabirds and sharks use chemical sensors to detect molecules as scent trails from injured prey on air and water currents. Primates and birds use chemical sensors in the same way to find ripening fruit.

Complex locomotor skills such as running, swimming and flying increase the search area for food and facilitate its capture. Predators exhibit the largest array of structural adaptations for the capture of food. Small prey is filtered from water by fine hairs or lamellae in dabbling ducks such as northern shovelers. Projectile tongues in chameleon lizards, toads (Family Bufonidae), and anteaters (Family Myremecophagidae) are used to seize larger prey. Protrusive jaws are likewise used by many fish to seize prey. Claws and talons secure prey for cats (Family Felidae) and raptors (Family Accipitridae) in a similar fashion to long rows of teeth and serrations along the beak of fish-eaters such as dolphins (Family Delphinidae), common mergansers and freshwater crocodiles. Immobilization of large prey may include physical impacts such as plummeting dives onto pigeons by peregrine falcons or tripping Thompson's gazelles by cheetahs. Large prey is incapacitated by a combination of suffocation (e.g., constriction by anaconda snakes or drowning by saltwater crocodiles) and blood loss (e.g., bites to the jugular vein with long canine teeth by lions and wolves). Small prey such as mice and rats may be incapacitated by nerve damage with venom from snakes (Family Viperidae, Family Elapidae), shaken to break the spinal cord by cats, or simply eaten whole. Herbivores capture plants, which are easier to apprehend than animal prey. Fruits, pollen and nectar that are located on the tops of tree or at the ends of branches are captured by agile climbing in primates and rodents, and dexterous flight in birds and bats. Elephants capture leaves on trees by first grasping the branches with their trunk before using their lips, tongue and teeth to strip the foliage from the branch.

The complex skills of search and capture often involve social behaviors that increase individual gains. Cooperative foraging by mammals and birds may include learning to find and handle foods with novel methods such as tools that confer large benefits to members of a group. Long-lived social animals such as common ravens, killer whales, wolves, meerkats, baboons and elephants increase the net gain from foraging activities as the individual matures and gains experience and as the size and experience of the group grows (Altmann 1998). Behaviors that involve manipulation of items to remove poor-quality components such as soil, bark and shell require an investment of time and energy, but may yield large gains once the skill is acquired. Ravens and gulls (Family Laridae) expend energy by carrying shelled foods such as tortoises and clams aloft and dropping them onto rocks to break them open. Shelling nuts by parrots (Family Psitaccidae) provides a rich source of energy as lipid and protein; stripping leaves from a tree branch provides a more concentrated source of protein and carbohydrate for moose than would the whole branch.

3.4 Form and Function of the Mouth

Captured food must be processed for swallowing. The main goal of handling food in the mouth is to prepare items for swallowing as a bolus or package that can be propelled through the digestive tract. Oral processes must contend with the physical and chemical characteristics of the food items generally selected by a species (Lucas 1994). The mouth of vertebrate animals is defined by mandibles at the side and front (jawbones or beak), a palate at the roof and a tongue at the floor (Fig. 3.8). All vertebrates can articulate mandibles in the medial plane (up and down) which allows the mouth to open and close and thus bite or crop food from a surface (Schwenk and Rubega 2005). Birds and mammals use the tongue to move food within the mouth and to mix and compress the solid and fluid phases of the food. Fish use the flow of water to move food by creating suction within the mouth and pharynx. Articulations of the maxillary bones above the mandible allow fish to expand the mouth by protruding the jaws. Mammals also can articulate the mandibles in the lateral plane (sideways) which allows solid food to be ground between the teeth. Many fish have arrays of teeth on their gill rakers and pharynx that allow them to grasp and also grind food with movements in all three planes (Clements and Raubenheimer 2006).

The form and function of the mouth is closely related to the range of foods selected by the species. Mouth volume affects the amount of food an animal can process in each feeding event. Sit-and-wait predators such as sculpin and grouper

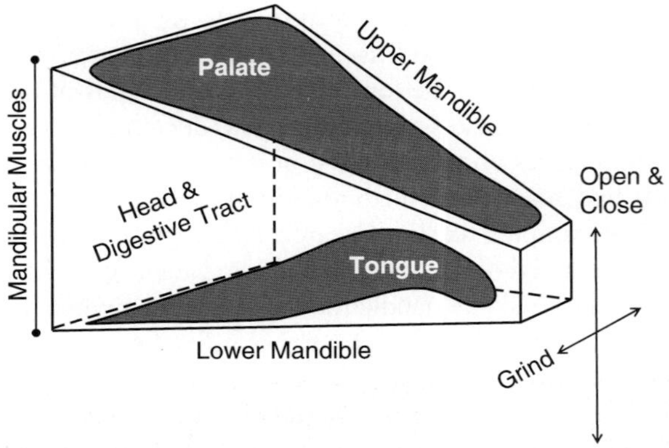

Fig. 3.8 Conceptual model of the vertebrate mouth. *Arrows* indicate movement in the medial plane (open and close) for all animals and in the lateral plane for grinding food in mammals

with large mouths engulf their meal in a single bite, whereas damsel fish process meals in a series of smaller bites. Bite size increases with body mass within groups of animals at allometric scalars that are typically less than 1.0 and similar to scalars for energy requirement (0.63–0.74 in ruminants) (Illius and Gordon 1999) (see Fig. 1.6 and Chapter 10). Food intake (grams per minute, $g{\cdot}min^{-1}$) is the outcome of bite size (grams per bite, $g{\cdot}bite^{-1}$) and bite rate (bites per minute, $bites{\cdot}min^{-1}$). Bite rate declines with bite size because increasing the fill of the mouth also increases the time required to process ingested food. The balance between bite size and bite rate depends on the form of the food and the function of the mouth in cropping and processing the item (Shipley and Spalinger 1992). Caribou may use small bites and high bite rates during selective feeding on the tips of forages such as sedges and lichens (Trudell and White 1981). Conversely, slower bite rates and larger bites provide greater gains for less selective consumption of forbs growing closely together. Bite size affects short-term intake rates and consequently may influence time spent feeding each day (Shipley 2007).

The form of food affects the function of the mouth and the time required for processing meals. The dimensions of the mouth provide an optimal range of function for size and shapes of food selected by a species (Fig. 3.9).

Omnivores may be considered as the generalist form of the mouth that allows an animal to consume the broadest diversity of foods. Examples include the bills of Japanese quail and mallard ducks in birds, and the mouths of rats and Virginia opossums in mammals. The mouth may be elongated or widened as a specialization of the generalist form. Long mandibles are well suited to small foods or elongated individual items. Elongated mandibles provide a large surface for lamellae of the baleen that traps plankton in bowhead whales, or long tooth rows that increase the surface for catching fish in gharial crocodiles. Narrow mandibles are also best suited as probes: butterfly fish glean prey from crevices; sandpipers (Family Scolopacidae) locate prey under water and sand; olive sunbirds and honey possums locate nectar and pollen in flowers. The ability to handle large items in long mouths depends on orienting prey items with the tongue or pharynx. Pelicans (Family

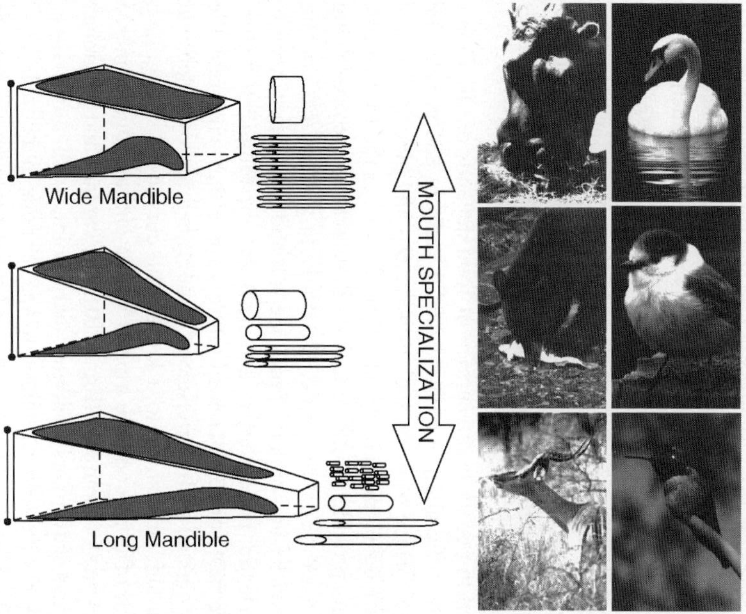

Fig. 3.9 Variation in the dimensions of the general model of the vertebrate mouth (see Fig. 3.8). *Cylinders* indicate the corresponding dimensions of foods for each model. Wide mandibles are specialized for handling bunches of fine foods or large items, whereas long mandibles are suited for fine particles and long items. Examples of specialization in the form and function of the mouth are shown for mammals and birds. Black bears and gray jays represent a generalist form of the mouth in omnivores. Cropping of vegetation is facilitated by a wide mouth in hippos and mute swans. Narrow mouths allow impala to select leaves from thorny branches and black-chinned hummingbirds to consume nectar from the base of flowers

Pelecanidae) and cranes (Family Gruidae) typically orient their prey to be swallowed head first. Wide mandibles and broad muzzles are best suited to broad or spherical items of fruit and prey, or bunches of narrow items such as grasses. Grazers such as hippos, plains bison and bighorn sheep have flat muzzles that facilitate close cropping and allow food intake to increase as the sward of vegetation gains height. The tongue is often used to secure plant leaves against the palate so that the lower mandible can cut the sward with incisors in ruminants or with the lower bill in geese. Both the lips and tongue provide flexibility for selection by animals with wide or narrow muzzles. Wild goats, giraffes and black rhinos use their lips and tongues to glean small leaves from trees with thorns.

Mandibular muscles provide the force required to bite and chew foods (Fig. 3.8). Short mandibles are better suited than long mandibles to transmit the force from the muscles to the food (Fig. 3.9). Long jaws are large levers that require much greater muscular force to hold open than to close. This mechanical constraint allows alligators to be safely handled by tying the mandibles together at the nose. Animals with proportionately large mandibular muscles may also have deeper mandibles to bear the compressive load of the bite. The strong bites of barracudas, turkey vultures, Tasmanian devils and spotted hyenas are reflected in their short broad mandibles and large cheek muscles.

The total force required to penetrate the surface of food depends on the food's toughness and elasticity, and the ability of the surface to deform and recoil. The force required to break the surface of fruits such as avocados and blueberries may be high when the tissues are developing, but gradually decreases as the fruit ripens. Although the skins of fruits provide an elastic tension that protects the underlying tissues, the surfaces of leaves are more elastic, especially in grasses that contain long flexible fibers. The muscles and organs of prey are soft and inelastic; vertebrate skins may be tough and elastic; bone, cuticle and shell may be tough but brittle. Pointed bills and teeth are best suited to concentrating a force that will break the surface of food that is not elastic, such as muscle or bone. Long, narrow ridges cut through elastic skins in a similar fashion to a knife. The curved bills of golden eagles and the canine teeth of lions combine a point with a cutting edge to pierce and cut the skin and soft tissues of their prey (Fig. 3.10). The canines and premolars of insectivorous rodents and scavengers combine sharp points with sturdy bases that can tolerate high compressive forces required to shatter bone and cuticle.

Milling of food into fine particles is best achieved by flatter surfaces such as tooth plates, as in parrot fish (Family Scaridae) that crush corals, and the molar teeth of mammals that crush plants. The composition and concentration of structural carbohydrates in plant cell walls influence the elasticity and toughness of the food. The leaves of woody plants are typically less elastic than those of grasses. Consequently, the molar teeth of browsing animals such as swamp wallabies have sharp and distinct ridges when compared to the molars of grazing kangaroos (Family Macropodidae) (Hume 1999). Increasing concentrations of cell walls in forages require more force and greater chewing effort, which may in turn select for broader molars and larger mandibular muscles in browsers such as moose that consume twigs, and in grazers such as zebra and African buffalo.

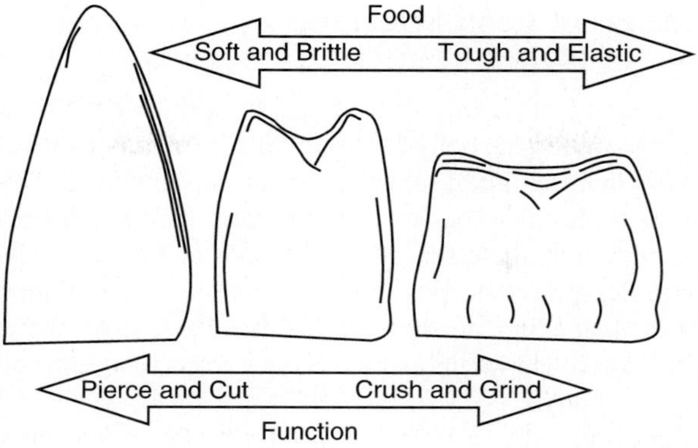

Fig. 3.10 General structure and function of teeth in relation to the physical characteristics of food

Chewing and mixing of food in the mouth increases wear on both soft and hard tissues as the food becomes tougher. Soft tissues are protected by lining the surface of the mouth with a protein called keratin that is also found in skin, beaks, claws, hair and horn sheaths. Lubrication by mucus secreted by salivary glands also reduces abrasion and injury from fine thorns and hairs in desert herbivores such as tortoises and goats. The repair of soft tissues is a continuous process that subtly reduces the net gain of energy and nutrients from an abrasive diet. Injuries to the soft tissues may increase the risks of both acute and chronic infections that ultimately lead to tooth loss and lower survival.

Repair of hard tissues may be limited by the developmental pattern of the animal and the ability to replace teeth as they are lost or worn. Tooth abrasion and wear limits the lifespan of carnivores and herbivores. Multiple tooth rows in requiem sharks (Family Charcarhinidae) maintain food processing when teeth are lost, but the loss of a canine in a tiger may dramatically impair its ability to immobilize prey. Reduction of the molar teeth with age or extreme wear in herbivores such as elk and koalas likewise limits their ability to mill plant cell walls and thus reduces their nutrient gain (Hume 1999). Teeth are comprised of a crystalline matrix of soft dentine and tough enamel at the surface. Grasses contain phytoliths of silica that are harder than tooth enamel and therefore abrade the surface more quickly than woody browse. Molar teeth wear more rapidly in grazers than in browsers even though enamel patterns are often more complex in grazers (Guthrie 1990). Tooth replacement can offset abrasion by grass, wood and sand particles in the diets of some animals such as rabbits (Family Leporidae), beavers (Family Castoridae), porcupines (Family Erethizontidae) and burrowing rodents (Families Geomyidae and Sciuridae). Tooth growth may be further combined with the pattern of enamel wear to maintain sharp cutting edges for grazers such as marsupial wombats (Family Vombatidae) (Hume 1999). The ability of wombats and rodents to process diets with high handling costs is partly related to their large diet breadth as well as their potential for exploiting new areas and new foods as populations grow.

3.5 Mechanisms of Food Selection

Animals must integrate net gains from food with short-term demands such as energy required to maintain the body and long-term demands such as the accumulation of fat for overwintering. Food selection is therefore modulated by cues for both short and long time scales of response. Ingestive cues facilitate acute responses that are stimulated by the immediate sensations of foods as they are captured and handled in the mouth. Responses to post-ingestive cues vary from short (seconds) to long (days) scales depending on the delay in feedback from the digestive tract, organs and central nervous system as metabolites from food enter into circulation and pass through the body.

Ingestive and post-ingestive cues may be used to guide preference or avoidance of food items as animals experience new foods (Provenza 1995). Ingestive responses involve somatic (voluntary) branches of the peripheral nervous system as well as autonomic responses such as secretion of saliva and swallowing. Post-ingestive responses are entirely autonomic and are neither under conscious control nor subject to learning. Animals may reject a bitter food as a learned response to the color, texture or smell (voluntary), or unconsciously regurgitate the item (involuntary) soon after a novel food containing a toxin is swallowed. Food selections and foraging behaviors that affect fitness through survival or reproduction ultimately select for the discrimination of ingestive cues such as the ability to detect and respond to scents and tastes. North American moose respond to scents from novel predators such as tigers (Berger et al. 2001b) by reducing foraging intensity. Similarly, urine from lynx and coyotes fed meat can be used as a repellent to discourage white-tailed deer from feeding in an area (Conover 2002).

3.5.1 Physical Characteristics of Foods

The physical sensation of food is perceived at the mouth as well as the digestive tract. Food texture is sensed by the force required from the mandibular muscles and the tongue to bite and handle the food, and by abrasion of soft tissues. Prey and plants with spines or thorns provide painful stimuli that serve as strong deterrents to consumers. Conversely, soft foods such as animal fat or the endosperm of seeds and nuts may be preferred by animals and are often used in baits for trapping wildlife. Post-ingestive cues of the structure of foods are mainly perceived by stretch receptors in the digestive tract that modulate mixing and propulsion of the digesta (see Chapter 5). Large meals of prey or bulky forage limit further ingestion by occupying the digestive tract for long periods of time. Foods that are impervious to chemicals or low in density of nutrients or energy (nutrient mass ÷ volume handled) require more time to degrade and more muscular effort to disaggregate and mix with secretions. Consequently, whole prey with thick skins or fibrous plants with thick cell walls delay chemical digestion and reduce the number of meals an animal can process in a day. Digesta flow and the process of digestion are discussed in further detail in Chapter 5.

Chemical cues during ingestion involve a broad range of small molecules that are perceived as tastes at the mouth and as scents at the nose. The flavor of a food is the combined sensation of all taste and scent molecules that provide many cues for food selection. Taste and smell overlap as molecules from food are chewed or mixed in the mouth and released into the air in the nasal passages of terrestrial vertebrates. Air-borne scents are described as volatile because they readily enter the gas phase, especially as temperature increases. A long warm day in the tropics provides scent trails to ripening fruit trees when fruit bats (Family Pteropodidae) leave their roosts at dusk to begin foraging.

3.5.2 Chemical Characteristics of Foods

Scents are typically soluble in fats or oils and are often located in glands or cells at the surface of plant tissues. However, all molecules must be dissolved in the watery fluids that bathe the mucous membranes of the nose and mouth before they can be received at sensory cells. Sensitivities to smells or flavors vary widely among species because the ability to detect a molecule depends on the number of receptor cells, their threshold for signaling a response and the perception of that response by the brain. Capsaicin from chili peppers is relatively benign to birds, but the same compound stimulates pain sensors and increases blood flow to raise temperatures in the mouths of mammals (Conover 2002; McGee 2004). Bears, dogs and rats can smell a much wider range of odors and foods than can humans, even though all three groups are omnivores and share a wide diet breadth. All animals probably share the broad categories of bitter, sour, salt and sweet tastes perceived by humans, albeit with differing acuity.

Scents and tastes attract animals to foods. The energy deposited as sugars in fruits and nectars is an attractant that ultimately disperses seeds and pollens when an animal carries plant propagules on its body or in its gut. Scents advertise feeding opportunities as fruits ripen and release odors (Harborne 1993). Many scents produced by degradation of muscles from vertebrate and invertebrate prey (Belitz et al. 2004) are similarly attractive to scavenging animals, from gulls to bears (Nevitt et al. 2004).

Plants and animals also may use molecules to repulse consumers. Many amphibians secrete neurotoxins that affect a broad range of vertebrate predators. This defense has facilitated the rapid expansion of the cane toad after its introduction into Australia. Australian snakes and other predators are vulnerable to the toxin, but South American species tolerate the toxin in the toad's native habitat. Insects may use toxins against vertebrate predators by incorporating the defensive compounds of plants. Insects and vertebrate species do not share the same sensitivities to plant defenses because ingestive and post-ingestive mechanisms differ between the taxa. The caterpillars of monarch butterflies can consume milkweed and other plants that contain cardiac glycosides (calotropin, oleandrin) even though these compounds are aversive and toxic to vertebrate herbivores and the birds that predate on caterpillars. The insect retains stores of the cardiac glycosides in its tissues after metamorphosis to a butterfly. Birds such as blue jays quickly learn to associate the color of the butterfly with the bitter taste of the toxin and a vomiting reflex (Harborne 1993). The feeding deterrent from milkweed therefore allows large aggregations of monarch butterflies to migrate with minimal predation by birds.

The chemical defenses of plants are derived from a variety of intermediary pathways that are secondary routes of C and N flow in the plant with no known plant primary metabolic function, and are thus called plant secondary metabolites (PSMs). All herbivores contend with some form of potentially aversive chemical in terrestrial and aquatic plants (Foley et al. 1999; Targett and Arnold 2001; Taylor and Steinberg 2005). Variations in the concentration and types of PSMs within a plant may reflect the growing conditions and the type of damage to the plant caused by herbivores (Bryant and Reichardt 1992). The time course of responses of vertebrate animals to PSMs ranges from seconds to days, depending on the mechanism of toxicity, the concentration in the food and the dose ingested by the animal. Bitter tastes are characteristic of several PSMs (e.g., glycosides, alkaloids and tannins) and result in low food intakes and thus low gains and production. Low levels of aversive taste do not necessarily prevent consumption of a plant if an animal can tolerate the load of toxin and still gain energy and nutrients. Canada geese can be discouraged but not prevented from grazing on golf courses by spraying the grass with bitter alkaloids. Plant toxins may also play a role in modifying wildlife food chains in response to global climate change if concentrations of PSMs (tannins, phenolics) increase in response to increasing carbon dioxide (CO_2) levels. Subsequently, insects that feed on these toxic plants may contain more of these PSMs and their vertebrate predators may well need to modify their prey choices (Müller et al. 2006).

Consumption of plants and their consumers varies with the concentration of the toxin, the energy requirements of the animal and its tolerance to the toxin. Toxins that are most likely to cause poisoning and sudden deaths in populations are those that are quickly absorbed and block common cellular processes (Fowler 1986). The likelihood of poisoning declines as animals gain experience of foods in an area and begin to use ingestive cues to avoid post-ingestive problems. However, poisonings increase as food choices decline and less defended foods become sparsely available. Woodrats avoid the leaves of creosote bush with its toxic resin unless the habitat provides few alternative food choices (Mangione et al. 2000). Vertebrate populations may tolerate poisonous plants and the insects that incorporate the toxins after prolonged exposure to poisons. Fluoroacetate is a potent poison from woody shrubs (*Gastrolobium* sp. and *Oxylobium* sp.) that are the base of the trophic hierarchy in southwestern Australia. The salt of fluoroacetate is purified as a commercial poison (Compound 1080). It is rapidly absorbed and blocks cellular oxidation; just 1 mg will kill 1,080 kg of carnivores such as red foxes or dingos. Insectivorous marsupials (Family Dasyuridae) are more tolerant of Compound 1080 in regions where fluoroacetate is produced by plants and consumed by the insect prey (Hume 1999).

Tolerance of toxins may involve reducing absorption or inducing an alternative route of metabolism that reduces the toxic effect. Species of woodrats that specialize on plants defended with resins have a lower absorption of toxin than those that consume a broader diet (Sorensen et al. 2004). In most vertebrates the absorbed toxins are first processed in the liver. Absorbed compounds are partly oxidized by enzymes called mixed-function oxidases and bound or conjugated to molecules such as glucose. Conjugated toxins of small molecular weight are excreted in urine from the kidney, whereas larger molecules are excreted in bile from the liver (Harborne 1993). Detoxification therefore uses C and N to produce enzymes and conjugates, as well as energy to produce those reactants and complete the pathways of oxidation and excretion (Foley et al. 1999). Detoxification reduces the net gain from the food, but if the potential nutrient gain is greater than the detoxification cost, animals will continue to consume plants that contain toxic PSMs (McArthur et al. 1993).

Detoxification also reduces the risk of injury to the body. Several common plants, from the leaves of white clover to the seeds of almonds, produce cyanogens that release hydrogen cyanide (HCN) when the tissues of the plant are broken by the bite of an animal. Cyanide (CN^-) can be lethal because it blocks respiration and thus energy metabolism. Although animals quickly learn to avoid cyanogenic foods, animals that cannot completely avoid cyanogens can develop a tolerance to cyanide. Tolerance to cyanide is increased with the activity of an enzyme (rhodanase) that converts cyanide to less toxic thiocyanate (CNS^-), which is excreted in the urine (Harborne 1993). Animals can modulate their intakes to control the concentration of toxin in the body. The diet of *Eucalyptus* leaves consumed by folivorous marsupials is high in phenols that are toxic to cells. One marsupial, the common ringtail possum, appears to control its food intake and food selection in relation to just one of these phenols by direct or indirect control of the part of the brain that initiates nausea and vomiting (Lawler et al. 1998; Lawler et al. 2000; Wiggins et al. 2006).

Handling of PSMs results in long-term responses such as reduced growth and reproduction by limiting food intake and body mass gain from digestion of the food. Tannins in the leaves of woody plants such as oak and willow and legumes such as trefoil are large molecules that bind to proteins and thereby reduce digestion of protein and the activities of enzymes. Phlorotannins are a similar group of compounds in algae that bind to proteins in the digestive tract of fish (Targett and Arnold 2001). Terrestrial animals may mitigate the effects of tannins by producing salivary proteins that preferentially bind to the tannins and thus minimize adverse effects on digestive enzymes from the animal and from microbes in the digestive tract (Hagerman and Robbins 1993). Species such as mule deer and black bears that commonly consume tannin-defended foods produce salivary tannin-binding proteins, in contrast to species such as sheep that select low-tannin forages. High levels of tannins reduce food intake and thus growth of domestic ruminants such as sheep (Min et al. 2003) with few adaptations to this plant defense. Conversely, low levels of tannins may be advantageous. In diets with high protein concentrations, tannins may bind and thereby protect dietary protein from microbial fermentation in the foregut of sheep. The protected protein is subsequently released for digestion in the intestine after acid treatment in the hindstomach (abomasum) (Barry and McNabb 1999). Similar positive associations between nutrients and PSMs may explain the surprising selection of potentially toxic plants by animals. Koalas prefer leaves of *Eucalyptus* that are high in moisture and protein, but also high in potentially toxic oils (Hume 1999). The effects of PSMs on microbial fermentation are discussed in Chapter 6.

Animal reproduction can be affected either positively or negatively by dietary PSMs. Isoflavones are a series of cyclic compounds that are similar in structure and action to some of the reproductive hormones of vertebrate animals. These phytoestrogens may enhance endogenous hormones to the benefit or detriment of the consumer. Phytoestrogens in subterranean clover and soybeans can reduce lamb production in domestic sheep and cub production in cheetahs by disrupting hormonal responses (Setchell et al. 1987). The correlation between PSM and the phases of plant production may also cue reproduction in the consumer. Reproduction of montane voles may be stimulated by phenolic compounds in plants (e.g., 6-methoxybenzoxazolinone) that correlate with increasing abundance of their food (vegetative growth) but may be suppressed by compounds (e.g., p-coumaric acid) that accompany senescence and signal declines in food production (Harborne 1993).

3.6 Summary: Feeding Dynamics

Feeding responses are studied at multiple scales of time and space to follow changes in the animal, its habitat and food supply. The rate at which animals can gain energy and nutrients is constrained by the relationship between food intake and food abundance. Decreases in food abundance reduce the absolute number of high-quality items available and shift selection to a broader range of foods that includes lower-quality items. Optimal foraging models predict the lowest investment of time or energy for the greatest benefit from a single behavior or method of foraging. Risk-sensitive foraging models solve for the best foraging behavior from a suite of responses with different probabilities of cost and benefit. The cost of foraging includes the time and energy to locate and capture each food item. The large diversity of structural adaptations used by animals during foraging reflects the strong selective pressure to minimize costs by maximizing the efficiency of finding and acquiring food.

Form and function of the mouth is closely related to the range of foods selected by the species. Long mandibles are suited to handling elongated foods or filtering small items whereas wide mandibles are suited to large items or to harvesting closely packed items. Food characteristics such as toughness, elasticity and abrasiveness correlate with differences among species in both hard and soft tissues in the mouth.

Ingestive and post-ingestive cues may be used to guide preference or avoidance of food items as animals experience new foods. Physical cues from foods include those associated with handling effort in the mouth as well as the digestive tract. Chemical cues include perceptions of smell, taste and metabolic feedback from attractants and deterrents. Plants produce a wide range of chemicals that may reduce the gain from food and increase the risk of injury to the body. Animals modulate food selection to minimize these costs of consumption.

Chapter 4
Measuring Food Consumption

The amount of food an animal consumes is a central concept of nutrition. Yet food intake is one of the most difficult parameters to measure in wildlife because feeding is influenced by multiple factors such as foraging behaviors, food abundance, food quality, and the requirements for survival and reproduction of the individual (Chapter 3). In this chapter, we discuss approaches to answering the deceptively simple question: *What and how much did an individual animal eat?*

Measures of food consumption are used to monitor the effects of external (e.g., temperature, food availability, food quality) or internal (e.g., body fat stores, reproductive demands) factors on individuals that ultimately affect fitness and dynamics of wild populations. Studies of both captive and wild animals are combined to define the many sources of variation in food intake of wild populations. For example, a study of wild voles might find that food intakes increase as plants senesce and the quality of the food declines. However, the large variation in food intake of the wild voles also is associated with changes in day length, ambient temperature and predation risk during the season of plant senescence. A subsequent captive study of the voles could be used to define the effect of food quality on food intake. Confounding effects such as predation, day length and temperature can be controlled in captivity so that most of the variation in food intake could be ascribed to changes in the diet. To predict food consumption of the wild population, the asymptotic food intakes of captive voles could be used to predict their maximal feeding rates in the wild at a given food quality and ambient temperature.

Food intake is used either as a dependent variable (response) or as an independent covariate in captive studies. Voluntary food intake is used as a response when food is available without limit (ad libitum). The study of captive voles would use ad libitum food intake as a measure of response to changing diet quality, ambient temperature or day length. Many studies of wild animals assume that food is available ad libitum, but that assumption may not hold when food availability declines with season or with unpredictable disturbances such as storms. Constraints on food intake in captive studies are used to mimic natural restriction of feeding time or foraging areas of wild animals. In studies that manipulate food availability, the dependent variable is an internal response such as body mass, body fat or litter size, whereas food intake is used as a covariate of the independent 'treatment' variable

P.S. Barboza et al., *Integrative Wildlife Nutrition*,

that ranges from 0–100% food availability (fasted to ad libitum). Responses to fasting are discussed further in Chapters 10 and 11.

4.1 Adjustment and Steady State

Variation in food intake is highest when animals are adjusting to novel conditions. Most studies that measure food intake assume that animals have entered a steady state of consumption in relation to internal or external conditions. The adjustment to steady state intakes is characterized by a decline in the variation of food intake over consecutive feeding cycles when environmental conditions such as location and food abundance are maintained (Fig. 4.1). Food intakes may remain uniform or

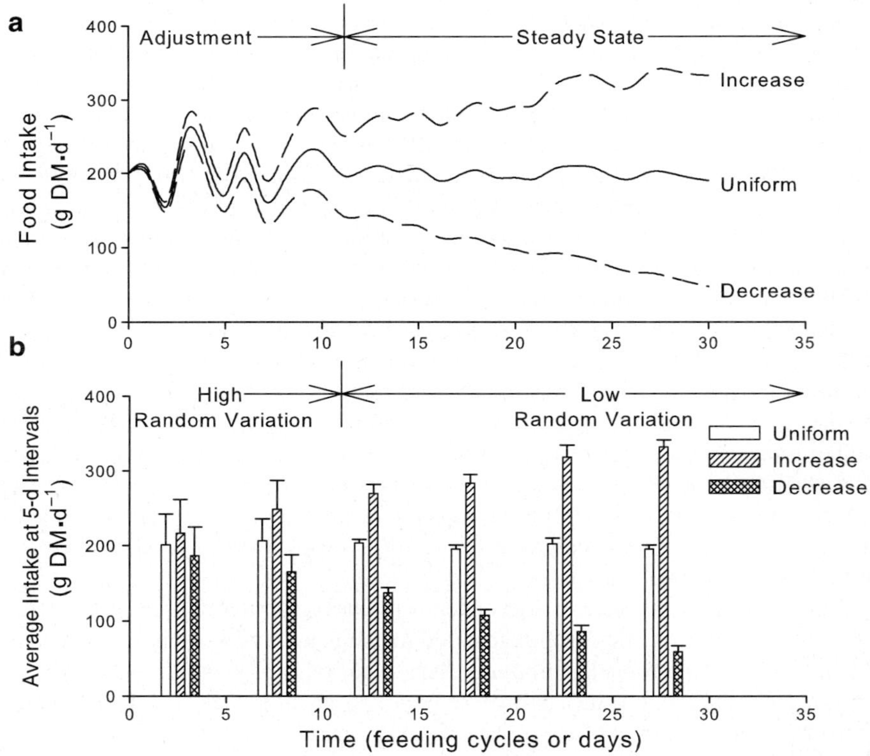

Fig. 4.1 Variation in food intake decreases as animals adjust to their diet and surroundings. **a** Food intake (dry matter, g DM•d^{-1}) of a hypothetical animal after introduction to a new diet or environmental condition in the wild or in captivity. **b** Average intake over consecutive intervals of 5 days. High random variation in the adjustment period masks differences between the three patterns of intake (uniform, increasing, decreasing). Steady state conditions allow different patterns of intake to be resolved

change slowly at steady state, but most variation is associated with predictable rather than random effects. Small amounts of random variation in food intake by individuals in steady state can be accommodated by assuming an average intake over multiple feeding cycles for one or more days. Average intakes are used to relate feeding activity to changes in the environment (e.g., temperature, light) or within the individual (body mass gain or loss, stage of incubation, pregnancy or lactation) over a series of intervals (Fig. 4.1).

The time required for adjustment depends on the behavior and physiology of the species and on the dissimilarity between old, familiar conditions and the new conditions of the experiment. Behavioral and physiological adjustments overlap because stressors can affect both consumption and metabolism of food. The time for behavioral adjustment can be minimized by familiarizing animals to handling procedures in captivity or to the presence of observers in the wild. Intensive studies of captive animals usually start training animals to standard procedures from a young age and may even imprint newly hatched birds on their human researchers. Preserves and parks that limit hunting and disturbance may have wild populations that are habituated to humans and vehicles and which therefore are more easily observed while feeding.

Although food intakes by individuals are most useful for comparisons with individual performance, animals often are studied in groups in the wild and in captivity. Aggregations of prey species into schools, flocks and herds increase the time that an individual can spend on feeding by reducing risk of predation with more eyes to watch (vigilance) and more bodies to distract the predator (predator swamping) (Caro 2005). However, the average intake of a group may mask large variation in the status and intake of individual members in the group. Dominant individuals and their relatives often consume a disproportionate amount of the available food and shelter by excluding subordinates from feeding sites and preferred roosts. Captive animals that are penned individually within sight, sound and smell of each other can still communicate their status and thus influence the feeding of subordinates even though food is provided ad libitum to all animals. Visual barriers and multiple feeding sites can reduce the effect of dominance within groups of captive animals. Exclusion of all behavioral effects is neither feasible nor desirable because isolation of individuals also can adversely affect feeding behavior. Consequently, individual food intake is often analyzed with covariates that account for status, including age and body size to resolve the main internal (e.g., growth) and external (e.g., temperature) physiological effects. Large variations in food intake attributed to behavioral interactions can be avoided by allowing groups enough time to form a stable hierarchy among individuals before measuring food intake responses. Aggressive encounters among animals and the movements of individuals can be useful indicators of the behavioral stability of a group.

Physiological adjustments of food intake to captive conditions (acclimation) or seasonal shifts in the wild (acclimatization) can vary from days to weeks depending on the time scale for ingestive and post-ingestive feedback in the species. The transition to a new diet is dependent on the rate at which gut contents (digesta) are replaced in the digestive tract and the rate at which absorbed metabolites are incorporated into tissues. As material flows into a discrete pool of mixing material in the body, the contents are gradually replaced. The rate of replacing the contents of a pool is dependent on turnover time (TT), which is the time required for inflow to equal the size of a pool. The time for feedback and equilibration of the body with a new diet decreases with short TT of a pool of digesta in the digestive tract or metabolites in tissue (Fig. 4.2). Turnover time of digesta is influenced by food intake and by the size, structure and function of the digestive tract (Chapter 5). The rate of tissue replacement is dependent on seasonal changes in maintenance, and production of new tissue (Chapters 10 and 11). High food intakes, simple digestive systems and high metabolic rates provide short TTs that allow small animals such as hummingbirds (Family Trochilidae) to adjust to simple diets of nectar within a few hours. Conversely, complex digestive systems that process a variety of plant parts may require many days for adjustment in large browsing ruminants such as giraffe. Times to equilibration decrease as the difference between the old and new diet diminishes; that is, small shifts in food composition may only require a few days for equilibration even in large animals. Acclimation times to equilibrate the contents of the digestive tract to a dietary shift range from 4–14 days in mammals and birds over 2 kg body mass. Complete equilibration of diet with metabolic pools in tissues may require as little as a few minutes for energy substrates to appear in the expired breath or as much as several weeks for isotopes of N to appear in muscle (Perga and Gerdeaux 2005; Podlesak et al. 2005).

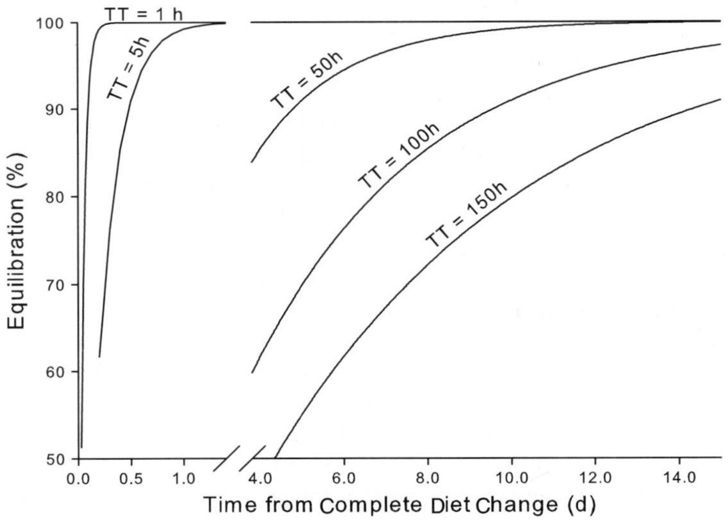

Fig. 4.2 Turnover time (*TT*) influences the time required to equilibrate material in the digestive tract or in tissues for animals on a new diet. High food intake, small digestive tracts and rapid rates of tissue replacement provide quick TT and thus rapid equilibration with a new diet. Times for equilibration depend on the difference between old and new diets. A small shift in one component of the diet may only displace the digestive tract to 90% of the equilibration and thus require only a small fraction of the TT to achieve equilibrium with the new diet

4.2 Direct Measures of Intake

Food intake can be measured directly by observations of feeding activity or by measuring the mass of food items removed through consumption. Indirect measures of food intake are based on the appearance of markers and metabolites from the diet in excreta or tissues. Direct observations are used to validate indirect approaches and to refine each application. A combination of indirect and direct approaches commonly is used to address the questions and logistics of studies on both captive and wild animals.

4.2.1 Behavioral Observations of Food Intake

The simplest approach to measuring food intake is to watch animals feeding in real time or on recording media (Jobling et al. 2001). This approach requires time and patience to systematically observe the animal through one or more cycles of activity. Direct visual observation is usually limited to applications that allow animals to be approached within line of sight. Large animals such as Dall's sheep in open terrain are much easier to observe than small animals such as American mink in dense cover. Observations from blinds of animals at waterholes, mineral licks and supplemental feeders may be used in some studies, but generally are not feasible for recording intakes over the entire period of activity each day. Continuous observations of large species such as geese, deer and moose can be achieved by following the animals (Hupp et al. 1996; Parker et al. 1999; Spaeth et al. 2004). Observations at close proximity require habituation of animals to minimize the effect of the observer on the movement and feeding habits of the animal subject. Animals that are imprinted or hand-reared allow close observation of food selection, even though they may not provide the full range of behaviors seen in wild animals.

Telemetry devices can relieve the observer of much of the tedium and difficulty of following an animal. The application is limited by the weight and size of the device, the method of attachment and any effects on the animal's movement or health. Many devices record geographic location and ambient conditions such as light, which may be used for measures of time spent feeding and to deduce the types of food available at a location. Similar devices attached to the head or neck can be used to estimate feeding movements: e.g., tilt sensors to record head down in foraging geese (Lane and Hassall 1996), and sounds of chewing or motion of the jaw in wallabies (Lentle et al. 2004). These technologies do not identify food items and are best applied when the food selection by the subject is known. However, small video cameras have been attached to the backs of seals to identify and count the number of prey consumed in foraging dives (Williams et al. 2004b). Surveillance devices and satellite transmitters will continue to improve our ability to remotely monitor feeding and the physiology of animals as the devices become smaller and as battery life and transmission range are extended (Block 2005).

Table 4.1 Continuous observation of feeding time and bite rate by an animal feeding for 300 min. Food intake rate (g•min^{-1}) is the product of bite rate and bite size. Bite size is typically measured by recording the number of bites used to consume a known mass of each item. Food intake is the product of intake rate and feeding time

Food item	Observed feeding time (min)	Observed bite rate (bites•min^{-1})	Estimated bite size (g•bite^{-1})	Food intake rate (g•min^{-1})	Food intake (g)
Calculation	A	B	C	D = B × C	E = D × A
I	15	4	0.2	0.8	12
II	45	6	1.4	8.4	378
III	60	6	1.6	9.6	576
IV	75	4	0.6	2.4	180
V	105	3	1.4	4.2	441
Total	300				1,587

Feeding observations often record the time (minutes) spent on a food item of an average mass (grams) with the corresponding bite rate (bites per minute, bites•min^{-1}) and processing rate (bites per item, bites•item^{-1}). These observations can be used to estimate the number of items consumed (harvest rate; items per minute, items•min^{-1}) and thus the total food intake (grams) (Table 4.1). Observation errors increase with the number of dietary items and reduced visibility of those items. Omnivores such as baboons may consume foliage, fruits and insects (Altmann 1998) throughout the day. Herbivores such as black-tailed deer may consume more than 70 species of plants in rainforest habitats (Parker et al. 1999). Large, discrete food items such as vertebrate prey or the leaves of skunk cabbage are easier to see during foraging and thus count than tiny ants or berries. Wide dietary selections also may increase the variation in food handling rates because the physical and chemical characteristics of food vary (Chapter 3).

Consequently, animals vary bite rate (bites per minute, bites•min^{-1}) or bite size (grams per bite, g•bite^{-1}) among foods to maintain intake rates (g•min^{-1}; Fig. 4.3). However, intake rates vary within a feeding bout as animals change their selection of foods during foraging (Gillingham et al. 1997). Foods that are optimal in size and structure for the mouth should provide the lowest variation in intake, whereas departures from the optimal range in size and quality of food will increase variation in rates of food intake (Fig. 4.3). Bite rates may best reflect intakes when bite sizes are constant and the quality of food within each bite is relatively consistent. Variation between bites is small when ruminants feed on emergent grass or forbs in spring, but their bite sizes become more variable as the forages within their ranges begin to senesce and the foods become more heterogeneous in form and quality. Estimates of food intake from bite rate for north-temperate species such as elk are thus more variable in late winter than in summer (Gedir and Hudson 2000a; Cook 2002).

The daily intake of a food item is the sum of all bouts of feeding on that item during the day. Intake of each food is calculated from feeding time (minutes per day, min•d^{-1}) and intake rate (grams per minute, g•min^{-1}). Intake rates may be measured separately for different foods by observing the time required for an animal to consume a known mass (grams, g) of each item. Daily food intake is then the sum

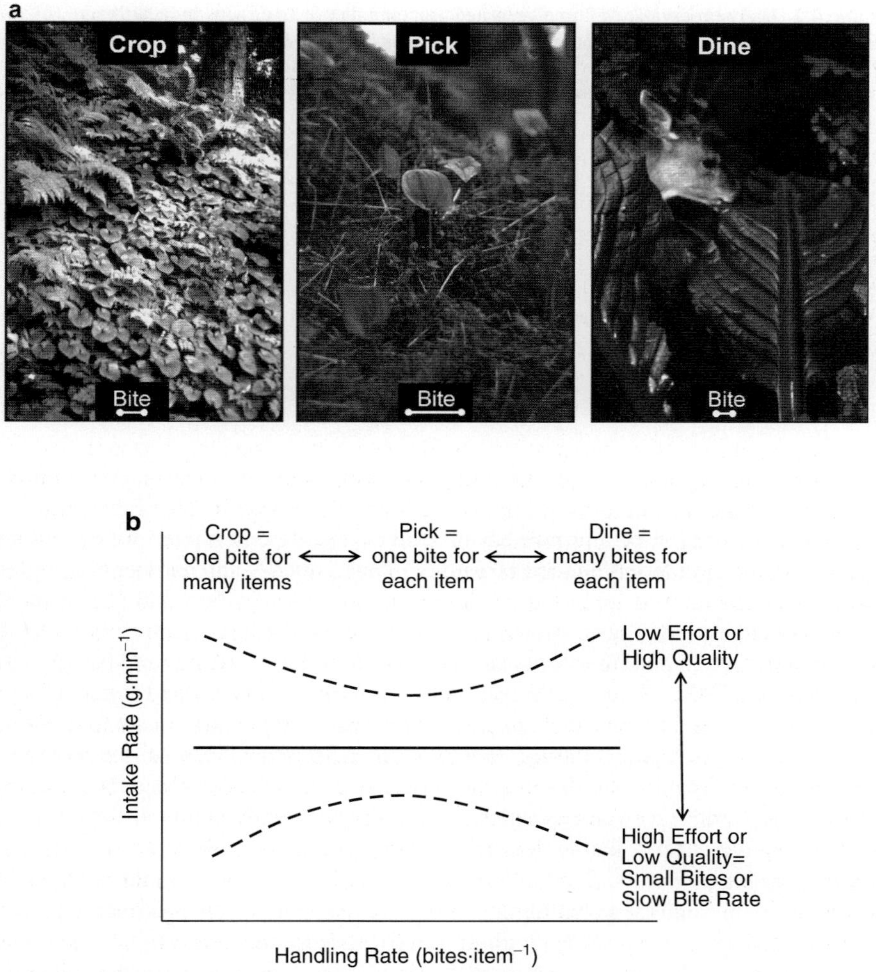

Fig. 4.3 Estimates of food intake that are based on bite rate and bite size may be affected by changes in the size and structure of the food. Food intakes are affected by the method of handling food. **a** Three examples of foods consumed by a medium-sized ungulate such as black-tailed deer. Small items such as forbs (false lily of the valley) and ferns (spreading woodfern) growing closely together can be cropped whereas discrete leaves and stems of low-density early-spring forbs are picked. Large items require dining with several bites per leaf (e.g., skunk cabbage). Intake rates usually are influenced more by bite rate than bite size for cropping, and vice versa for dining. **b** Food intake rate ($g \cdot min^{-1}$) can be maintained (*solid horizontal line*) by balancing bite rate ($bites \cdot min^{-1}$) and bite size ($g \cdot bite^{-1}$), and by varying the number of $bites \cdot item^{-1}$. *Broken lines* indicate variation in food intake across a range of handling rates. Variation in food intake rate is predicted to increase with heterogeneity in food quality for each bite. Shifts in the method of handling also increase variation in intake rate as bite rate or bite size become more important in determining intake rate for crop and dine methods, respectively

Table 4.2 Daily observation of time spent feeding on a diet of five foods by a herbivore. Intake rates of dry matter (g DM•min^{-1}) for each food were measured in a separate experiment (see Table 4.1). Food intake (g•d^{-1}) is the product of feeding time and food intake rate. Corresponding intakes of nutrients such as protein are calculated as the product of nutrient content (g•g^{-1} DM) and intake of each food

Food item	Feeding time (min•d^{-1})	Food intake rate (g DM•min^{-1})	Food intake (g•d^{-1})	Protein content (g•g^{-1} DM)	Protein intake (g•d^{-1})
Calculation	A	D	E = D × A	F	G = E × F
I	27	0.8	21.6	0.10	2.2
II	81	8.4	680.4	0.15	102.1
III	108	9.6	1,036.8	0.09	93.3
IV	135	2.4	324	0.05	16.2
V	189	4.2	793.8	0.08	63.5
Total	540		2,856.6		277.2

of all foods consumed. The corresponding intake of nutrients can be calculated from the concentration of nutrients in each food item and the food intake (Table 4.2).

The concentration of any nutrient in food is usually determined by collecting representative samples of the food consumed. The frequency and number of samples depend on spatial and temporal variation in food composition and the nutrient under consideration. Light, moisture and nutrients affect plant growth and the variation in protein and fiber contents of forages for elk in clearings and old-growth forests (Cook 2002). Studies of selection by herbivores for water and sugar content of leaves may need to consider sampling at sunrise, midday and sunset to monitor the changes in food plants during transpiration and photosynthesis throughout the day. However, weekly samples may be sufficient for studies of seasonal variation in food consumption (Van Soest 1994). Foods are usually sampled in the units selected by the animal, that is, leaves vs. whole plants for herbivores or tissue vs. whole prey for predators. Sample preservation varies with the type and number of nutrients to be analyzed. Nutrients are best conserved when food samples are stored at freezing temperatures with minimal exposure to air and light as soon as possible after collection. Less exhaustive methods of field preservation such as drying forages or preserving prey in solvent may be sufficient to conserve many important components of the food. The analyses of different nutrients are discussed in Chapters 6–10.

4.2.2 Food Intake by Mass Balance

The most direct approach to measuring food intake is to measure the disappearance of food by weighing available food before and after an animal has completed a feeding cycle. Captive animals are typically held in individual pens, but individual intakes also may be determined in animals housed together by using feeding gates that open only to a unique key frequency emitted by the collar of each animal

(Parker et al. 2005). Similar exclusion systems may be used to restrict access to feeders by animals based on body size; young animals can be provided exclusive use of a feeder by setting a 'creep' or hurdle in front of the feeder to exclude adults. Conversely, calves can be excluded from adult foods by elevating feeding stations. Balance approaches assume that disappearance of food is due to ingestion by the animal. Thus feeders must be constructed to minimize errors from other losses such as food spillage and any extraneous consumption by pests.

Residual or refused food (orts) should be collected and replaced with fresh food at the end of the feeding cycle to provide a consistent amount and quality of food for the animal throughout the experiment. Unfed or control batches of food are used to monitor changes in composition of food offered each day, especially if the diet is moist and ambient conditions are warm and conducive to bacterial spoilage. Most importantly, animals may select within the offered food, leaving a portion of food that differs in composition from the average composition of the food offered. Balance experiments should therefore collect daily samples of the food offered as well as the food refused by each animal. A hypothetical balance experiment on a mammal is described in Table 4.3 with the calculations for food intake. Foods offered and refused are expressed on the basis of dry matter (DM) to remove the effect of changes in moisture during each day. The equivalent energy and protein content of each food is also provided for calculation of energy and protein intake.

4.2.3 *Digestible and Metabolizable Food Intake*

Experiments that confine animals to metabolism cages allow for the collection and measurement of all excreta produced during each day of feeding. Animals may require considerable training to acclimate to metabolism cages, but the effort can

Table 4.3 Direct measurement of food intake by mass balance is combined with excreta collection to determine intakes and efficiencies of digesting and retaining energy and nutrients such as protein

Parameter	Calculation	Dry matter balance (g)	Gross energy content ($kJ \cdot g^{-1}$ DM)	Energy balance (kJ)	Crude protein content ($g \cdot g^{-1}$ DM)	Crude protein balance (g)
Food offered	A	1,000	20	20,000	0.19	190
Food refused	B	400	18	7,200	0.15	60
Food intake	C = A − B	600		12,800		130
Feces	D	120	21	2,520	0.15	18
Urine	E	20	20	400	0.50	10
Digestible intake	F = C − D	480		10,280		112
Digestibility	G = F ÷ C	0.80		0.80		0.86
Metabolizable intake	H = F − E	460		9,880		102
Metabolizability	I = H ÷ C	0.77		0.77		0.78

provide a wealth of information about responses to diet, levels of food intake and the validity of indirect methods (Fig. 4.4). The mass balance between food intake and fecal output is used to calculate the apparent absorption of food from the digestive tract (Table 4.3). The disappearance of mass between food and feces is the digestible intake. We refer to the fraction of food intake that is digested as the digestibility coefficient, but this term is also known as digestive efficiency or assimilation efficiency (Hume 2005). The animal apparently retains most of the digested material for metabolism once urinary losses are deducted from digestible intake. Digestible intake that is corrected for urinary loss is called metabolizable intake (Table 4.3). Metabolizable intakes approximate the food retained by the animal if losses from respiration and the skin are small. The fraction of food intake metabolized by the animal is the metabolizability coefficient. The role of endogenous losses in these apparent intakes and efficiencies is further discussed in Chapters 8 and 10.

The efficiencies of digestion and metabolism in captive animals can be used to predict the value of foods for animals in the wild (Table 4.4). Digestible and metabolizable contents of a food are calculated from the concentration of the nutrient in the food and the corresponding efficiency. These digestible and metabolizable contents are used to convert observed intakes of food into the amount of nutrients

Fig. 4.4 Pacific black brant in a metabolism cage for measurement of food and water ingestion as well as total excretion. Metabolism cages are designed to provide easy access by animals to food (*A*) and water (*B*) with minimal spillage during feeding. The cage floor is constructed of mesh that allows excreta to fall through the floor into a pan (*C*) for collection. Cages for mammals also include screens beneath the floor to separate feces and urine for collection. Mesh on the floor must be large enough to allow excreta to fall through, but small enough to support the animal and minimize damage to the feet. Perches, roosts and nest boxes may be used to provide resting areas within each cage

Table 4.4 Food intake measured by behavioral observation or mass balance can be converted to absorbed (digested) and retained (metabolized) intakes of nutrients with coefficients for digestibility and metabolizability (see Table 4.3)

Parameter	Calculation	Energy	Crude protein
Dry matter intake	A	850 g	850 g
Nutrient content	B	$19\,\text{kJ}\cdot\text{g}^{-1}$	$0.20\,\text{g}\cdot\text{g}^{-1}$
Digestibility (digested/ingested)	C	$0.80\,\text{kJ}\cdot\text{kJ}^{-1}$	$0.86\,\text{g}\cdot\text{g}^{-1}$
Metabolizability (retained/ingested)	D	$0.77\,\text{kJ}\cdot\text{kJ}^{-1}$	$0.78\,\text{g}\cdot\text{g}^{-1}$
Digestible content	$E = B \times C$	$15.2\,\text{kJ}\cdot\text{g}^{-1}$	$0.17\,\text{g}\cdot\text{g}^{-1}$
Digestible intake (digested)	$F = E \times A$	12,920 kJ	146.2 g
Metabolizable content	$G = B \times D$	$14.6\,\text{kJ}\cdot\text{g}^{-1}$	$0.16\,\text{g}\cdot\text{g}^{-1}$
Metabolizable intake (retained)	$H = G \times A$	12,436 kJ	132.6 g

or energy digested and retained by the animal for maintenance and production. Such calculations have been used to evaluate the quality of winter habitats for waterfowl such as snow geese (Alisauskas et al. 1988; Raveling 2004).

4.3 Indirect Measures of Intake

Markers allow the estimation of food intake and digestion when logistical constraints preclude one or more direct determinations on either captive or wild animals. Many estimates of digestion in captive fish use markers because total collection of excreta is more difficult with aquatic than with terrestrial animals (Guillaume and Choubert 2001; Jobling et al. 2001). Markers and other indirect methods make assumptions that must be tested by direct approaches or corroborated by other indirect approaches. The principal assumption is that a marker represents a food item as it is processed by the consumer. Markers of food items belong to one of two general fractions: indigestible or digestible (Fig. 4.5). Completely indigestible markers remain with food residues throughout the digestive tract and emerge in the feces. Completely digestible markers are absorbed from the digestive tract and are assimilated into blood and tissues.

Indigestible markers represent a fraction of the diet that is not absorbed by the animal. The indigestible fraction is therefore most representative of foods with a large proportion of refractory material such as plant fiber consumed by grazers or the cuticle in exoskeletons consumed by insectivores. Indigestible and digestible fractions of the diet usually have different rates of flow and turnover in the digestive tract. These disparities between diet fractions are small when food intakes and flow rates are high, but increase with the complexity of the diet and the digestive system (see Chapter 5). Markers that are absorbed best represent the portions of food used by an animal's tissues. The relationship between assimilated food and the markers incorporated in tissues, however, is complicated by the processes of digestion, absorption, excretion, intermediary metabolism and cell turnover (Fig. 4.5).

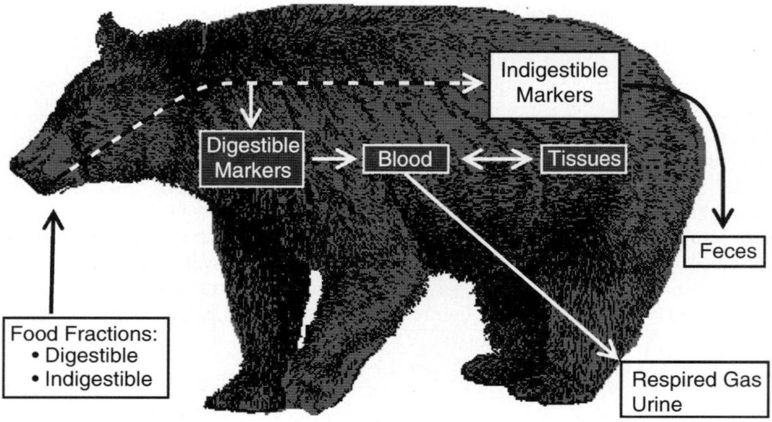

Fig. 4.5 Markers can be used to track both indigestible and digestible fractions of the food in an animal. Indigestible markers are subject to digestive processes, but nonetheless appear in refractory residues of the feces. Digestible markers are absorbed into the circulation of the animal, deposited in tissues and excreted. *White arrows* indicate processes that can contribute to metabolic discrimination of absorbed markers

4.3.1 Measuring Intake with Indigestible Markers

Markers for the refractory portion of foods can be intrinsic components of the diet or extrinsic materials that are usually synthetic additions to the diet. Intrinsic markers are most likely to behave like the food, but their concentrations may be low and their range of concentrations small, both of which can limit application and detection. Extrinsic markers can be applied to foods to measure both digestibility and flow of material through the digestive tract. Intrinsic markers provide additional information on the composition of the diet (Table 4.5).

Counting fragments of food in feces or digesta is the most common method for estimating the proportions of the diet in a wide variety of carnivores and herbivores. For example, hair, feathers, insect cuticle and bones can be used to identify species of prey in feces (scats) from foxes and in regurgitated pellets from owls. Some prey items have well-defined structures such as the otoliths in fish or beaks of squids that can be used to identify the species and size of prey consumed by marine carnivores (Moyle and Cech 2004). Similarly, the structure of plant cell walls is used to identify fragments of plants by 'microhistology' of feces or digesta from ungulates to rodents. The biggest concern about these methods is the differential detection of fragments in a mixed diet (McCullough 1985). The contributions of highly digestible components to the diet are underestimated because they leave few fragments, whereas large indigestible or highly fragmental items are most easily detected and best represented in fecal samples. Such complexities arise in accurately quantifying the winter diets of reindeer and caribou, which include both highly digestible lichens and highly fragmentable and less digestible mosses (Russell et al. 1993).

Table 4.5 Some intrinsic (natural) and extrinsic (synthetic) indigestible markers and their application to measures of food intake in animals

Marker class	Marker group	Application
Intrinsic	Fragments – plant cell walls, fish otoliths, silicates	Diet proportions
Intrinsic	Plant cuticular waxes and alcohols	Diet proportions
Intrinsic	Minerals – Mn, Ti, Sr	Digestibility
Intrinsic	Lignin in plant cell walls	Digestibility
Extrinsic	General – beads, food dyes, chromic oxide, alkanes	Digestibility, digesta flow and digestive capacity
Extrinsic	Phase specific – fluids (polyethelyne glycol, Co-EDTA), lipids (labeled wax), particles (beads, rare earths and mordanted fiber)	Digesta flow and digestive capacity

The number of fecal samples required to define the diet increases with the number of dietary items and the difficulty of detecting fragments. Items that are infrequently consumed (e.g., $< 5\%$ of the diet) are most difficult to detect and therefore require more samples of feces. Many samples are required to define the diets of pelagic predators such as sea turtles because they frequently include a few species of prey but also a much larger number of incidental prey items (Bjorndal 1997). Diverse diets improve comparisons among groups of animals because differences in feeding are more easily detected when there are more dietary items on which to base the comparison (Trites and Joy 2005).

A simple method for estimating the average food intake of individuals in a group of fish uses the rate of disappearance of food from the stomach to calculate the net rate of filling the stomach each day (Diana 2004). Indigestible markers such as plastic beads (extrinsic) or the skeletons of prey (intrinsic) can be used to time the disappearance of a meal from the stomach (Jobling et al. 2001). Stomach contents can be visualized with X-ray or removed by gastric lavage (flushing the stomach with a tube) or by dissection (post mortem). Serial measures of stomach digesta in captive fish provide the average amount of marker remaining from the dosed meal. Stomach contents are sampled from individuals over intervals during each day of feeding (Table 4.6). The fractional rate of disappearance is used to estimate the mass of digesta that remains in the stomach from the start of each feeding interval. Food intake is calculated as the net change in mass of digesta during the interval and converted over the total number of feeding intervals in the day to the corresponding daily food intake (Table 4.6). This estimate of food consumption is known as the 'food ration' for fish populations (Diana 2004).

Chemical methods for detecting food fragments often allow better detection of small dietary components than visual methods of identification. The indigestible waxes and alkanes of the plant cuticle provide a signature of lipid-soluble compounds that can be used as a 'fingerprint' for identifying plant species and plant parts in a diet (Dove and Mayes 2003). The number of food items that can be resolved by this method depends on the uniqueness of the patterns in each mixture and thus the statistical discrimination of multiple alkane concentrations. Alkanes have been

Table 4.6 Estimating the amount of food consumed by fish from average changes in the mass of marker and digesta dry mass in the stomach

Component	Calculation	Measure
Stomach clearance		
Marker dose (mg)	A	20
Marker remaining (mg)	B	6
Time since dosing (h)	C	8
Fractional rate of disappearance [1] (h^{-1})	D = (Ln A – Ln B) ÷ C	0.15
Feeding		
Start of feeding interval (g digesta)	E	8.0
End of feeding interval (g digesta)	F	15.0
Feeding interval (h)	G	6.0
Residue from start [1] (g digesta)	$H = E \times e^{(-D \times G)}$	3.2
Food intake over interval (g)	I = F – H	11.8
Daily feeding window ($h \cdot d^{-1}$)	K	16.0
Daily food intake ($g \cdot d^{-1}$)	L = I × (K ÷ G)	31.4

[1]Markers and digesta are assumed to leave the stomach in an exponential manner. The amount of marker (mg) in the stomach declines from e^A to e^B over the period C (h).

used to measure seasonal changes in diets of a wide range of herbivores including red deer (Bugalho et al. 2005), Przewalski's horses (Kuntz et al. 2006) and mountain hares (Hulbert et al. 2001).

Chemical markers that are widely distributed in a diet can be used to measure digestibility. This application assumes that absorption of the markers is known or negligible and that concentrations in digesta or feces are directly related to that of the marker in the diet. Chemical analysis of minerals provides high sensitivities for measures of digestibility with poorly absorbed minerals such as manganese and titanium (Peddie et al. 1982; Van Soest 1994). Plants that are high in fiber often contain high concentrations of lignin, a plant secondary metabolite that can be used to measure digestibility in grazing animals (Fahey and Jung 1983). One assumption of this and other chemical methods is that the marker is unchanged by the process of digestion so that its behavior and detectability do not change between ingestion and defecation. Organic markers such as lignin may be vulnerable to fermentation, acidity and alkalinity as digesta pass through the herbivore digestive tract (Susmel and Stefanon 1993). Inorganic markers such as minerals are probably also vulnerable to digestion because they are associated with the organic matrix of the plant. Consequently, these indirect measures of digestibility should be validated against direct measures of total intake and output when either the diet or the digestive response is variable.

Extrinsic markers must be applied to foods evenly and consistently to avoid changes in concentration that are unrelated to changes in food intake or digestion. A single dose of a general marker such as food dye can be used to label one food item or an entire meal to measure digesta flow and digestive capacity of animals (see Chapter 5). Alternatively, repeated doses or incorporation of the marker into the diet of captive animals can be used to indirectly measure food intake from digestibility and fecal output. Large animals including ruminants can be implanted with controlled-release devices (CRD) that continuously release indigestible markers such as chromic oxide or odd-chain alkanes into the gut for up to 30 days. These devices have been used to measure food intakes of animals at pasture with minimal disturbance because the method only requires the collection of spontaneously voided feces once the marker has equilibrated in the digestive tract (Furnival et al. 1990a, 1990b). Food intakes of wapiti that were measured with CRDs demonstrated that food intakes of females are highest during lactation and lowest during late pregnancy (Gedir and Hudson 2000b).

Phase-specific markers are used to measure digesta flow and capacity of the digestive tract for important components of the diet. For example, labeled wax has been used to study the dynamics of stomach oils, a large energy depot in nesting seabirds (Roby et al. 1989). When measuring digesta kinetics, we assume that markers stick with their respective food residue or food fraction. Absorption of solute markers such as cobalt-EDTA and chromium-EDTA can reduce the concentration in the feces leading to the perception that the marker dose was diluted by food or digesta.

These fluid markers have been used to measure the response of the digestive tract to large volumes of water when desert goats, camels and donkeys return to watering holes and drink a volume equivalent to 25% of their body mass (Shkolnik et al. 1980). Particulate markers are rarely absorbed but may migrate between large and small particles, leading to erroneous conclusions on the rate of digesta flow if small particles move through the digestive tract faster or slower than large particles.

Fidelity of markers to particles is probably highest for mordants with minerals such as chromium. Large errors can be introduced by the use of markers with low fidelity. Foley and Hume (1987) found in the marsupial greater glider that the mean retention time (MRT) of large food particles marked with ruthenium-phenanthroline (which has been found to migrate from large to small particles that are selectively retained in the large cecum of this folivorous species) was 50h, but only 23h when marked with a chromium mordant. High fidelity of a mineral label may also increase the density of the particle and remove sites for microbial attachment, rendering the particle indigestible (Van Soest 1994). Because any digesta marker can fail to completely mimic the respective phase of the diet or digesta (Faichney 1975), it is important to understand the potential errors of each marker system in the context of its application.

An application of indigestible markers is demonstrated in Table 4.7. We calculate the food intake of a hypothetical ruminant with three indigestible markers: dietary fragments, an intrinsic digestibility marker and a CRD delivering an extrinsic marker at a constant rate into the rumen. The regular release of an extrinsic marker such as an alkane is used to calculate the amount of feces produced each day. This calculation relies on even and regular dosing with the alkane so that the concentration in the feces reaches a plateau when the marker is equilibrated with the digesta. Dose rates are determined in a subset of animals by recovering the CRD from the rumen. Equilibration of the marker can be confirmed by sampling feces at regular intervals to verify the rise to a plateau from dosing to 20 days. The time to equilibration depends on the TT of digesta (Fig. 4.2).

Fecal output is used to calculate food intake from the indigestibility of the diet determined using an intrinsic marker such as lignin. Indigestibility is calculated as Feces DM ÷ Food DM when total input and output are measured. The indigestible marker (X) serves as a reference for comparing discrete samples of feces with the diet. Indigestibility also can be calculated from the marker concentrations ($X \cdot DM^{-1}$) as $X \cdot Food\ DM^{-1} \div X \cdot Feces\ DM^{-1}$ which can be rearranged as Feces DM ÷ Food DM. Fecal output (grams per feces dry matter per day, g Feces $DM \cdot d^{-1}$) is divided by indigestibility to calculate the corresponding food intake (grams of food per day, g $Food \cdot d^{-1}$) (Table 4.7). We assume that the digestibility marker equally represents all food in the diet. This assumption holds when animals are fed a common diet with an extrinsic marker such as chromic oxide. If an intrinsic marker is used, the assumption is met when the marker concentration is similar between food items. If the marker varies with food item, then the dietary concentration of the marker is the weighted average of dietary items. Weighted concentrations are products of marker concentration and the proportion of that item in the diet: $X \cdot g\ Item^{-1} \times g\ Item \cdot g\ Diet\ DM^{-1}$.

The proportions of the diet can be determined by microhistology of plant fragments in feces. Fragment analysis calculates the proportional count. We assume that proportional counts and proportional mass are equal if the density of fragments is similar (grams per item, $g \cdot item^{-1}$). If the assumption of equal density of fragments does not hold, the proportions should be expressed on the basis of mass: ($count \cdot total\ count^{-1} \times g \cdot item\ count^{-1}$) ÷ Σ ($count \cdot total\ count^{-1} \times g \cdot item\ count^{-1}$). The proportions may be further corrected for the detectability of rare or highly degradable fragments produced from some items. The consumption of each item is calculated as the product of its proportion and the total intake of food DM. Microhistology measures have been used to assess the competition for food between herbivores in habitats with low plant production; between feral goats and the endangered yellow-footed rock wallaby

Table 4.7 Predicting food intakes from indigestible markers in a ruminant

Component	Calculation	Measure
Extrinsic marker (e.g., alkane)		
Daily marker dose rate	A	$100\,mg \cdot d^{-1}$
Fecal marker concentration	B	$0.20\,mg \cdot g^{-1}$
Fecal marker output	C = A ÷ B	$500\,g\ feces \cdot d^{-1}$
Intrinsic marker (e.g., lignin)		
Dietary concentration	D	$50\,mg \cdot g^{-1}$ food
Fecal concentration	E	$125\,mg \cdot g^{-1}$ feces
Indigestibility	F = D ÷ E	$0.40\,g \cdot g^{-1}$
Digestibility [1]	G = 1 – F	$0.60\,g \cdot g^{-1}$
Total food intake	H = C ÷ F	$1{,}250\,g \cdot d^{-1}$
Diet proportions (e.g., plant fragments)		
Food item I	I	$0.50\,g \cdot g^{-1}$diet
Food item II	J	$0.40\,g \cdot g^{-1}$diet
Food item III	K	$0.10\,g \cdot g^{-1}$diet
Intake of food item I	L = I × H	$625\,g \cdot d^{-1}$
Intake of food item II	M = J × H	$500\,g \cdot d^{-1}$
Intake of food item III	N = K × H	$125\,g \cdot d^{-1}$

[1]Digestibility = (Food DM– Feces DM) ÷ Food DM = 1 – (Feces DM ÷ Food DM) = 1 – Indigestibility.

during drought in Australia (Hume 1999; Hume et al. 2004); and between Arctic hares and muskoxen in the high arctic of Greenland (Klein and Bay 1994).

4.3.2 *Measuring Intake with Digestible Markers*

Materials that have been absorbed from the digestive tract appear in circulating blood plasma or lymph. The circulating materials are then incorporated into cells of muscle, adipose, liver and kidney or are deposited in products such as hair, eggs and milk. Circulating materials are also oxidized and then excreted from the body in expired air and urine. Many absorbed materials, even pollutants and toxins, can be used to track food consumption (Moriarty 1999). Radioactive fallout of cesium isotopes from nuclear testing has been used to estimate the consumption of lichens by reindeer and the subsequent consumption of those herbivores by wolves and humans (Holleman et al. 1979; Staaland et al. 1995). Substrates for energy metabolism (carbon compounds) or cellular constituents (e.g., nitrogen compounds) best represent the diet and are therefore used as digestible markers of foods. The relationship between the concentrations of a marker in tissues and the diet depends on food intake and the metabolic route of the marker. The metabolic routes of dietary carbohydrates, lipids and proteins are discussed in subsequent chapters, but the general patterns of exchanges are shown in Fig. 4.5. The concentration of the marker depends on the number of metabolic paths, the size of metabolic pools and the TTs of those pools.

Discrimination factors, which are metabolic corrections, are used to adjust for changes in concentration during routing through the animal. These discrimination factors are most effective when animals are at steady state, that is, when their metabolic response is not changing rapidly. Discrimination factors may vary with life stage, tissue and the composition of the diet because metabolism changes with those factors in all species. The concentration of a marker in a tissue (M_{tissue}) is expressed as the sum of multiple dietary contributions:

$$M_{tissue} = D_{tissue} + \sum P_i \times M_i$$

where 'D' is the discrimination factor between diet and tissue, 'P_i' is the dietary proportion of item 'i', and 'M_i' is the marker concentration of that item. The proportion of each item (P_i) in the diet is estimated by using the corresponding marker concentration in each item (M_i) to solve for M_{tissue}. These calculations rely on the accuracy of measuring marker concentrations in all dietary items and on applying the appropriate discrimination factors for tissues and the species. The number of items that can be resolved by this algebraic approach is one less than the number of markers.

Stable isotopes are used as digestible markers in a wide array of organisms for large-scale comparisons of food webs (Kelly 2000). They can be powerful tools in dietary reconstructions and in defining geographic, temporal and age-specific variation in diets. Isotope ratio analysis by mass spectrometry measures small changes in the ratio (< 0.001) of rare to abundant isotopes of nitrogen ($^{15}N/^{14}N$), carbon ($^{13}C/^{12}C$), hydrogen ($^{2}H/^{1}H$), oxygen ($^{18}O/^{16}O$), sulfur ($^{34}S/^{32}S$) and strontium ($^{87}Sr/^{86}Sr$) (Ehleringer and Rundel 1988). These analyses have contributed to surprising findings of significant amounts of whitebark pine seeds with elevated $^{34}S/^{32}S$ signatures in the diets of grizzly bears (Felicetti et al. 2003b) and of the use of different-sized streams with different geochemical signatures ($^{87}Sr/^{86}Sr$) by fish of different ages (Kennedy et al. 2005).

Although both rare and abundant isotopes follow the same metabolic paths, some reactions favor one isotope over the other which changes the isotope ratios in the pools of reactants and products. Ratios of $^{15}N/^{14}N$ in animal proteins increase as N is incorporated by each trophic level from primary producers to herbivores and carnivores (Martinez del Rio and Wolf 2005). Consequently, gray wolves would have higher muscle $^{15}N/^{14}N$ ratios than their ungulate prey (Urton and Hobson 2005). Conversely, synthesis of fatty acids from carbohydrate decreases the ratio of $^{13}C/^{12}C$ from the substrate to the lipid product (Stott et al. 1997). Isotope ratios can be used to track substrates from diet to tissue and among tissues by analyzing general classes of molecules (e.g., lipids, proteins) or specific molecules (e.g., amino acids in protein).

Dietary isotopes of N are found in the crude protein fraction of tissues and include protein, nucleic acids and their components (e.g., amino acids and bases). Dietary isotopes of H and O are associated with water (H_2O), carbon dioxide (CO_2) and exchangeable H in lipid and protein. Dietary isotopes of C can be deposited in carbohydrate, protein and lipid in tissues, whereas oxidation of C produces CO_2 in exhaled air. Isotope ratios of plants and animals vary with species, tissue, location and time of year. Seasonal changes in diets, therefore, can be reflected across antler and hoof tissues in ungulates and vibrissae in marine mammals and canids (Kurle and Worthy 2002). It is also possible that the tissues of an animal could vary isotopically even if it consumed the same types of plants and animals during the year (Barboza and Parker 2006). Food items that are collected for isotopic analysis should correspond with times and locations of tissues collected from animals, especially if the differences between items are small and the seasonal shifts in metabolism are large. It also is particularly important to use an appropriate discrimination factor, which may significantly influence estimates of dietary composition. Discrimination factors are used with isotopic values for foods and the consumer in models such as Isosource to quantify the diets of numerous species including furbearers, bears, wolves and waterfowl (Phillips and Gregg 2001; Phillips and Eldridge 2006).

The composition of fatty acids in the lipids of plants and animals is crucial to energy storage as well as membrane function. The pattern of fatty acids in the tissues of an animal therefore can indicate the source of dietary lipids. Fatty acid profiles in blood fractions, adipose tissues and milk have been used to determine the diets of several carnivores, frequently using Quantitative Fatty Acid Signature Analysis (QFASA) (Iverson et al. 2001, 2004). The technique works most effectively for simple-stomached species in which the dietary fatty acids pass into the circulation intact and are not modified by bacterial metabolism which can occur in foregut fermenters. Fatty acid analysis provides a larger number of markers for partitioning diets than isotope analysis because two groups of essential fatty acids (see Chapter 7) can be used as potential markers.

Discrimination of fatty acid patterns between food items is similar to the statistical methods used for the analysis of indigestible alkanes in feces (Cooper et al. 2005). Fatty acid analysis is best suited to diets that are high in lipid and to animals that store that lipid in tissues, such as the subcutaneous blubber in seals and sea lions. The differences between diet and the tissue profiles of fatty acids (i.e., discrimination factors or calibration coefficients) increase as the animal uses the dietary lipid for energy, synthesizes new lipids and modifies the dietary fatty acids before depositing them in tissues. Food items that are similar in fatty acid profiles are difficult to resolve, as are items that are similar in isotopic signatures. Consequently, the use of isotopes and fatty acids in concert may discriminate foods that are chemically similar but derived from different locations (Pond et al. 1995a). In both methodologies, samples from animals are analyzed using mixing models that optimize the solution to estimate the proportions of different items in the diet.

An application of digestible markers is demonstrated in Table 4.8. We calculate the intake of three foods by a hypothetical carnivore such as a seal with an energy

Table 4.8 Predicting food intakes from energy demand and dietary proportions

Component	Calculation	Measure
Energy demand	A	20,000 $kJ \cdot d^{-1}$
Dietary proportions (e.g., fatty acids)		
Proportion of dietary dry mass		
from item I	B	0.50 $g \cdot g^{-1}$
from item II	C	0.40 $g \cdot g^{-1}$
from item III	D	0.10 $g \cdot g^{-1}$
Metabolizable energy content		
of item I	E	20 $kJ \cdot g^{-1}$
of item II	F	15 $kJ \cdot g^{-1}$
of item III	G	30 $kJ \cdot g^{-1}$
Weighted energy density		
of item I	$H = E \times B$	10 $kJ \cdot g^{-1}$
of item II	$I = F \times C$	6 $kJ \cdot g^{-1}$
of item III	$J = G \times D$	3 $kJ \cdot g^{-1}$
Average energy density of diet	$K = H + I + J$	19 $kJ \cdot g^{-1}$
Proportion of dietary energy		
from item I	$L = H \div K$	0.53 $kJ \cdot kJ^{-1}$
from item II	$M = I \div K$	0.32 $kJ \cdot kJ^{-1}$
from item III	$N = J \div K$	0.16 $kJ \cdot kJ^{-1}$
Energy intake[1]		
from item I	$O = L \times A$	10,526 $kJ \cdot d^{-1}$
from item II	$P = M \times A$	6,316 $kJ \cdot d^{-1}$
from item III	$Q = N \times A$	3,158 $kJ \cdot d^{-1}$
Dry matter intake		
of item I	$R = O \div E$	526 $g \cdot d^{-1}$
of item II	$S = P \div F$	421 $g \cdot d^{-1}$
of item III	$T = Q \div G$	105 $g \cdot d^{-1}$
Total dry matter intake	$U = R + S + T$	1,053 $g \cdot d^{-1}$

[1]Assumes animal feeds to meet energy demands.

demand of 20,000 kJ•d^{-1}. This type of calculation has been used to determine the potential impact of grey seals on commercial fisheries (Bowen et al. 2006a) and of killer whales on Steller sea lions and sea otters (Williams et al. 2004a). Dietary proportions can be derived from the analysis of fatty acids or isotopes. Energy equivalents for each food are derived from metabolizable contents. The energetic value of each food is estimated from the composition of the food and the fuel values for carbohydrate, protein and lipid (see Chapter 1), or determined directly by holding animals in a metabolism cage to measure total energy ingested and excreted for each food item (see Table 4.3). The amount of energy contributed to each unit of dry mass in the diet is calculated as the weighted density of each item. The contribution of each item to the average energy density of the diet is therefore the proportion of dietary energy. Energy consumed from each prey item is the product of the proportion of dietary energy and the energy demand. These energy intakes can be converted to DM of each item using the metabolizable energy content of the food. The total DM intake on this diet is the sum of the DM from each item.

4.4 Summary: Food Consumption

Food intake is a central concept in animal nutrition, but it is one of the most difficult parameters to measure in wildlife. Measurements of food intake must be made with the animals in steady state. The time to reach steady state when transferred to a new diet or environment depends on several factors, including the complexity of the digestive tract, and rates of feeding and digesta turnover.

Techniques to measure food intake can be either direct or indirect. Direct measurements include behavioral observations of feeding animals (best suited to wild animals) or mass balance between food offered and food left at the end of the feeding cycle (used in captive animal studies). Indirect measurements of food intake use markers, which may be either digestible (such as stable isotopes or fatty acids that are incorporated into animal tissues) or indigestible (either a natural fraction of the diet that is not absorbed by the animal or a chemical that can be incorporated into the diet). A series of calculations show how the various approaches are used to establish food intake in wild and captive animals.

Chapter 5
Digestive Function

The common purpose of digestion is to supply the nutrients and energy required by the animal for maintenance and production. Physical and chemical structures of prey and plants must be disrupted because most dietary substrates are too large to be absorbed into the circulation of the animal. Digestion commonly includes both mechanical (e.g., chewing, grinding) and chemical (e.g., acidic and enzymatic) disruptions; these are part of the cost of handling food once it is captured. Handling costs are therefore investments of time, energy and nutrients in processing food.

Enzymes that are produced either by the animal or by microbes resident in the animal's digestive tract break down food particles to nutrient molecules small enough to diffuse or be transported across the digestive tract wall. Absorbed molecules are metabolized in the body by incorporation into tissues, oxidation to energy or excretion as waste. Often the process of digestion is multi-stepped, and the various steps take place sequentially along the tract. The morphology of the digestive tract of many animals is regionally specialized to store, mechanically degrade and/or selectively delay the passage of either solid particles or solutes (Stevens and Hume 1995). These various specializations enhance the overall digestive performance of animals by maximizing the rate of absorption of nutrients.

This chapter introduces general aspects of digestive function. Subsequent chapters focus on the digestion and metabolism of the main dietary substrates (carbohydrates, lipids and proteins).

5.1 Food Intake, Digestive Efficiency and Digestive Tract Capacity

The conversion of food to absorbable products depends on both digestive function and diet structure. Digestive efficiency is expressed as the digestibility coefficient, which is the ratio of nutrient or energy absorbed from the digestive tract to that ingested in the food (grams per gram or kilojoules per kilojoule, $g \cdot g^{-1}$ or $kJ \cdot kJ^{-1}$; Table 4.3) (Hume 2005). Digestibilities are high for foods that lack protective barriers such as tough skins, exoskeletons and cell walls. Bird and turtle nests are depredated

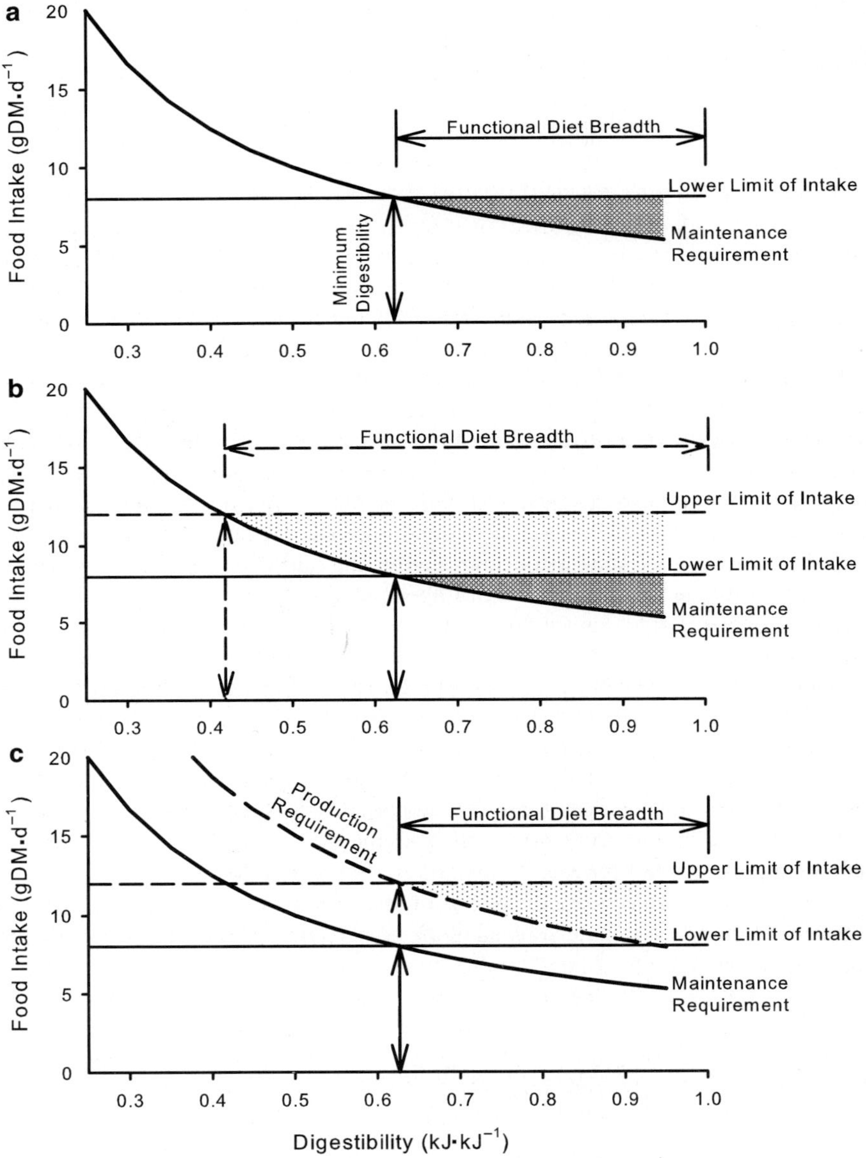

Fig. 5.1 Model of the relationship between food intake (g DM·d^{-1}) and energy digestibility (kJ·kJ^{-1}) in a small mammal (100 g body mass) consuming a diet containing 20 kJ·g^{-1} DM. *Curved lines* represent minimal food intakes for maintenance (100 kJ·d^{-1}; *solid*) and production (150 kJ·d^{-1}; *broken*). *Horizontal lines* represent the lower (*solid*) and upper (*broken*) limits to food intake due to environmental or physical/physiological constraints. **a** Minimum digestibility of the diet (0.625 kJ·kJ^{-1}; *vertical arrow*) is predicted at the intersection between the limit to intake and the requirement curve. The *shaded area* is the range of intakes and digestibilities that will meet or exceed the requirement within the limit of food intake. Functional diet breadth (*horizontal arrow*)

by foxes, raccoons, lizards and even the occasional herbivore because eggs are highly desirable foods. Egg proteins and lipids are easily digested by a wide range of animals once the egg shell is breached. Highly digestible foods are often high quality because they require minimum time and energy to convert them into the tissues of the animal. Foods of lower quality are more difficult to digest, but are usually more abundant in the environment than high-quality items (see Chapter 3). Diet breadth is the range of food qualities that an animal can use when foods vary in abundance (Fig. 3.3). Digestive function by the animal determines its ability to handle the breadth of physical and chemical characteristics of the diet, which is the functional diet breadth. The broad diet of white sharks, for example, is accommodated by a large, muscular and acidic stomach that can degrade skins, scales and fur surrounding the more digestible muscle tissue in small fish as well as large seals.

Digestibility determines the amount of food that an animal will need to meet its nutrient requirements for maintenance and production (Fig. 5.1). Food intake therefore increases exponentially as digestibility declines, but it is limited by the time available for foraging (Chapter 3), the capacity of the animal's digestive tract and the time for digestion. The minimum average digestibility that an animal can tolerate coincides with the highest food intake that will meet its requirements (Fig. 5.1a). Its functional diet breadth is the range of foods that can be digested at or above the minimum digestibility. The range of intakes that meets or exceeds the requirement increases as digestibility exceeds this minimum (Fig. 5.1a). By increasing the limit to food intake (Fig. 5.1b) (e.g., increasing foraging time or digestive capacity), the animal is able to use a greater range of digestibilities and thus tolerate lower-quality foods to meet the same requirement. The capacity of the digestive tract is usually larger in herbivores than in carnivores (Stevens and Hume 1995; Clements and Raubenheimer 2006). The large digestive capacity of herbivores relaxes the limit to food intake and allows these animals to process more food of lower digestibility.

With higher nutrient requirements for growth or reproduction (Fig. 5.1c), animals may be forced to use a narrower range of highly digestible foods unless the limit to food intake is increased by increasing foraging time, digestive capacity or the efficiency of the digestive system. Prairie voles increase the size of their digestive tract to accommodate greater food intakes when energy requirements are increased by low ambient temperatures in autumn and winter (Hammond 1993; Hammond and Wunder 1995). They also increase the capacity of their digestive tract (i.e., gut capacity) when the quality of the food is decreased by dilution with dietary fiber (Hammond and Wunder 1991). Other species such as American badgers that are

Fig. 5.1 (continued) is the range of digestibilities that the animal can use (0.625–1.00) to meet the requirement. **b** Increasing the limit to food intake increases the diet breadth of the animal and the range of intakes and digestibilities (now 0.415–1.00) that will satisfy the requirement. **c** Increasing the energy requirements of the animal decreases diet breadth. A 50% increase in requirements cannot be satisfied at the lower limit of food intake. An increase in the limit of food intake by 50% offsets the added requirement for production and conserves the same functional diet breadth required for maintenance exposed to periods of prey abundance followed by prey scarcity show increased retention time of food in the digestive tract that leads to more efficient extraction of nutrients when food supply is limited (Harlow 1981).

The gut capacity required to process food increases with higher energy requirements of the animal and decreases with higher digestibility of the food (Fig. 5.2). Increased food intake at the same digestibility increases the amount of both digestible and indigestible components of food in the digestive tract. The indigestible residues occupy more space as the digestibility of the food declines. Differences in gut capacity between levels of food intake are small when retention time in the tract is short and the contents of the digestive tract turn over rapidly. However, increasing the retention time rapidly increases the volume of digestive tract required to accommodate indigestible bulk. The maximum volume of the digestive tract rarely exceeds 25% of body mass in herbivores and is often less than 10% of body mass in most carnivores and omnivores. The maximum capacity of the digestive tract is determined by the space within the body cavity taken up by other organs (especially for egg production in fish, reptiles and birds, or for fetal development in placental mammals) and by the movement of the animal. The mass of the digestive tract alters the center of gravity and thus influences the musculature and movement of large herbivores such as gazelles and buffalo (Grand 1997). Consequently, increasing mass of the digestive tract may increase the cost of locomotion. Sedentary sloths (Family Bradypodidae) can hold more of a leafy diet than agile colobus and leaf

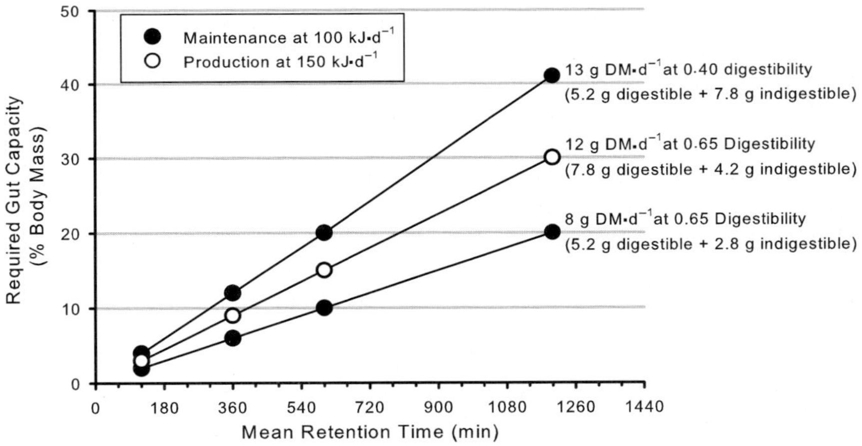

Fig. 5.2 Capacity of the digestive tract required for a small mammal (100 g body mass) using 100 kJ • d^{-1} for maintenance or 150 kJ • d^{-1} for growth or reproduction. The capacity required by the animal (% body mass) increases with food intake and with the load of digestible and indigestible digesta. Gut capacity also increases as food is retained in the digestive tract (min) for longer times each day (1,440 min). Food intakes and digestibilities are derived from the model in Fig. 5.1. Capacity was calculated by the method of Holleman and White (1989) from the daily output of DM in feces (indigestible residue) and mean retention time under the assumption that digestion follows an exponential decay. Total gut fill was calculated on the basis of 0.2 g DM • g^{-1} whole digesta and expressed as a proportion of whole body mass (100 g)

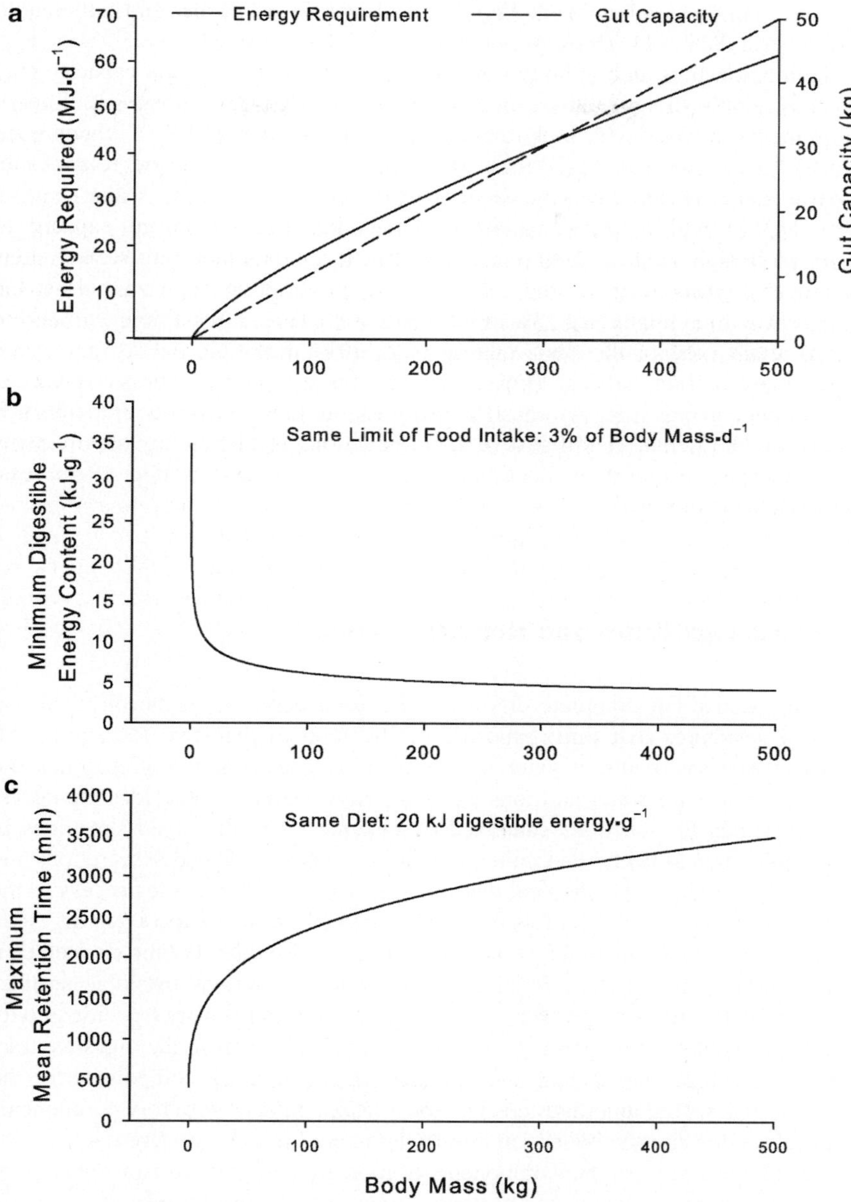

Fig. 5.3 Model of energy requirements and digestive tract capacity in mammals from 100 g (0.1 kg) to 500,000 g (500 kg). **a** Energy requirement was calculated as 2 × the average basal metabolic rate for mammals (2 × 293 kJ • $g^{-0.75}$ • d^{-1}) (Kleiber 1947). Digestive tract (gut) capacity was calculated as 10% of body mass with a dry matter content of 0.2 g DM • g^{-1} digesta. **b** The minimum energy content (kJ • g^{-1}) of the diet that meets the animal's requirements declines with increasing body mass when food intake is set to the same proportion of body mass (3% • d^{-1}). **c** Time (min) available for retention of the food increases with body mass when digestible energy content is set at 20 kJ • g^{-1} food

monkeys (Family Colobinae) and flightless herbivores such as rhea and ostrich hold more digesta than geese (Stevens and Hume 1995).

The relationships among body size, energy requirements and gut capacity may also constrain the diet breadth of small animals such as voles and songbirds. Energy requirements increase with body mass at a scalar less than 1 ($kg^{0.75}$) whereas gut capacity is isometric to body mass ($kg^{1.0}$) (Parra 1978). Thus the relationship between energy required and the capacity of the gut has a negative scalar ($kg^{0.75} \div kg^{1.0} = kg^{-0.25}$), which means that small animals must use a small gut capacity to satisfy their high requirements (Fig. 5.3a). Small animals therefore require diets with much higher concentrations of digestible energy than large animals at the same limit of food intake (e.g., 3% of body mass$\cdot d^{-1}$) (Fig. 5.3b). Large animals are able to retain food of the same quality (e.g., 20 kJ digestible energy$\cdot g^{-1}$) much longer than small animals can (Fig. 5.3c). Lemmings, voles, possums and koalas, for example, avoid some of these constraints on food retention time through increases in the size of their digestive tract, reductions in their energy requirements and/or changes in digestive function which maximize the digestion of the least fibrous fraction of the diet.

5.2 Reaction Rates and Retention Time

The time needed for complete digestion of a food depends on the physical and chemical structures that limit exposure of the food to digestive secretions. All digestive reactions require a water-soluble interface for the action of enzymes and acid. Solvation with water and ions from digestive secretions provides the necessary conditions for reactions catalyzed by enzymes. The dissolved substrates in liquid foods such as nectar and milk are rapidly exposed to digestive secretions and are quickly degraded. Mechanical disruption of solid foods provides access to the soluble components and increases the surface area of particles exposed to digestive secretions. The starch in a seed is more rapidly degraded than the fibrous seed coat because starch granules can be dissolved in these watery secretions. Insoluble particles such as the cell walls of plants vary in size, surface area and density (specific gravity or mass per volume, e.g., $g\cdot mL^{-1}$). The array of particle sizes in the digestive tract changes with both mechanical and chemical degradation as components of the particle are dissolved and digested. The most insoluble and refractory components of particles comprise the bulk of indigestible material in the digestive tract.

The soluble and insoluble components of food are a mixture of substrates, principally carbohydrates, lipids and proteins. Consequently, most foods are degraded by a series of reactions and enzymes rather than by just one process. All animals use endogenous enzymes produced by mucosal cells or secreted by the glands of the digestive system. Microbes in the digestive tract provide an additional source of exogenous enzymes that augment the activities of endogenous secretions. Foods comprised of readily soluble substrates that require only one or two enzymatic reactions are rapidly digested. Decreasing solubility and increasing complexity of digestion sequences with multiple enzymes slow down digestion rates (Fig. 5.4). Digestion of the soluble starch in a seed involves fewer steps than digestion of the complex fiber matrix in the seed coat. Fiber degradation requires fermentation with enzymes from several microbes as well as the secretion of buffers by the animal. Herbivores that utilize the energy in plant fiber must therefore accommodate slow digestion rates and a slow release of the abundant energy from structural carbohydrates.

Retention time is the average time available for digestion of a food. It is measured as 'mean retention time' (MRT), which is the average time that an oral pulse dose of an inert, indigestible marker takes to be eliminated in the feces. Retention time depends on the size of the animal's digestive tract and its food intake (Fig. 5.2). Reactions that proceed slowly require more time to reach the average digestibility required by the animal (Fig. 5.4). Animals with short retention time and small gut capacity must rely on high-quality foods with rapidly degraded substrates when their requirements and minimum digestibilities are high (Fig. 5.1). Consequently, they degrade only a small fraction of the most slowly degraded substrates such as structural carbohydrates in plant fiber. Animals do not need to completely digest all the substrates in their food so long as they digest enough for their requirements. Black and brown bears are classified as omnivores but they can be herbivorous; bears digest the soluble starch and protein in the cell contents of plants and relatively little of the fiber in the plant cell walls (Pritchard and Robbins 1990).

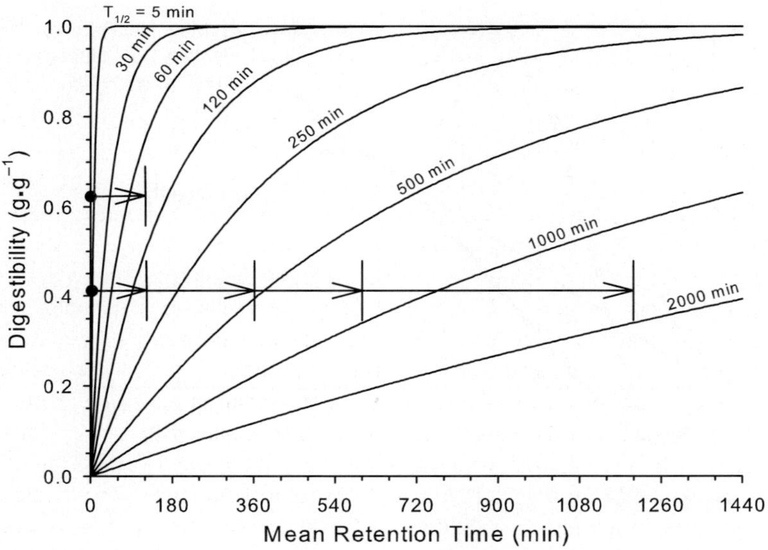

Fig. 5.4 Digestibility ($g \cdot g^{-1}$) of a substrate in relation to the available mean retention time (min) in an animal. *Solid lines* are reactions with different rates or half lives. The half life ($T_{1/2}$) of a reaction is the time required to convert 50% of the substrate in the food to products that can be absorbed from the digestive tract. *Arrows* indicate the range of reaction rates that correspond to the model of a small mammal in Figs 5.1 and 5.2. Increasing digesta retention time allows animals to use a broader range of substrates that involve slower rates of degradation and thus lower-quality foods

Finding and processing food both entail costs that must be met by the animal before energy and nutrients can be used to satisfy other requirements. The time, energy and nutrients invested in digestion are handling costs that can be optimized in a fashion similar to the selection of foraging patches by the marginal value theorem (see Chapter 3). The model in Fig. 5.5 predicts that retention time increases as diet quality declines because more time is required to recover handling costs and thus maximize the net rate of return from the food. Net gain from digestion meets or exceeds an animal's requirements when foods fall within its functional diet breadth (Fig. 5.1). Net gain decreases if the diet is diluted with slowly degraded substrates and if handling costs are higher. When the amount of fiber in the diet of a bear or a vole increases, food intake often increases as do the costs of maintaining the digestive tract. Added costs of high food intake include abrasion of the tract, higher secretory and absorptive activity as well as new tissue for enhancing digestive capacity. The metabolism of excess nutrients or toxins also contributes to handling costs. Substrates that are degraded too rapidly may similarly reduce the net gain to the animal by disproportionately increasing the handling costs for metabolism of absorbed nutrients. Digestive and metabolic functions therefore can impose an upper limit to diet breadth for some species best suited to slow rates of net gain (Fig. 5.1).

Readily digested substrates such as soluble carbohydrates can cause severe disturbances to fermentation in the digestive tract. Microbial enzymes convert these substrates to acids and gas that overwhelm the secretory and absorptive capability

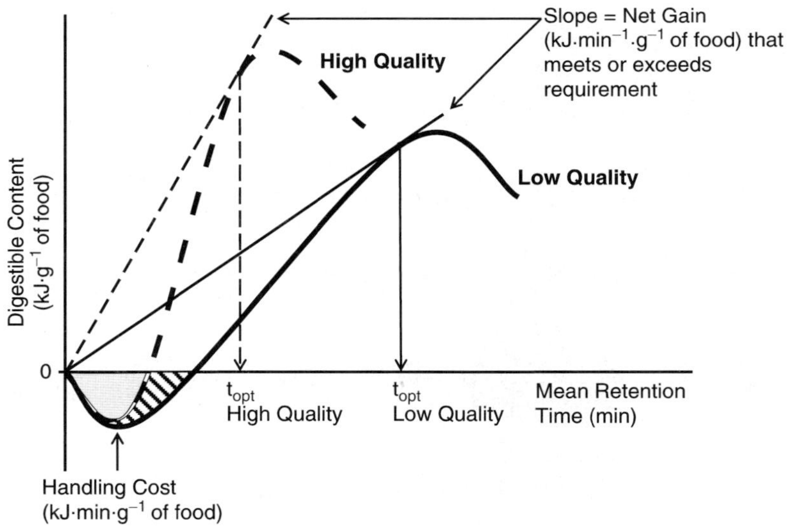

Fig. 5.5 Optimal mean retention time (t_{opt}; min) of two foods containing high and low concentrations of digestible energy. The amount of digestible energy released from the food increases to a maximum as substrates are degraded and falls as only the most slowly degraded substrates remain. The area below the *X axis* is the cost of handling the food in terms of energy and time (kJ•min•g^{-1}) for digestion and metabolism. The slope of the *line* from the origin to the tangent of the *curve* is the net gain of energy that is available to the animal. Adapted from Sibly (1981) and Hume (1989)

of the animal's digestive system. Foods that are high in fermentable starch can cause acidosis and bloat in captive ruminants (Owens et al. 1998), or in wild ruminants that have access to grain fields or are supplemented with concentrate rations during severe winters. The form and function of each digestive system influences the cost of handling foods and determines the range of foods that a species can use in its habitat.

5.3 Common Functions of Digestive Systems

Dissection of the digestive tract of most wildlife species reveals a complex and convoluted series of tubes and chambers that may be several times the length of the animal's body. The complexity and diversity of the vertebrate digestive system reflect the wide range of solutions to the common task of handling food. The harder it is to extract the required nutrients, the more complex is the digestive tract. Among carnivores, complexity of the digestive system increases from the simplest straight tube in parasitic lampreys that consume blood, to longer coiled tubes and the large stomachs of large carnivores and insectivores. Herbivores show even more complexity. However, in all vertebrates the digestive tract can be divided into four segments from mouth to anus: headgut (mouth and pharynx or throat), foregut (esophagus and stomach), midgut (small intestine) and hindgut (large intestine or cecum, colon and rectum) (Fig. 5.6). Mechanical degradation and solvation are

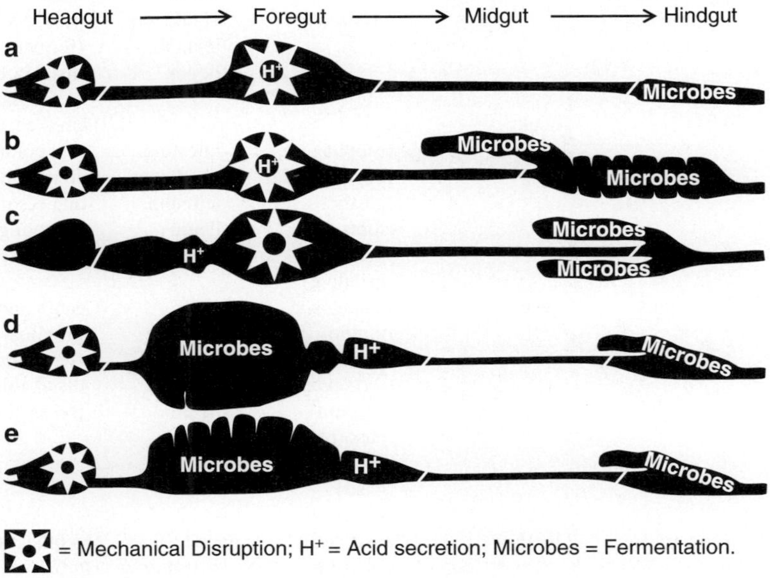

Fig. 5.6 Five general models of the vertebrate digestive tract indicating the sites of mechanical disruption, gastric acid secretion, and microbial fermentation. **a** Carnivore; **b** hindgut fermenter (e.g., rodent); **c** birds; **d** foregut fermenter (ruminant); **e** foregut fermenter (kangaroo)

emphasized in the headgut and foregut whereas solvation, absorption and retention are emphasized in the midgut and hindgut (Tables 5.1 and 5.2).

The headgut combines elements of the skeletal system with the soft tissues of the digestive tract. Mandibles, teeth and beaks seize food and begin mechanical disruption. Articulation of the lower mandible in mammals allows the teeth to function as a mill that provides the most thorough mechanical disruption among vertebrates. The absence of teeth in birds shifts the emphasis on grinding food particles to the gizzard (or ventriculus) in the foregut (Fig. 5.6). Grinding is enhanced by solvation with saliva in mammals and by acid treatment from the proventriculus in birds (Tables 5.1 and 5.2). Gastroliths (stomach stones) and grit in the gizzard of birds and in the stomach of crocodiles also enhance the disruption of food particles (Stevens and Hume 1995). Similar foods can be handled equally efficiently by different

Table 5.1 Representative animals for each model of the vertebrate digestive system, with brief descriptions of structures in each segment of the digestive tract

Model	Representatives	Headgut	Foregut	Midgut	Hindgut
A	Fish, reptiles, amphibians, carnivorous mammals	Teeth or beak (turtles) or pharyngeal plates and gill rakers (fish)	Esophagus, gastric (acid) stomach (except some fish)	Small intestine: duodenum, jejunum, ileum	Large intestine: small cecum and colon (resorption)
B	Mammalian and reptilian omnivores to herbivores	Teeth	Esophagus, gastric stomach	Small intestine: duodenum, jejunum, ileum	Large intestine: cecum and proximal colon (fermentation), distal colon (resorption)
C	Birds	Beak	Crop (esophagus), proventriculus (gastric stomach), gizzard (grinding)	Small intestine: duodenum, jejunum, ileum	Paired ceca and colon (fermentation and resorption) emptying into a cloaca
D	Ruminants	Teeth	Esophagus, rumino-reticulum (fermentation), omasum (absorption), abomasum (gastric stomach)	Small intestine: duodenum, jejunum, ileum	Cecum and proximal colon (fermentation), distal colon (resorption)
E	Kangaroos	Teeth	Haustrated colon-like forestomach, gastric stomach	Small intestine: duodenum, jejunum, ileum	Cecum and proximal colon (fermentation), distal colon (resorption)

Table 5.2 Sequences of digestive and metabolic functions for models of digestive systems (A–E) described in Fig. 5.6 and Table 5.1. Intensity of shading in each cell indicates the contribution of each segment to a particular function over the whole digestive tract. For example, the darkest cell in the row for absorption indicates that most absorption occurs in the midgut for all models

Function	Headgut	Foregut	Midgut	Hindgut
Mechanical disruption	A,B,D,E	A–C	All	All
Solvation	All	All	All	All
Chemical disruption – endogenous enzymes	All	All	All	All
Chemical disruption – microbial enzymes	All	D and E	All	All
Absorption	All	All	All	All

digestive mills: rats (teeth) and finches (gizzard; Family Fringillidae) both use seeds efficiently, and rabbits (teeth) and grouse (gizzard) are equally able to use the leaves of many plants. Mechanical disruption becomes more important to the animal with increasing insolubility and toughness of the food and with higher energy requirements (Schwenk and Rubega 2005). In nectarivorous birds such as hummingbirds (Family Trochilidae), the gizzard is greatly reduced because their liquid diets do not require mechanical disruption. In tortoises, low energy requirements facilitate long retention times that allow them to rely on chemical digestion of tough grasses in the absence of both teeth and gizzard (Barboza 1995).

The soft tissues of the vertebrate digestive tract are a series of variations on a long tube that is best represented by the midgut or small intestine (Fig. 5.7). Connective tissues of the serosa support the tract and allow the convoluted system to retain its organization as an animal moves. The serosa holds the gut in place when monkeys brachiate through the trees of a forest and when songbirds fly through dense thickets. The digestive tract is integrated with the rest of the body by vessels and nerves that run through the serosa to the submucosa. Connective tissue and muscle layers (muscularis) provide most of the tissue mass and thus the general appearance of the digestive tract on dissection (Fig. 5.6). Increasing the complexity of the serosa and muscularis expands the tube into fermentation chambers in the foregut of ruminants and kangaroos and in the proximal colon of zebras (Fig. 5.6, Table 5.1).

The digestive tract can be closed by contraction of a band of circular muscle (sphincter) or by opposition of muscle bands (valve). In mammals, closure of the gastric stomach confines the potentially harmful reactions with acid and permits vigorous mixing for mechanical disruption. Circular and longitudinal muscles act in opposition to move the digesta in waves of peristalsis (away from the mouth) or antiperistalsis (back towards the mouth) (Fig. 5.7). Peristaltic flow predominates in the midgut, but antiperistaltic flow is commonly used to regurgitate indigestible parts of prey in owls, deliver partially digested food to offspring in wolves and seabirds, and to eject potentially toxic foods in primates. In the hindgut, antiperistaltic flow often predominates, resulting in long retention times of digesta for microbial fermentation (Stevens and Hume 1995).

Muscle activity in the digestive tract is coordinated by nerves that convey information from stretch receptors in the muscularis to the central nervous system (CNS), and

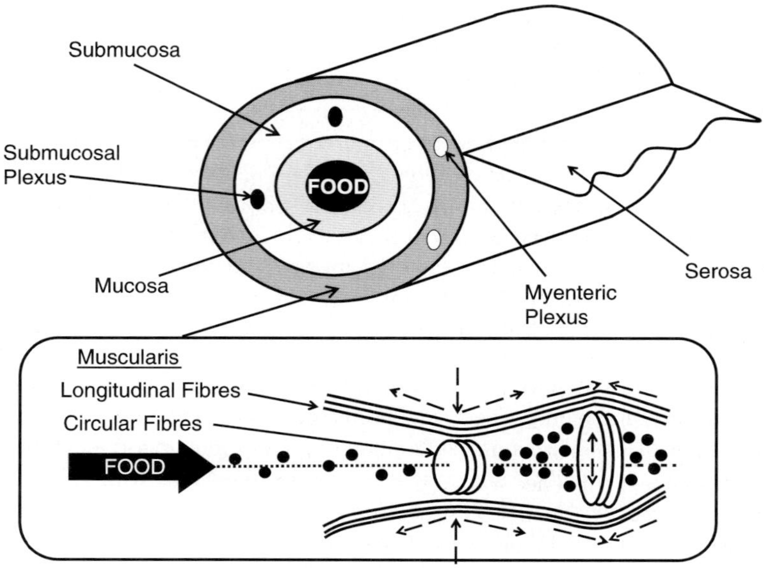

Fig. 5.7 Basic structure of the vertebrate midgut and the propulsion of food by the muscularis. Contraction of circular muscles (*vertical broken arrow*) coordinated with relaxation of longitudinal fibers narrows the tract and squeezes the contents. Conversely, longitudinal contraction and circular relaxation open the tract. Food is propelled by waves of closing and opening of the digestive tract

transmit responses back to muscle fibers (Holmgren and Holmberg 2005). Bundles of nerves (myenteric plexi) in the muscularis send messages to and from the gut (Fig. 5.7). Stretch receptors convey information on digesta fill in the tract and on the tension between muscle layers that stimulate mixing and provide feedback for peristalsis. Motility is controlled by autonomic systems that can be modulated by voluntary (conscious) control at the ends of the digestive tract (mouth, esophagus and anus). The neurotransmitters that effect motility of the gut have complementary actions throughout the body, especially at the heart and at the liver. Gut motility is therefore integrated with systemic energy demands that initiate foraging behavior as well as substrate utilization in the liver. Immobilizing drugs that are used to capture or restrain animals can affect gut motility and feeding because they act on receptors for common neurochemicals. Sedation with ketamine HCl increases motility and mixing of the digesta in the foregut of rufous rat-kangaroos for several hours (Hume et al. 1988).

The effects of anesthetics on gut function vary with the site of action and the time for clearance from the body. Gaseous anesthetics usually have the lowest effect on motility because the drugs are rapidly cleared from the body. Systemic immobilants that are reversed with antagonists may continue to affect the gut after the animal gains consciousness. Consequently, movements of animals that have been anesthetized may not represent normal foraging behavior for 1–2 days after release.

The mucosa is a thin interface between the body and the digesta that flow through the tract (Fig. 5.7). The underlying submucosa provides the infrastructure for secretion and absorption. Foregut regions that emphasize storage or fermentation must be able to stretch to accommodate a variable load of digesta. Consequently, the rumen of deer and the crop of geese are lined with a stratified squamous epithelium that is similar to the external skin of the animal. Foregut regions that are subject to abrasion such as the esophagus and the gizzard incorporate a resistant coat of protein (keratin) at the surface. The submucosa in the foregut includes goblet cells and glands that secrete mucus to lubricate the digesta and reduce abrasion. Regions such as the gizzard that are heavily fortified to resist grinding are too thick for effective absorption. However, small molecules such as short-chain fatty acids (SCFAs) produced from fermentation are absorbed across lightly keratinized epithelia in the rumen. This allows ruminants to mix a fibrous and potentially abrasive diet and still absorb most of the energy produced from fermentative digestion.

The absorptive segments of the tract in the midgut and the hindgut (cecum, colon and rectum) are typically lined with a columnar epithelium. These large columnar cells synthesize enzymes and transporters at the plasma membrane exposed to the digesta (the apical membrane). Midgut and hindgut absorption transports digestive products and also recovers digestive secretions. The absorptive surface of the digestive tract is the combination of nominal surface (calculated from the apparent length and surface area of segments at dissection) and the microscopic enhancement within each segment. Nominal areas are increased by ceca or diverticula of the midgut in rainbow trout, spiral midgut valves in sharks (Subclass Elasmobranchii) and by elongation of the distal colon in goats and wallaroos (Freudenberger 1992; Stevens and Hume 1995; Rust 2002). Micro-structural enhancement of surface area is found at three levels: folding of the submucosa and mucosa (mucosal folds), folding of the mucosa (villi) and folding of the plasma membrane on each cell (microvilli). Total surface areas for absorption may be 800 times the nominal surface area (Karasov and Hume 1997). Small seasonal changes in the length of the small intestine, therefore, rapidly increase the total surface area for absorption in small birds and mammals that increase food intakes to meet added requirements for thermoregulation in the cold (Karasov and McWilliams 2005; McWilliams and Karasov 2005).

Abrasion and chemical damage require regular replacement of the mucosa, especially the fragile columnar epithelia. The digestive tract therefore requires a steady supply of oxygen and substrates for maintenance, and is one of the most expensive organs in the body to maintain. The gut receives oxygenated blood from arteries in the serosa (Fig. 5.8). Substrates absorbed from the digesta may be used extensively for maintenance of the mucosa. However, most absorbed molecules are transported to either the blood or the lymph in the submucosa. Lipids and lipid-soluble molecules are transported by the lymph to tissues, whereas blood-borne molecules from the digestion of protein and carbohydrates are first transported to the liver by the portal vein. This difference in circulation reduces the number of metabolic conversions for dietary fatty acids and makes fatty acids useful dietary markers in studies of feeding behavior of marine mammals (see Chapter 4). Dietary lipids that are not immediately incorporated into tissues such as adipose tissue or egg yolk return in the venous

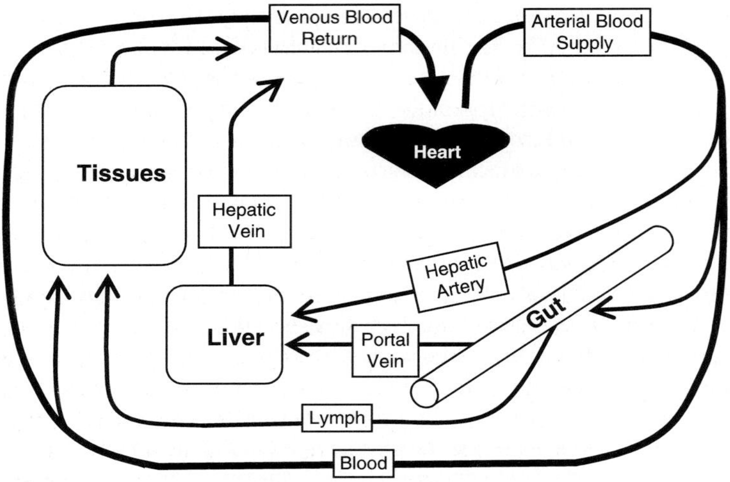

Fig. 5.8 Circulation of blood and lymph from the digestive tract (gut) to liver and tissues. Entero-hepatic circulation through the portal vein conveys water-soluble nutrients as well as resorbed secretions from the digestive tract to the liver. Lipid-soluble fractions are transported in the lymph from the digestive tract to tissues without passing through the liver. In birds, both lipid and water-soluble fractions enter the portal vein

blood and ultimately reach the liver via the hepatic artery. Entero-hepatic circulation also returns resorbed secretions from the digestive tract to the liver. The liver, therefore, plays a central role in modulating the metabolic cost of handling food by regulating intermediary metabolism of carbohydrates, lipids and proteins as well as some minerals and vitamins (Chapters 6–9).

Digestive secretions comprise a large proportion of metabolic handling costs because they involve the commitment of both energy and substrates. Secretions are produced by glands in the submucosa throughout the digestive tract and by glandular organs. The glandular organs include the salivary glands that empty into the headgut, and the liver and pancreas that empty into the midgut at the duodenum. Digestive secretions serve three functions: protection of the wall of the tract, maintenance of a suitable reactive medium and provision of digestive enzymes. Mucus coats the digesta and protects soft tissues against abrasion during mechanical disruption in the headgut and foregut, against chemical damage from acid in the gastric stomach and the duodenum, and against abrasion in the distal colon as water is removed from the digesta.

The largest component of digestive secretions is the watery mixes of ions (e.g., Na^+, Cl^- and HCO_3^-) that allow the animal to adjust the solute concentration (osmolality) and the pH of the digesta for reactions with enzymes. Reaction conditions are controlled by secretions and by absorption of the digestion products during the reaction. Control of secretions by the CNS is transmitted through submucosal nerve bundles (submucosal plexi; Fig. 5.7) as the chemosensors in the mucosa monitor the acidity or hydration of digesta. This control of reaction conditions by the animal maximizes the gain from a broad range of foods with the minimum investment in a set of catalytic tools (enzymes).

Endogenous enzymes are expressed from the genes of the animal, whereas exogenous enzymes are produced by a greater genetic diversity of microbes. Enzyme reactions run at either acidic (< 2–4) or near-neutral (7 ± 1) pH. Neutral conditions predominate in the headgut and from the midgut to the hindgut, whereas acid conditions are confined to the gastric stomach. Neutral pH is the least damaging to the mucosa because intracellular pH is near 7. Most endogenous enzymes are secreted into the midgut (Tables 5.1 and 5.2) at neutral pH. Similarly, neutral conditions favor microbial enzymes in both foregut and hindgut fermentations.

Secretion of HCl is costly because the digestive tract must produce enough acid to lower the pH of digesta, secrete mucus to protect the wall of the tract from the acid, and then produce enough alkali to return the digesta to neutral conditions in the midgut (Fig. 5.6). Gastric acids disrupt the proteins in skins of prey and cell walls of plants, which enhances mechanical disruption and therefore increases access to softer, more soluble tissues for further digestion in the midgut. In ruminants, the cost of acidifying digesta is reduced by resorbing water from the omasum to reduce the volume of digesta that is passed into the gastric stomach (the abomasum; Fig. 5.6, Tables 5.1 and 5.2). Acid is not always required for digestion, however; several abundant species of fish such as minnows and carp (Family Cyprinidae) as well as the monotremes (platypus and echidna) do not secrete acid in the foregut (Stevens and Hume 1995; Guillaume and Choubert 2001). The energetic cost of acidifying food may be prohibitively high for some marine fish that consume large amounts of alkaline calcium carbonate ($CaCO_3$) from corals and shelled molluscs.

5.4 Digesta Flow

The morphology and specialization of an animal's digestive tract affect the flow of food particles and solutes. Patterns of elimination of indigestible markers in the feces provide estimates of the mean retention time (MRT) and the role of structures such as tubes and chambers in controlling the movement of digesta. These measures can be used to assess changes in digestive function with different foods and levels of food intake. Indigestible markers are usually administered by dosing in a small amount of the usual food before presentation of the daily meal and subsequently collecting feces (see Chapter 4). Feces are collected at intervals determined by the body size and rate of food intake of the animal (Fig. 5.3). Sparrows (Family Passeridae) and ducks eliminate digesta markers within a few minutes of dosing (Karasov 1990), whereas markers only emerge after several hours in elephants (Foose 1982; Clauss et al. 2007) and in tortoises (Family Testudinidae) (Bjorndal 1989). These procedures are most easily conducted with captive animals. It is only feasible to measure marker elimination in wild animals if they are relatively sedentary (e.g., koalas) or return to a site such as dens or nests from which feces can be collected (Krockenberger and Hume 2007).

The simplest digestive tracts, including those of carnivores such as seals (Fig. 5.6a), have simple tubular systems through which digesta flow in a pulsatile fashion. That is, following an oral pulse dose of indigestible markers, the concentration of

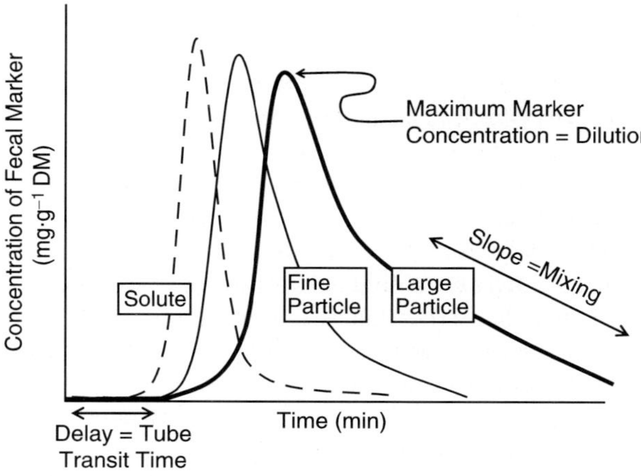

Fig. 5.9 Patterns of elimination of markers in solutes and on particles of different size from the digestive system of a carnivore typical of model **a** in Fig. 5.6

the markers in the feces rises and falls rapidly, forming distinct peaks after a delay (Fig. 5.9) (Trumble et al. 2003). Retention times in tubular segments are calculated from the first appearance of the marker in the feces. The declining slopes of marker concentration provide estimates of turnover times in mixing chambers. Mean retention time is calculated from the weighted average of the marker concentration in feces over time (Warner 1981) (Fig. 5.9).

Solutes emerge first, followed by particles of increasing size and density. This pattern indicates that the most digestible fraction of the diet in solutes is retained for the shortest time, which is consistent with predictions of optimal retention time (Fig. 5.5). Less digestible particulate fractions are delayed by the gastric mill, and by sedimentation in the intestine as the contents are propelled through the gut. Mixing in chambers such as the stomach broadens the marker elimination peak and decreases the slope of declining marker concentration. Animals that retain little digesta between meals provide the highest concentrations of fecal marker because the dose is minimally diluted by digesta in the tract. Herbivores with slower turnovers of gut contents and large digesta volumes dilute the marker dose, especially in the mixing chambers of the foregut and the hindgut (Fig. 5.6). Herbivores must handle increasing loads of insoluble particles as the quality and thus the digestibility of their food declines. Large fibrous particles occupy more space in the digestive tract than do dissolved substrates, so large particles have the highest handling costs.

5.4.1 Digesta Flow in the Foregut of Ruminants and Kangaroos

The cost of handling particles depends on the structure and function of the fermentation chambers. In ruminants, particles that enter the foregut fermentation are mixed in a large vat, the reticulo-rumen (Fig. 5.6, Table 5.1). Particles are degraded by rumination (regurgitation and rechewing) and by microbial fermentation in the rumen (Fig. 5.10a). Mixing of the vat turns over the entire contents and thus mixes both solutes and particles from one meal with residues from prior meals. Consequently, in ruminants, solute and particulate markers are eliminated as broad overlapping peaks (Fig. 5.11a).

The distribution of particles in the rumen depends on the structure of the plant cell walls. In large grazing ruminants such as muskoxen, long fibers of grass form a buoyant layer at the top of a sedimentation column, with fine, dense particles at the bottom of the rumen (Fig. 5.10a). Leaves of woody plants (browse species) and forbs have less recoil than those of grass (see Chapter 3), and are chewed into smaller, denser particles. Consequently, the ruminal contents of reindeer and other browsing ruminants tend to be less stratified in particle size (Clauss et al. 2003).

Particles can leave the rumen when they are reduced to a size that can pass through the orifice between the reticulum and the omasum. Large particles are therefore selectively delayed in the reticulo-rumen. However, on poor-quality diets, particle delay may exceed the functional capacity of the rumen, resulting in impaction and inhibition of food intake. These conditions also occur when starving ungulates that are supplemented with hay during winter still starve to death because of large particle size and lack of digestive adaptations to a new food source. The stomach of kangaroos does not restrict the size of particles passing from the forestomach fermentation because the system is not constrained by an omasal filter. Consequently, kangaroos can increase digesta flow to maintain food intake and avoid impaction

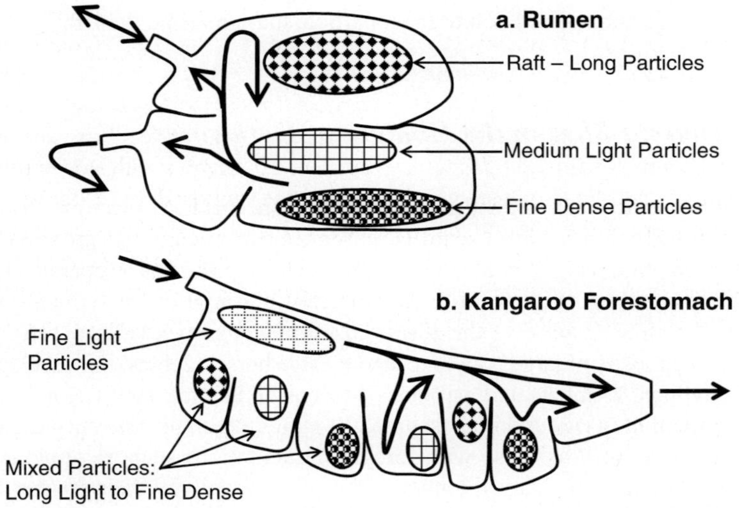

Fig. 5.10 Flow of solutes and particles through fermentation systems in the foregut of two grazing mammals: the rumen of muskoxen (**a**) and the forestomach of kangaroo (**b**)

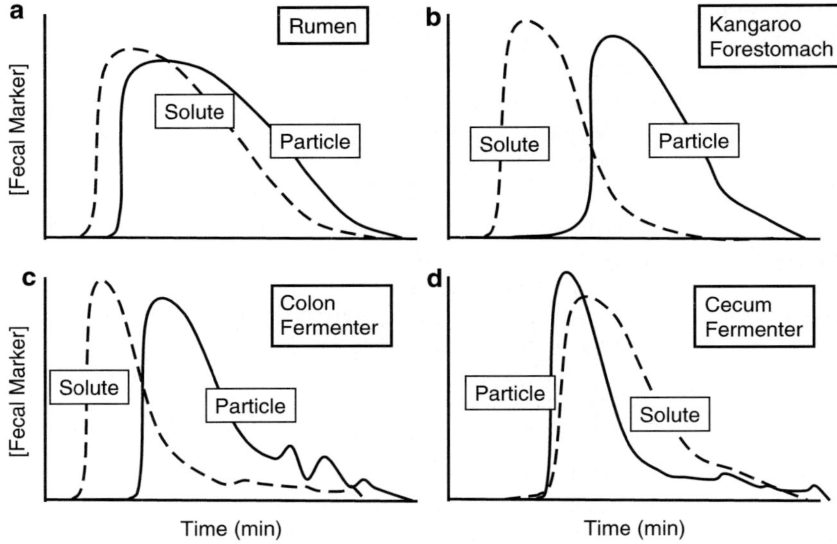

Fig. 5.11 Patterns of marker elimination from the digestive systems of herbivores with contrasting fermentation chambers. Foregut fermentations are represented by the rumen of elk or sheep (**a**) and the forestomach of kangaroos and wallabies (**b**). Hindgut fermentations are represented by the proximal colon of horses and wombats (colon fermenters, **c**) and by the cecum of rabbits and possums (cecum fermenters, **d**)

on low-quality food (Hume 1999). Because the kangaroo forestomach is an expanded tube (Fig. 5.10b), digesta are mixed locally rather than through the whole fermentation chamber. The predominant flow of digesta along the axis of the chamber results in clear separation of solute from particle markers (Fig. 5.11b).

5.4.2 Digesta Flow in the Hindgut of Herbivores

Hindgut fermentations can be divided into two types: one that takes place predominantly in the proximal colon or, if a cecum is present, in the cecum and proximal colon (colon fermentation) (Fig. 5.12a), and one that uses the cecum as a separate chamber (cecum fermentation) (Fig. 5.12b). Markers flow through colon fermenters (Fig. 5.11c) in a similar fashion to the forestomach of kangaroos (Fig. 5.11b). Colon fermenters are mostly of large body size, and include horses, zebras, rhinos, elephants, pandas, dugongs, wombats and ostriches (Stevens and Hume 1995). Some form of colonic fermentation is found in most animals, including omnivores such as javelina and red foxes, as well as those with foregut fermentations such as ruminants and kangaroos. Particles are delayed longer in the colon simply by their greater density compared with solutes (Figs. 5.11c and 5.12a). Small peaks of markers that follow the main peak of fecal appearance after an oral pulse dose indicate that digesta may be mixed in discontinuous pockets. As with the forestomach of kangaroos, colon

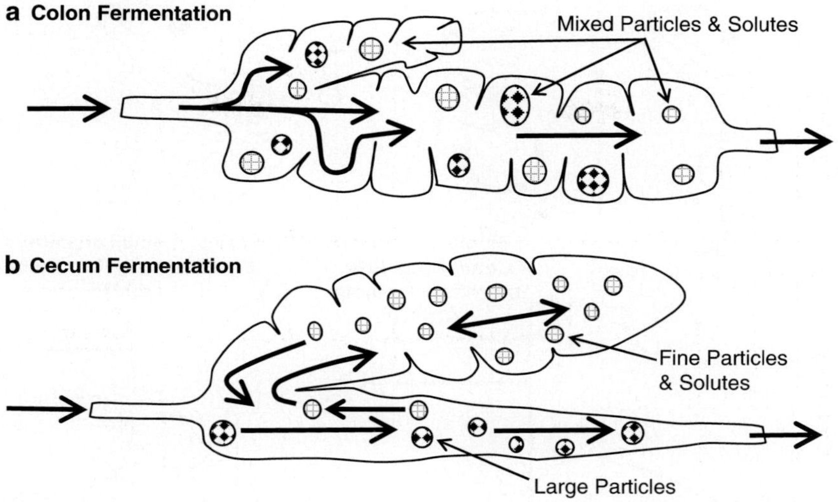

Fig. 5.12 Flow of solutes and particles through two fermentation systems in the mammalian hindgut: the colon fermentation of horses, elephants and javelina (**a**), and the cecum fermentation of rabbits, lemmings and possums (**b**)

fermenters can increase the flow of particles as indigestible fill increases to the limit of the proximal colon. Consequently, this system allows animals such as horses to increase food intake and thus maintain digestible intakes when food becomes more indigestible as plants senesce.

Cecum fermenters include a variety of small herbivores that can eat surprisingly poor-quality foods, such as coarse senescent grass and browse, in spite of their relatively small gut capacity (Fig. 5.3a). Examples of cecum fermenters are voles, lemmings, rabbits, marmots and possums as well as ptarmigan and geese. Cecum fermenters use retrograde flow to selectively retain solutes and fine particles in the cecum longer than large particles. These animals therefore reduce the cost of handling their fibrous diet by minimizing the amount of space allocated to large particles. Selective retention of solutes and small particles with large surface areas also increases their rate of gain because these are the most digestible components of the diet. Retrograde flow is facilitated by antiperistaltic motility and in some species, such as rabbits, hares and pikas, by fluid reflux that results from net secretion of water into the proximal colon and net absorption from the cecum (Cork et al. 1999). The solute marker emerges as a broad peak after the particle marker, indicating prolonged mixing and retention of solutes (and fine particles) in the cecum (Fig. 5.11d).

5.5 Optimizing Digestive Systems

The efficiency of tubes and chambers in a digestive system can be explored through chemical reactor theory (Penry and Jumars 1987) (Fig. 5.13). Three types of continuous flow reactors can be related to the flow of markers in Figs 5.10 and 5.11.

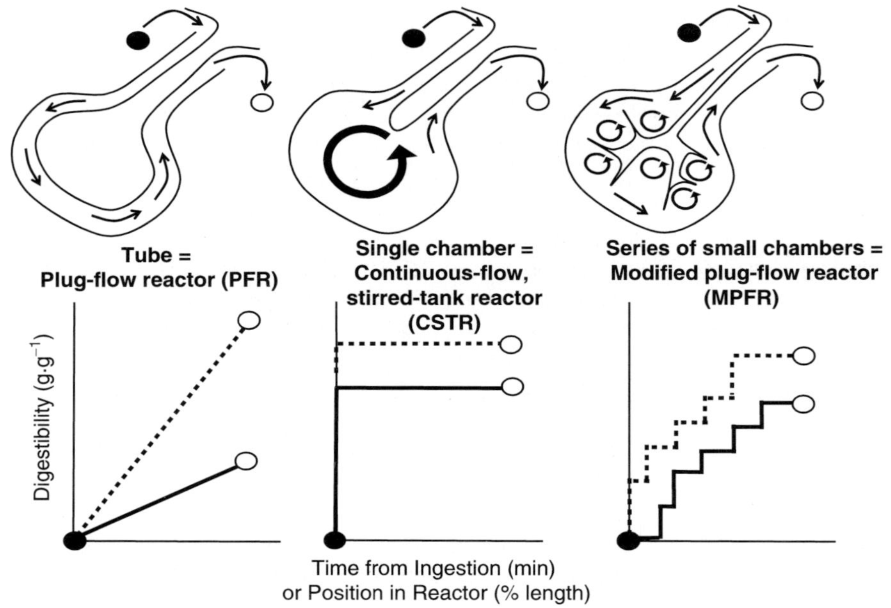

Fig. 5.13 Digestibility ($g \cdot g^{-1}$) of high-quality (*broken line*) and low-quality (*solid line*) foods in continuous-flow chemical reactors. (Adapted from Hume (1989) and Penry and Jumars (1987).)

Plug-flow reactors (PFRs) are analogous to the tubular system of the midgut in all vertebrate animals (Hume 1989; Horn and Messer 1992). Continuous-flow, stirred-tank reactors (CSTRs) are analogous to the single vat of the rumen, and to the solute fraction of the diet in the cecum of cecum fermenters. Modified plug-flow reactors (MPFRs) are a series of stirred-tank reactors analogous to the forestomach of kangaroos and the proximal colon of horses (Hume 1989).

Plug-flow reactors are best suited to high-quality diets and simple reactions with one or two enzymes. Absorption is also favored because they provide the maximum surface area for the uptake of products. Simple diets with substrates digested by endogenous enzymes therefore are handled most efficiently by the midgut, which dominates the digestive systems of carnivores and omnivores (Fig. 5.6). In rodents and birds, increases in the daily intake of simple foods are accommodated by elongation of the small intestine. Plug-flow reactors permit the highest flows and thus the highest food intakes. Ducks can attain daily food intakes that are several times higher than their own body mass on prey with indigestible shells (Jorde et al. 1995). Plug-flow reactors achieve the greatest yield of energy and nutrients of any system on simple foods, but more complex substrates such as plant fiber are better digested when reactants are retained for longer and are better mixed (Fig. 5.13).

Single chamber reactors (the CSTRs) can achieve higher digestibilities of low-quality plant diets because reactants are mixed continuously in a common pool and retention times can be long. However, incoming meals are diluted with undigested residues of previous meals in the rumen, which slows the rate of fermentation. Therefore long retention times are needed for maximum digestibility. In serial reaction chambers (the MPFRs), incoming meals are diluted with only a part of the indigestible residues from previous meals, so that fermentation rate is usually higher than in a CSTR but, because of faster flow of the digesta, extent of digestion may be lower. The plant fiber of low-quality diets therefore is digested better in ruminants than in kangaroos or horses, but kangaroos and horses can maintain food intake levels better as diet quality falls. Wildebeest (ruminants) and zebra (colon fermenters) share the same grasslands in the Serengeti by selecting different plant parts (Bell 1971). Zebras, as MPFRs, are better able to use the poorer-quality grass stems with low retention and high daily food intake. The higher-quality grass sheaths and blades are used by wildebeest, as CSTRs, at lower intakes and longer retention times (Bell 1971).

Ruminants are the most abundant herbivores between 15 and 1,000 kg because ruminal CSTRs deliver more energy from plant fiber at the lowest handling cost. Very large herbivores with serial chambers (MPFRs) retain food long enough to achieve the same digestibility as ruminants (CSTRs), and thus elephants and rhinos both handle plant fiber at low cost. Many herbivores with small body size use the cecum as a CSTR because that reactor provides the highest return from the minimal space and retention time, albeit at a lower digestibility of fiber than larger herbivores.

5.6 Summary: Digestive Function

The common purpose of digestion is to convert food into smaller nutrient molecules that can be absorbed across the wall of the digestive tract. The efficiency of digesting food (i.e., digestibility) depends on both the quality of the food and the digestive function of the consumer. Digestibility determines how much of a food will be needed to meet the animal's energy and nutrient requirements for maintenance and production. Functional diet breadth is determined by the lowest digestibility at which an animal can meet its requirements. Increases in gut capacity allow animals to process more food of lower digestibility, whereas increases in energy requirements (e.g., for growth or reproduction) may force an animal to select a narrow range of highly digestible foods.

The headgut (the mouth and pharynx) prepares food for digestion by mechanical disruption (chewing) and solvation. The foregut (esophagus and stomach) is also involved in mechanical disruption of food, and the initial stages of digestion. Digestion of simple foods by endogenous enzymes occurs mainly in the tubular midgut (small intestine). Digestion of plant material occurs by microbial enzymes in fermentation chambers that are expansions of either the foregut (forestomach) or hindgut (cecum and colon). Digesta flow affects the rates of digestion in the tubular midgut as well as in the fermentation chambers. A single mixing vat in the rumen provides more thorough digestion of fibrous particles, but a series of chambers in the colon allows faster flows of particles when food quality declines.

Part II
Substrates and Tissue Constituents

Chapter 6
Carbohydrates: Sugars, Fiber and Fermentation

6.1 Complementary Substrates for Metabolism

Dietary carbohydrates, proteins and lipids interact during both digestion and metabolism. Digestion is a complementary process that depends on mechanical and chemical degradation of all three substrates (Chapter 5). Digestibility of algae in fish, for example, is enhanced by degradation of proteins in cell membranes, which ruptures the cell and exposes the lipids and the carbohydrates in cell contents to other enzymes. Digestibility of the diet also can be reduced by waxy lipids on the surface of plants, fibrous carbohydrates in cell walls and tough proteins in connective tissue, which limit the access of digestive enzymes to their substrates.

The carbon in carbohydrates, lipids and proteins can be oxidized to release chemical energy through enzyme pathways such as glycolysis, the tricarboxylic acid (TCA) cycle and oxidative phosphorylation (Fig. 6.1). Some of this chemical energy is trapped in the form of energy currencies such as NADH, $FADH_2$ and ATP, which support physical activities such as locomotion and thermoregulation, and synthetic activities such as the production of feathers, eggs or milk (Fig. 6.1). The metabolic pathways involved in these transactions are complementary in that they allow protein to be converted to carbohydrate, and both carbohydrate and protein to be converted to lipid. These pathways allow species such as bears to convert the starch and sugar in berries to body fat in autumn, and to convert body protein to glucose for brain function during the winter fast. Metabolic routing allows animals to partition the use of substrates between oxidation for energy and incorporation into tissue; fat can be oxidized for energy while protein is reserved for cell maintenance in bears during the winter fast.

Oxidation of carbohydrates, protein and fat also produces water (Fig. 6.1). For example, fat oxidation can provide all the water required by bears during their winter fast. All animals require water as a medium or solvent for metabolism. Metabolic pathways are affected by the concentrations of substrates as well as other dissolved materials such as hydrogen and sodium ions. Membranes with embedded enzymes control the exchanges of both water and solutes between the animal and its environment and among the cells of the body to maintain homeostasis, a narrow range of internal conditions required for metabolism. For example, the release of

P.S. Barboza et al., *Integrative Wildlife Nutrition*,

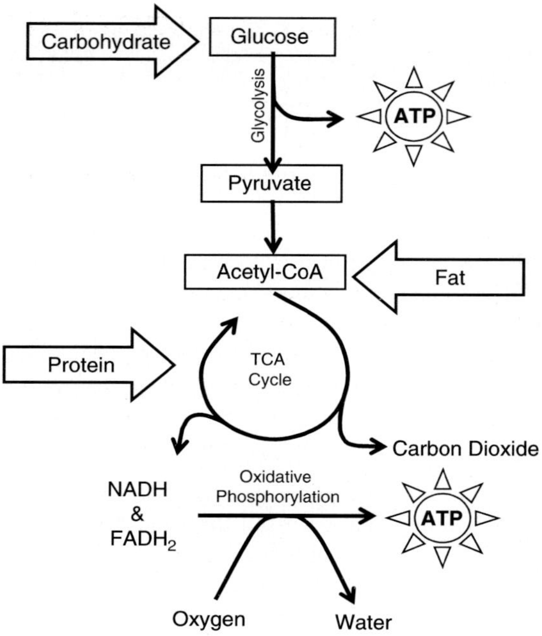

Fig. 6.1 Overview of pathways for converting carbon from carbohydrate, protein and fat into energy as ATP in animals

calcium from bone maintains the concentrations of calcium ions for muscle contraction in wintering bears. Trace nutrients participate in reactions as either an integral component of enzymes or as substrates. Bears that restore lost body mass by consuming salmon and berries in summer incorporate dietary protein and fat as well as vitamin C for building connective tissues and iron for building oxygen stores in blood and muscle.

Digestion, absorption and metabolism of carbohydrates, lipids and proteins are discussed in this chapter and the two chapters that follow. Water, and the minerals and vitamins required for metabolism are discussed in Chapter 9. The use of energy substrates for maintaining the function of animals and for the production of new tissues for growth and reproduction are discussed in Chapters 10 and 11 (Part III).

6.2 Functions of Carbohydrates

Most of the carbon that flows from plants to animals is in the form of carbohydrate because carbohydrates comprise most of the organic matter in plants (Chapter 1). The fibrous cell walls of terrestrial and aquatic plants are an abundant source of energy for herbivores such as moose, hippos and brant geese. Sugars in flowers and fruit entice animals to visit plants and to inadvertently disperse pollen and seeds.

Foraging activities of birds such as the nectarivorous rainbow lorikeet and the frugivorous king parrot distribute pollen and seeds, which ultimately sustains the population of trees from which the birds collect their sugary rewards (Jordano 1992). Starches in tubers, seeds and fruit may provide most of the dietary energy for a wide variety of omnivores including turkeys, raccoons and boars. Some species rely on carbohydrates in seeds and fruit for reproduction. The seasonal availability of acorns and beech nuts cues reproduction in dormice, increases the body mass of fawns born to roe deer, and affects the milk composition of black bears (Bieber 1998; McDonald and Fuller 2005; Kjellander et al. 2006).

Carbohydrates are involved in both the structure and function of animal tissues. Glucose and its metabolites provide the central axis for energy flow in animal cells. Carbon is stored as a readily usable form of carbohydrate in glycogen, which is released through oxidation of glucose in glycolysis and the TCA cycle (Fig. 6.1). Predators use the energy from glycogen in muscle to pursue their prey. Sugar chains also form the structure of chitin, which provides the framework for cuticle in invertebrate prey such as insects and shrimp. Genetic records of both predator and prey are based on the structural functions of sugars that link bases in DNA and RNA. The sequences of sugars in glycoproteins also identify cell surfaces for immune function and for exchanges of metabolites.

6.3 Functional Chemistry of Carbohydrates

Sugars are the smallest carbohydrates; compounds that contain three or four carbons are readily dissolved in water and are usually only found in metabolic pathways. Simple sugars with five carbons (e.g., xylose, ribose) and six carbons (e.g., glucose and fructose) are found in diets and in tissues as monomers (single units or monosaccharides) or dimers (two units or disaccharides). Complex sugars are formed by linking simple sugars to form polymers (>10 units or polysaccharides) that can be deposited as energy stores (e.g., amylose in starch granules) or structures for support (e.g., cellulose in plant cell walls).

Simple sugars are distinguished by small differences in the orientation and position of functional groups such as hydroxyls ($-H_2C-OH$) and aldehydes ($-HC=O$) that ultimately affect the structure and function of the whole molecule. A small shift in the position of the aldehyde group produces a six-sided ring in glucose and a five-sided ring in fructose (Fig. 6.2). Subtle differences in the structure alter the attraction of sugars to animals. In humans, among the monosaccharides fructose is sweeter than glucose, while among the disaccharides sucrose is sweeter than either maltose or lactose (McGee 2004). These differences in sweetness are not related to the energy content of the molecule because the number of oxidizable carbons is similar in each group (Blaxter 1989). Changes in the structure of carbohydrates in fruits favors dispersal of seeds by animals; ripening increases the palatability of fruits by altering the concentration and chemical structure of sugars once the seeds have completed development (McGee 2004).

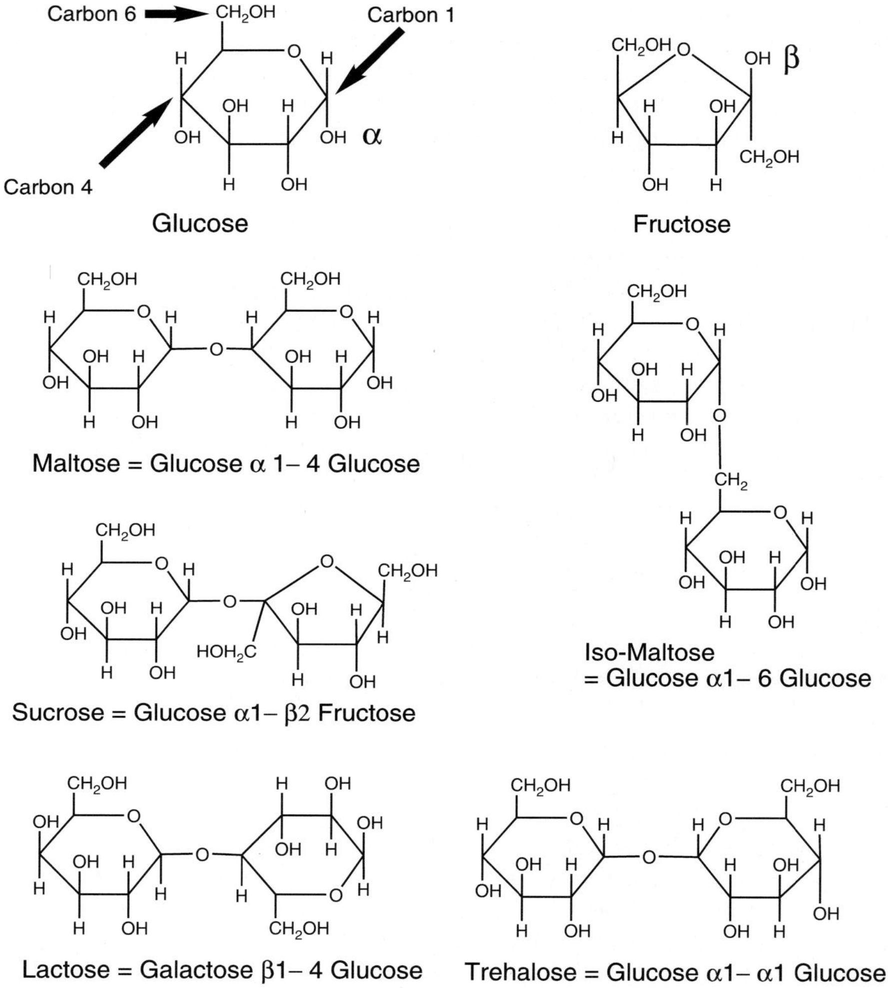

Fig. 6.2 Structures of simple sugars based on glucose and fructose found in foods. *Arrows* indicate carbon atoms that form glycosidic bonds with glucose. Linkages between carbons 1 and 4 can form repeating units for long chains, such as amylose in starch, which is a polymer of maltose. The orientation of the linking hydroxyl group (α or β) affects that of the next sugar; linkages with β hydroxyls invert the next molecule in lactose

Hydroxyl groups form glycosidic bonds between sugars; linkages are started from carbon 1 in glucose and carbon 2 in fructose. Glycosidic linkages between these starting carbons form stable disaccharides that are used to transport carbohydrate in a soluble form between tissues of both animals and plants: trehalose is a blood sugar in insects; lactose is the principal sugar in milk; sucrose is a sugar in the phloem of trees (maples), grasses (sugar canes), and tubers (beets) (Fig. 6.2). Linkages between carbon 1 and carbons 4 or 6 in glucose allow further linkages and the formation of polymers from maltose and iso-maltose (Fig. 6.2). Repeated linkages between maltose units form amylose chains, which are connected with iso-maltose residues to form amylopectin in starch and glycogen (Fig. 6.3a). The solubility of amylopectin affects the speed with which enzymes can release the glucose for energy. Long chains of amylose favor a solid granule of starch that can be packaged

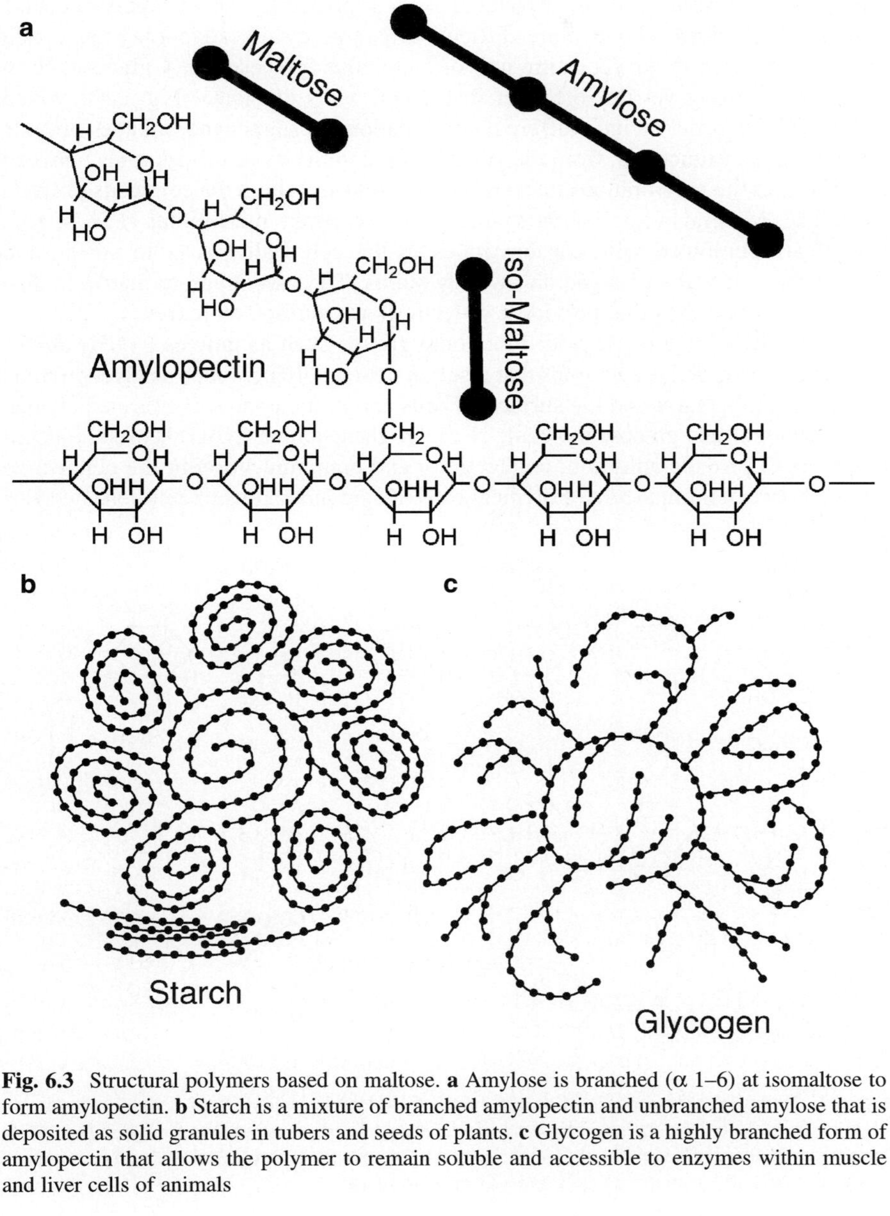

Fig. 6.3 Structural polymers based on maltose. **a** Amylose is branched (α 1–6) at isomaltose to form amylopectin. **b** Starch is a mixture of branched amylopectin and unbranched amylose that is deposited as solid granules in tubers and seeds of plants. **c** Glycogen is a highly branched form of amylopectin that allows the polymer to remain soluble and accessible to enzymes within muscle and liver cells of animals

in a coat of protein for storage until the seed germinates. Increasing the number of iso-maltose branches in amylopectin produces glycogen, which is more water-soluble than starch and readily available to enzymes in muscle for exercise (Fig. 6.3b).

The orientation of the hydroxyl on carbon 1 affects the glycosidic linkage; glycosidic bonds that start with a β hydroxyl create a tension that inverts the next sugar molecule (Fig. 6.2). Polymers with β1−4 linkages form chains that interdigitate and stack into insoluble fibers or plates. The close proximity between polymers also makes the glycosidic bond more difficult to access by digestive enzymes. Long chains of glucose in the repeating unit of cellobiose (glucose β1−4 glucose) form cellulose, a strong yet flexible fiber in the cell walls of plants (Fig. 6.4). Mixed polysaccharides that include different combinations of sugars and linkages provide a wide range of functions, from the gel of pectins in the connective tissues between plant cells to the membranous fibers of hemicelluloses within the cell walls of fruits and algae. Minerals (e.g., calcium) and plant secondary metabolites (PSMs; e.g., lignin) are combined with carbohydrates in the cell wall matrix to stiffen and defend the cell walls of algae and woody stems. The carbohydrate matrix is also combined with waxes that provide a water-resistant surface on leaves.

Modification of glucose with functional groups such as amines (−NH) allows connections to proteins and lipids that result in mixed polymers such as glycoproteins and glycolipids that lie on the surface of cells and in the connective tissue between cells. The modified glucose molecule N-acetyl glucosamine (NAG) is used in mixed polymers to provide adhesion, conductivity and communication in the connective tissues between animal cells. Polymers of NAG are analogous to cellulose because

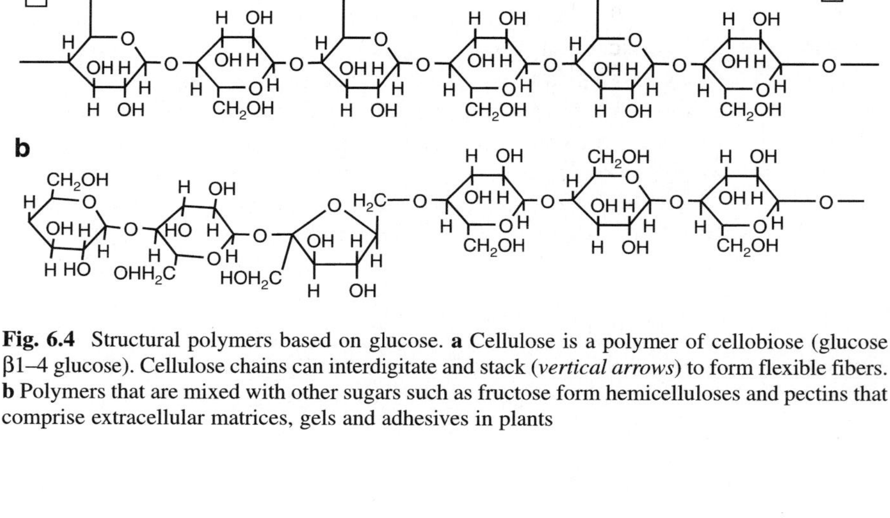

Fig. 6.4 Structural polymers based on glucose. **a** Cellulose is a polymer of cellobiose (glucose β1–4 glucose). Cellulose chains can interdigitate and stack (*vertical arrows*) to form flexible fibers. **b** Polymers that are mixed with other sugars such as fructose form hemicelluloses and pectins that comprise extracellular matrices, gels and adhesives in plants

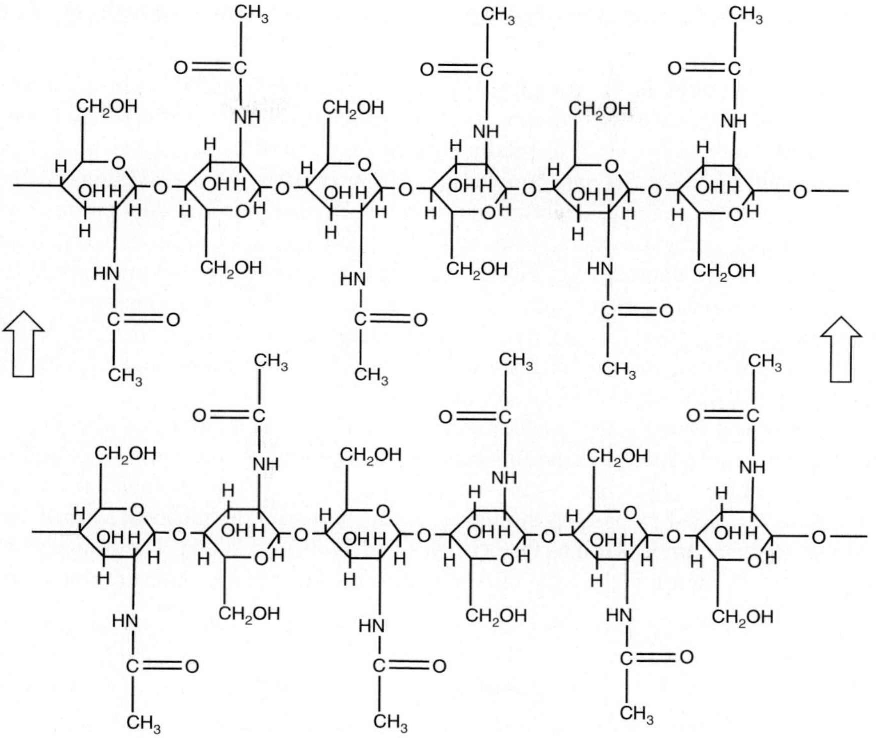

Fig. 6.5 Chitin is a polymer of chitobiose (N-acetyl glucosamine β1–4 N-acetyl glucosamine) that is found in the cuticle of invertebrates and the hyphae of fungi. *Vertical arrows* indicate stacking and interdigitation of adjacent polymers

they are based on the same β1–4 glycosidic linkage (Fig. 6.5). The repeating unit of chitobiose (NAG β1–4 NAG) produces chitin, which forms the exoskeletons of arthropods and the cell walls of fungi. Plates of chitin are coated with wax to provide a hard water-resistant cuticle in insects.

Non-structural sugars and starches provide most of the digestible energy in many seeds, tubers and fruits. Soluble sugars in plant secretions are easily accessible to digestive enzymes and are therefore quickly digested by a wide variety of animals from honey possums to hummingbirds (Family Trochilidae). However, most sugars and starches are contained within the cell contents of plants and are protected by the fibrous cell wall. Disruption of the plant cell wall by chewing, grinding and acid secretion usually precedes digestion of the sugars and starches in the cell contents. Protein coats on starch granules must also be ruptured before the starch can be dissolved and digested. These physical barriers often reduce the digestibility of non-structural carbohydrates, which ultimately allows the seeds of many plants to survive passage through the digestive tract and be dispersed in the animal's feces (Jordano 1992).

6.4 Digestion and Absorption of Non-Structural Carbohydrates

Digestive secretions in the foregut begin dissolving starch granules; in mammals this process is enhanced by salivary amylase. Amylases break the glycosidic bonds in amylose and amylopectin to produce shorter polymers as well as maltose, iso-maltose and glucose. Salivary amylases are destroyed by acid conditions in the stomach, but the pancreas secretes much more amylase into the duodenum (Fig. 6.6). Pancreatic amylase degrades the available amylopectin to maltose, iso-maltose and glucose. The maltoses are further degraded to glucose by enzyme complexes attached to the plasma membranes of the mucosal cells. The mucosal enzymes have multiple binding sites for substrates, which allow them to cleave the glycosidic bonds of more than one sugar. For example, maltose, iso-maltose and sucrose are all degraded by the same enzyme complex (Fig. 6.6).

Glucose and fructose are too large to diffuse through the membranes of cells, but they can be transported into the mucosal cell by specialized proteins embedded in each membrane. The sugar is moved through the cell (transcellular) from the digesta to the blood by passing through one transporter on the mucosal surface and another at the submucosal interface (Breves and Wolffram 2006). The transport of sugar requires energy in the form of ATP (active transport) or a concentration gradi-

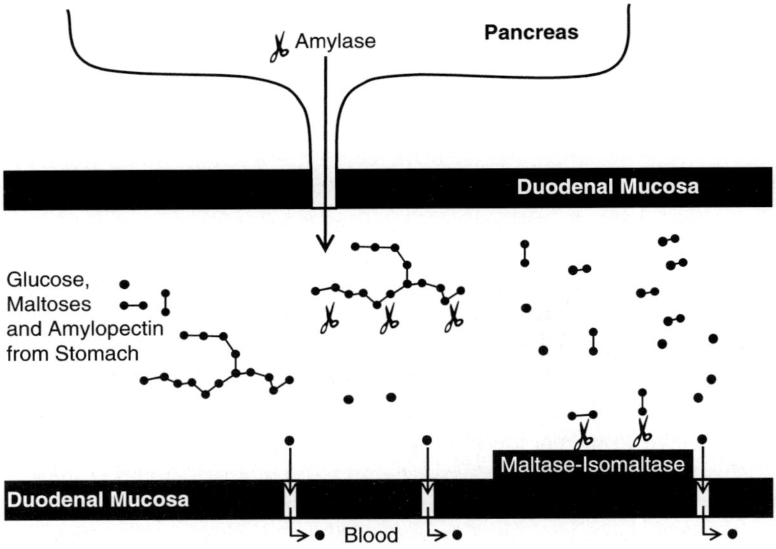

Fig. 6.6 Digestion of starch (amylopectin) releases the component sugars maltose (glucose α1–4 glucose), isomaltose (glucose α1–6 glucose) and glucose. In the small intestine, amylopectin chains are broken with amylase from the pancreas. Maltose and isomaltose are degraded to glucose by the maltase–isomaltase enzyme complex on the surface of the intestinal mucosa. Glucose is absorbed by either active (SGLT) or facilitated (GLUT) transporters at the mucosal surface and exported to the blood stream by facilitated transport

ent (facilitated diffusion). The SGLT transporters can absorb glucose at very low concentrations by using ATP that is linked to the sodium (Na) and potassium (K) exchange of the cell. Active transport allows the animal to absorb glucose from the digesta even when the concentration of glucose is lower than that of the blood; a grazing zebra can absorb glucose by active transport even though grass leaves contain little glucose. The GLUT family of transporters can only absorb glucose by facilitated diffusion along a concentration gradient. GLUT transporters can absorb fructose in mammals and birds because the concentration of fructose in blood is relatively low. The absorption of glucose with GLUT transporters relies on high concentrations of glucose in the digesta, such as during the digestion of starchy tubers by wild boars.

Facilitated diffusion cannot remove very low concentrations of glucose from the digesta, but it does provide a greater gain of energy than active transport because ATP is not involved. A third diffusion pathway (paracellular) is even cheaper because it moves glucose between mucosal cells (Karasov and Cork 1994; Chang et al. 2004) (Fig. 6.7). Paracellular transport of glucose through specialized junctions between the mucosal cells depends on the flow of water down a steep concentration gradient between the digesta and the blood (McWhorter 2005). Paracellular absorption of glucose has been reported in birds such as parrots, quail and sparrows that feed on nectar or starchy foods but has not been demonstrated in other vertebrates such as mammals or fish (Levey and Cipollini 1996; Clements and Raubenheimer 2006). This passive form of diffusion allows 70–80% of the glucose to be transported at the lowest cost before returning to the common transcellular routes of active and facilitated transport at lower concentrations of glucose in the digesta.

Fig. 6.7 Paracellular absorption can account for 80% of the glucose absorbed in the small intestine of the nectarivorous rainbow lorikeet (Karasov and Cork 1996)

Very little glucose or fructose escapes absorption along the small intestine because transporters are located on the same mucosal membranes as the enzymes that cleave disaccharides such as sucrose and maltose. Evolutionary selection for complete uptake of simple sugars is favored by the value of sugar as an energy substrate and also by the adverse consequences of allowing sugar to flow into the hindgut. Simple sugars such as sucrose and lactose can draw water from the blood into the digestive tract and thus prevent the normal resorption of water and ions, resulting in osmotic diarrhea. Rapid fermentation of the sugar in the hindgut can produce gas and acid that can further alter gut motility and absorption. Malabsorption of sugar may be caused by the genetic loss of enzyme activity during evolution or development. The omnivorous thrushes in the Family Muscicapidae do not express the gene for sucrase; common birds such as the American robin minimize their intake of fruits that are high in sucrose and thus minimize the consequences of malabsorption (Brugger 1992; Levey and del Rio 2001). All mammals express the enzyme lactase for digestion of milk sugar at birth (lactose; Fig. 6.2) but many species decrease the expression of the gene after weaning, when lactose is normally no longer encountered in the diet. Lactase activity of young mammals also varies with the concentration of lactose in milk among species. Consequently, insufficient lactase activity causes diarrhea and debility when orphaned wildlife is reared on artificial milk that does not match the lactose concentration of the natural milk.

Dietary sugars can modulate the expression of genes for both digestion and absorption in concert with structural changes in the digestive tract (Starck 2005). The induction of digestive changes varies widely among species, which probably reflects both diet breadth and genetic history. Increases in dietary starch and sugar decrease the activities of the corresponding digestive enzymes in carnivorous trout and have little effect on the enzymes of omnivorous catfish, but they increase those activities in omnivorous chickens, mice and rats (Karasov and Hume 1997). In migratory birds, the length of the small intestine as well as the number of enzymes and transporters on each mucosal cell can increase to augment digestive and absorptive capacities when fruit intake increases before migration (McWilliams and Karasov 2005). These digestive changes allow birds to minimize the time required to build body stores at the start of migration and to rebuild those stores at each staging ground. The flexibility of digestive and absorptive responses to changes in dietary substrates and food intake are discussed further in Chapter 11.

6.5 Glucose Metabolism and Homeostasis

Glucose is the focus of carbohydrate metabolism because most of the carbohydrates absorbed by animals are in the form of glucose from amylose, maltose, sucrose and lactose. Other sugars such as galactose and fructose also can be converted to intermediates of glucose metabolism (Fig. 6.8). The carbon in all these sugars can be oxidized to yield energy equivalents (ATP, NADH and NADPH) or stored as glycogen or fat.

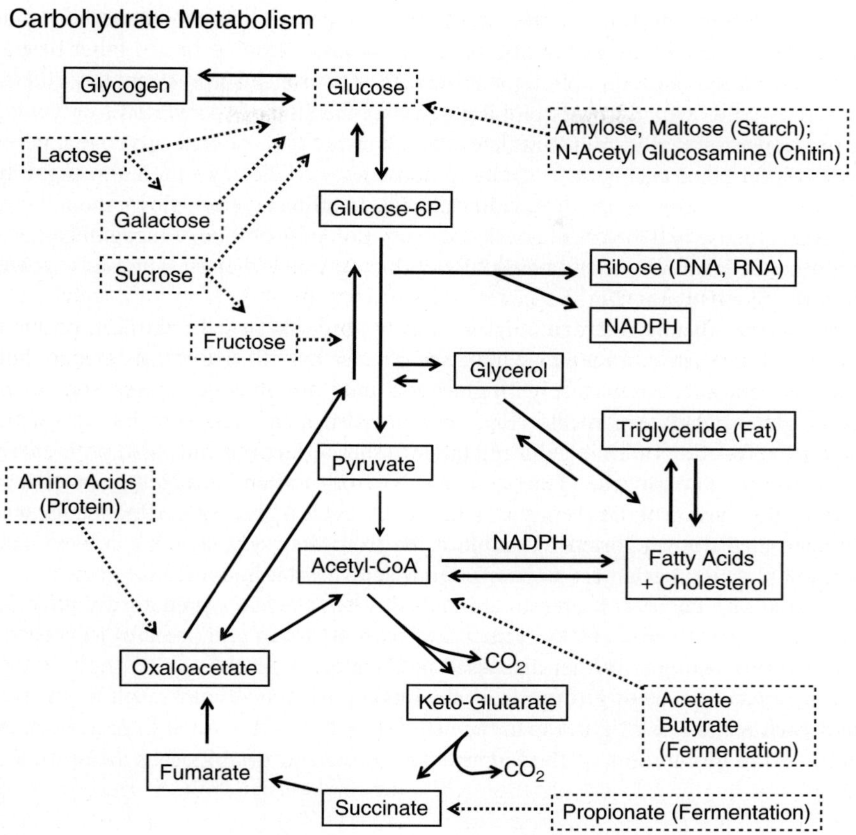

Fig. 6.8 Metabolic pathways for utilizing carbohydrates for energy or for storing the carbon from carbohydrate in fat (triglyceride) or glycogen. Carbon may be converted back to glucose from glycerol in fat or from amino acids via intermediates in the tricarboxylic acid (TCA) cycle

Absorbed glucose is an immediate source of energy for the gut mucosa because the sugar can quickly enter glycolysis and the TCA cycle (Figs 6.1 and 6.8). Mucosal cells export sugars to the blood by facilitated diffusion (GLUT transporters) at the submucosa. Dietary sugars are received at the liver (see Fig. 5.8), where they are converted to glucose. The glucose is used for three purposes: re-export to other tissues via the blood, oxidation of the carbon for current energy demands, and storage as either glycogen or fat. Circulating glucose is used for the energy demands of all cells, especially those of the brain and the rest of the nervous system as well as red blood cells. Glucose is stored in muscle and liver cells as both glycogen and fat, and in adipose cells throughout the body as fat. More carbon is stored as fat than as glycogen because fats form compact droplets of oil whereas glycogen is associated with water and occupies a larger volume within the cell (Fig. 6.3). Glucose is used to synthesize both the glycerol backbone and the fatty acid tails of triglycerides (triacylglyerols) as well as to produce NADPH that is used in the synthesis of fatty acids (Fig. 6.8).

Glucose concentrations in the blood are regulated by the pancreatic hormones insulin and glucagon to ensure adequate fuel supply to the brain and other tissues, and to avoid the toxic complications of excess glucose. Insulin stimulates glucose uptake by cells, which lowers blood glucose as the sugar is converted to glycogen or fat. Conversely, glucagon stimulates the release of glucose from glycogen, which raises blood glucose (Fig. 6.9). Blood glucose rises as dietary starch and sugar are digested and glucose is absorbed after a meal. Insulin secretion slows the increase in blood glucose and lowers the peak concentration in the blood. Glucagon balances the effect of insulin by maintaining blood glucose as the dietary glucose is cleared from the bloodstream (Fig. 6.9).

Anomalies of glucose regulation are rarely observed in wild animals, probably because inadequate storage or clearance of glucose has adverse effects that are both acute and chronic. Animals rely on glucagon and their glycogen stores to maintain blood glucose between meals (Fig. 6.9) because acute drops in blood glucose impair the function of brain cells and other tissues. Glucagon may also promote the formation of new glucose (gluconeogenesis) from the amino acids in protein by routing the carbon through parts of the TCA cycle (Fig. 6.8). Gluconeogenesis between meals is supported by amino acids from the body, which can eventually degrade muscle, the largest store of amino acids in the body. Consequently, poor regulation of glucose can alter an animal's ability to evade a predator by affecting both its behavior and mobility. The use of carbohydrates and proteins as fuels for exercise and fasting is further discussed in Chapters 8 and 10.

Inadequate uptake of glucose from the blood after a meal may result in glucosuria, which is the loss of glucose in the urine (Fig. 6.9). Glucose is filtered from the blood at the glomerulus of the kidney but returned to the blood by resorption in

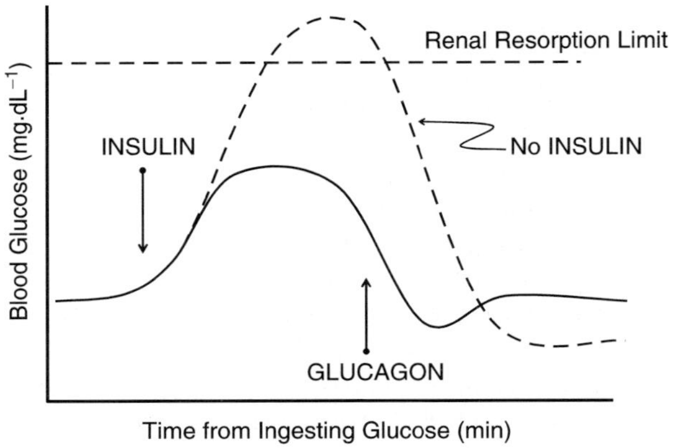

Fig. 6.9 Homeostatic control of blood glucose after a meal of carbohydrates. *Vertical arrows* indicate increases in the secretion of hormones from the pancreas. The *dotted line* is the response in blood glucose when insulin is not secreted. Insulin attenuates the rise in blood glucose whereas glucagon attenuates the fall in blood glucose. Glucose may be lost in the urine if insulin secretion is not sufficient to reduce blood glucose concentrations below the capacity for resorption at the renal tubules of the kidney

the renal tubules. Tubular resorption cannot return all the glucose to the blood when the concentration of the filtrate is too high or when the transporters are incapacitated by disease or by immobilants such as ketamine. Glucosuria may also increase the water lost in the urine (polyuria) because glucose can draw water from the blood into the filtrate and eventually dehydrate the animal if water is not freely available. Low secretion of insulin increases the frequency of glucosuria, which is a common symptom of diabetes mellitus in humans and domestic animals. Repeated exposure to high concentrations of glucose also increases the likelihood of glycosylation, which is the binding of glucose to proteins. Glycated proteins impair the functions of membranes, receptors, transporters and fine vessels, resulting in progressive damage to sensitive organs such as the eyes. Glycosylation of hemoglobin in blood cells indicates frequent exposure to high glucose in mammals and birds. Glycated hemoglobin in hummingbirds is lower than many mammals even though they consume large amounts of sugar that result in extremely high concentrations of blood glucose after a meal (Beuchat and Chong 1998). These long-lived birds must therefore avoid the chronic consequences of high blood glucose by rapidly using glucose for high energy demands and for storage as glycogen (Gass et al. 1999; Welch et al. 2006).

6.6 Digestion of Structural Carbohydrates

Structural carbohydrates form the matrix of plant cell walls that is commonly known as dietary fiber. Tubers, seeds and fruit may be high in carbohydrates such as starch or sugar but are usually low in fiber. Conversely, most of the carbohydrate in grasses and woody browse is fibrous (Fig. 6.10). Fiber can affect maximal food intake, digestibility and digesta flow of herbivores because the cell walls are slow to dissolve and degrade (Chapter 5). Plant cell walls are typically divided into three chemical entities using a system that sequentially removes fiber components with hot detergents at pH 7 and pH 2, and with strong acid (Van Soest et al. 1991) (Fig. 6.11). Neutral detergent dissolves the proteins, lipids, starch and simple sugars in the cell contents to leave a residue (neutral-detergent fiber or NDF) that approximates total cell walls. Neutral detergent may also remove some mixed polysaccharides such as pectin and β-glucan that form the gel-like connections between plant cells. Acid detergent removes mixed polysaccharides such as hemicellulose from the cell walls, which leaves a dense residue (acid-detergent fiber or ADF) of cellulose, lignin and mineral. Strong sulfuric acid dissolves the cellulose from the acid-detergent residue. Lignin is measured as the loss of organic matter from the acid residue during combustion. Lignin is not a carbohydrate but a polyphenolic complex that imparts rigidity to the fiber of stalks, stems and wood. The pliable fiber of tubers, seeds and fruits is lower in lignin than the rigid fibers of twigs in winter browse (Fig. 6.10). Plant senescence increases the lignin content of fiber, which can increase the time and energy required for chewing and digestion, and usually reduces the digestibility of the fiber (Chapter 3) (Van Soest 1994).

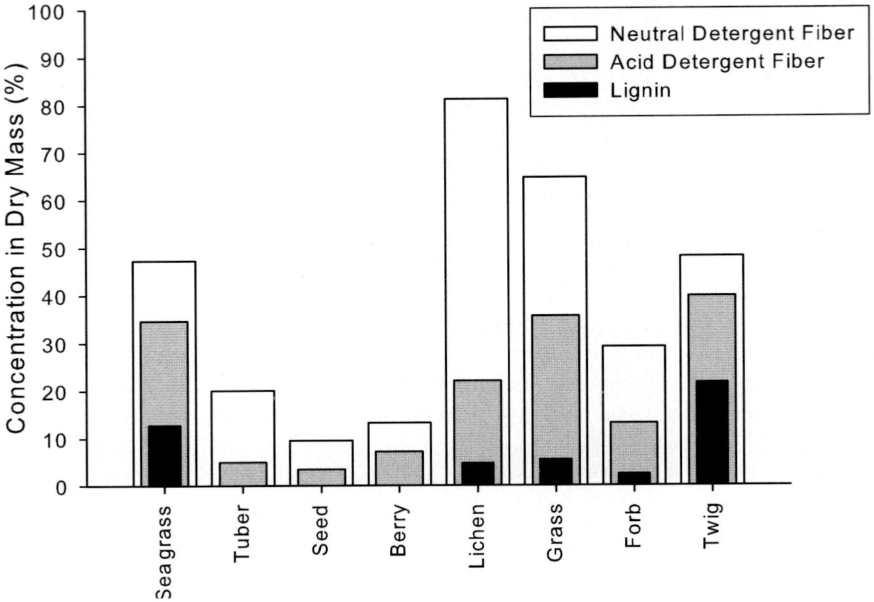

Fig. 6.10 Cell wall components of selected plants that were sequentially extracted with neutral detergent, acid detergent and strong acid (lignin) to measure fiber content (% DM). Plants are the same as those characterized in Figs 1.2–1.5: leaves of sea grass (Mason et al. 2006); tubers of sweet potato (National Research Council 2003); seeds of corn (National Research Council 1996); blueberries (National Research Council 2003); terricolous lichen (Barboza, unpublished); aerial parts of *Schismus* grass and the leaves of the forb globemallow in spring (Barboza 1995, 1996); twigs of Barclay willow in winter (Spaeth et al. 2002)

The content and composition of detergent fiber in temperate grasses and legumes are used to predict digestibility for domestic ruminants (Van Soest 1994; Huhtanen et al. 2006). Crude fiber is also used to evaluate the digestive responses of herbivores. Crude fiber is extracted with both acid and alkaline treatments, but the resulting residues are not the same as the detergent extracts of the plant cell walls. Fiber from woody plants, algae and fungi also may include resins and chitin in addition to several polysaccharides other than cellulose and hemicellulose. Detergent extracts of lichen, a combination of algae and fungi, are comprised mainly of mixed polymers of lichenans rather than the cellulose found in vascular plants (Elix 1996) (Figs 6.10 and 6.11). The same amount of detergent fiber from two different foods therefore may provide very different digestible intakes for an animal because of differences in the functional chemistry of the cell walls.

The chemistry of the fiber affects both solubility and flow through the digestive tract, which in turn affects the time available for enzymatic digestion in the animal. Soluble polymers (e.g., pectins from fruit or β-glucans from seeds) are more readily digested than solid polymers (e.g., crystalline cellulose and hemicellulose), and are more easily moved and mixed with the fluid phase. Soluble fibers and fine particles therefore provide the greatest rate of return of nutrients and energy within the limited

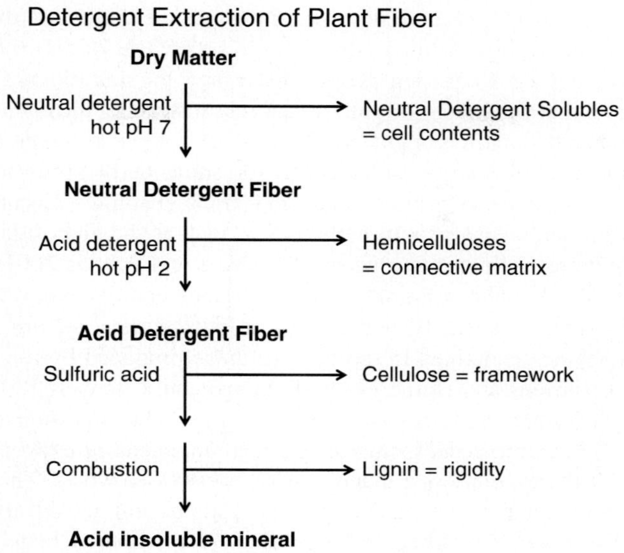

Fig. 6.11 Sequential extraction of plant fiber by the detergent method (Van Soest et al. 1991). Neutral detergent dissolves protein, lipid, simple sugars and starch within the plant cell, leaving the fibrous cell wall. Mixed polymers of carbohydrate such as alginates and hemicelluloses in the plant cell wall are dissolved with acid detergent to leave the structural framework of cellulose and lignin

volume of the digestive tract in small herbivores such as rabbits and voles (Chapter 5). Digestion of bulky fiber requires a large digestive tract to retain the fiber for two processes: exposure of the surface of molecules by chemical or mechanical degradation, and enzymatic digestion. Hemicelluloses have been found to be more digestible than cellulose in several birds including waterfowl (Australian wood duck) and ratites (ostrich), probably because the combination of acid from the proventriculus and grinding in the gizzard dissolves these components of neutral-detergent fiber before they are fermented in the ceca (Dawson et al. 1989; Swart et al. 1993a; Vispo and Karasov 1997; Dawson et al. 2000).

Complete digestion of structural polysaccharides is similar to that of starch because the polymers are degraded by a sequence of enzymes (Fig. 6.6). Long polymers are first cleaved by endo-enzymes; mixed polymers are broken by laminarinase, lichenanse and xylanase; cellulose is broken down by cellulase, and chitin by chitinase. The shorter chains are further degraded to simple sugars: cellulose fragments are degraded to cellobiose (glucose β1–4 glucose) by cellobiohydrase, and then to glucose by β-glucosidase; chitin fragments are similarly degraded to N-acetyl-glucosamine by chitobiase.

Secretion of endo-enzymes allows the animal to disrupt the structural barriers of plant cell walls or insect cuticle. Chitinase is produced in the pancreas and/or the stomach mucosa in a wide variety of insectivorous and omnivorous fish, amphibians, reptiles, mammals and birds (Stevens and Hume 1995; Gutowska et al. 2004).

Animals that express chitinase do not always express the chitobiase that would be required for complete degradation of the chitin (Karasov and Hume 1997; Clements and Raubenheimer 2006), which means that many animals do not completely degrade the polymer for energy or glucose but rely instead on the cell contents that are released when the chitin is degraded. Cellulose digestion, however, may provide significant amounts of energy and glucose for some herbivorous invertebrates. Edible snails, subterranean termites and land crabs (Families Gecarcinidae and Coenobitidae) produce endogenous cellulases as well as the glucosidases required to completely digest cellulose to glucose (Watanabe and Tokuda 2001; Linton and Greenaway 2007). Vertebrate animals do not produce endo-enzymes for cellulose and mixed polymers. Rather, fiber digestion in fish, reptiles, mammals and birds depends on enzymes contained in the food or those produced by microbes in the digestive tract (Stevens and Hume 1995; Guillaume and Choubert 2001).

Microbial fermentation is responsible for most of the digestion of structural carbohydrates in wildlife because microbes can complement the enzymes expressed by the animal and also express a much wider variety of enzymes. This mutualism expands the nutritional niche of the microbe and the animal by allowing both partners to use a larger number of foods that are distributed in more habitats. Consumption of invertebrates by seabirds and whales favors digestion with both endogenous and microbial sources of chitinase and chitobiase because both microbe and host benefit from the abundant source of carbon (Jackson et al. 1992; Olsen et al. 2000). Digestion of plant fiber is also a complementary process because it is achieved by a community of microbes rather than any one organism. The microbial community colonizes the surface of particles and soluble polymers to form a bio-film. Bacterial degradation of cellulose involves multifunction enzyme complexes called cellulosomes that bind to the fiber surface and catalyze a series of reactions (Forsberg et al. 2000). Large numbers of cellulolytic bacteria interact with fungi that open the cell wall matrix and with other bacteria that produce branched chain acids, which are essential for growth of the cellulolytic bacteria (Russell 2002) (Table 6.1). The digestion of plant fiber therefore depends on a suitable environment to sustain a microbial community that expresses the enzymes for converting cellulose and mixed polymers to short-chain fatty acids (SCFA) that can be absorbed by the host.

Table 6.1 Abundance and common functions of microbial groups in anaerobic fermentations of ruminants (Hespell et al. 1997; Mackie et al. 2000)

Microbe	Abundance (counts·g^{-1} digesta)	Common functions
Eubacteria	10^9–10^{11}	Carbohydrate, lipid and protein digestion
Archaebacteria	10^8–10^9	Hydrogen consumption and methane production
Protozoa	10^4–10^6	Bacterial predation, starch digestion
Fungi (zoospores)	10^3–10^5	Invasion of ligno-cellulose matrix
Viruses (bacteriophages)	10^7–10^9	Infection of bacteria and protozoa

6.7 Microbial Fermentation

Microbial fermentation is best suited to regions in the gut where conditions of pH, hydration and temperature are relatively constant. Secretions of buffers and absorption of acids maintain a pH close to 7 for fermentation in both the foregut and the hindgut. Acidic conditions in the gastric stomach inhibit the activities of all but a few bacteria; the microbial community that digests cellulose is inhibited when digesta pH falls below 5 (Russell and Wilson 1996). Microbial numbers are reduced in the duodenum by the addition of bile and alkali, and in the distal colon by removal of water and ions. Changes in body temperature of the host affect the rate of fermentation, especially in fish and reptiles. Marine iguanas feed on seaweed in the cold water around the Galapagos islands but bask in the sun after each foraging bout to raise body temperatures to levels that favor fermentation (Wikelski and Carbone 2004).

Digesta flow affects the substrates exposed to fermentation and the distribution of microbes. Particle retention increases the time available to colonize and degrade bulky fiber in the capacious rumen of bison or finer fiber with large surface areas in the smaller cecal volume of voles. Microbes associate with both fluid and particulate phases of the food as well as with the mucosal surface of the tract. Bacteria that colonize the surface of large particles are responsible for most of the cellulose digestion whereas the solute-associated bacteria digest colloidal polymers such as pectin and starch. Some microbes can move independently of the digesta flow: ciliated protozoa can move from the mucosal surface to the fluid phase during each meal; fungal spores are also motile until they affix to particles of food to become sporangia. Rapid digesta flow through the small intestine usually limits the microbial population to those microbes associated with the mucosal surface. Retrograde flow (antiperistalsis) and expansion of the intestine reduces the wash-out of microbes and substrates to allow emus and red-bellied turtles to gain energy from the fermentation of fiber in the distal small intestine (ileum) (Herd and Dawson 1984; Bjorndal and Bolten 1990).

Each population of microbes expands to fill a niche, which varies with the food selected by the animal and with conditions in the digestive tract (Russell and Rychlik 2001). Microbial populations persist in a fermentation chamber when the rate of replacement exceeds the rate of loss, just as rates of birth and death balance an animal population at equilibrium (Chapter 2; Eq. 2.1). Bacteria are the most abundant microbes in the fermentation chamber and are the largest proportion of the microbial biomass at the base of the trophic hierarchy within the microbial ecosystem (Table 6.1). Bacterial populations in the rumen are predated by protozoa, infected by viruses and inhibited by bacteriocins from competing bacteria (Hespell et al. 1997; Whitford et al. 2001). Fortunately, bacteria replicate within a few minutes when energy and nutrient sources are abundant; intrinsic growth rates (R_{max}) of ruminal bacteria range from 7.17 to 0.22 h^{-1}. These growth rates would allow a population to grow from 10^2 to 10^{12} cells·mL^{-1} in 11 to 115 h (Theodorou and France 1993).

High rates of replication allow bacteria to respond quickly to changes in the food intake of the animal; bacterial counts in the rumen fluid of muskoxen increase after each meal and decrease with daily food intakes that decline from autumn to winter (Barboza et al. 2006; Crater et al. 2007). Rapid replication also allows microbes to respond quickly to new substrates because the gene for an enzyme that provides more energy or nutrients for replication will rapidly spread through the population after a delay (lag time) for inducing the gene. Ruminants may require several days of exposure to a new plant before microbes can induce and express sufficient enzymes for effective digestion. The ruminal bacteria in white-tailed deer and reindeer may require 5–10 days to adjust to changes in fiber chemistry as their hosts switch from vascular plants in summer to lichens that contain mixed polymers in winter (Garmo 1986; Jenks and Leslie 1988). The inclusion of plant secondary metabolites (PSMs) in the diet may likewise select for microbes that can use those compounds for energy, which may in turn reduce any toxic effect on the animal. Goats that consume tropical plants in the genus *Leucaena* harbor the bacterium *Synergistes jonesii*, which degrades mimosine, a toxic analog of the amino acid tyrosine (Allison et al. 1992).

However, microbial enzymes are not completely under the control of the animal. An energetically favorable reaction for a microbe can also convert PSMs to even more toxic compounds (e.g., release of cyanide from cyanogenic glycosides; Chapter 3) (McSweeney and Mackie 1997). Some PSMs are also toxic to the microbes and can therefore impair the use of fiber by the community and its host. Koalas consume leaves high in tannins, but bacteria in their hindgut can disrupt complexes between tannins and proteins that impair fiber digestion (Osawa et al. 1993). Pronghorn antelope, pygmy rabbits and sage grouse all consume diets that are more than 90% sagebrush, a forage that contains high concentrations of terpenoids and volatile oils that are known to be antimicrobial. The ability to use these toxic diets probably relies on selection for detoxification pathways in the microbes as well as the mucosa, liver and kidney of the host over many generations (Foley et al. 1999).

6.7.1 Host–Microbe Relationships

The mucosal surfaces of the digestive tract are colonized by microbes soon after animals are born or hatched (Conway 1997; Mead 1997). Arctic ruminants such as muskoxen and caribou are born with sterile digestive tracts but soon develop a fermentation in the rumen and begin digesting the same foods as adults within 4 weeks of birth (Knott et al. 2005). Young animals are inoculated with microbes or spores that enter the digestive tract in food and by contact with the saliva, skin and feces of other animals. The relationship between animal host and microbe ranges from parasitic (microbe benefits at the expense of the animal) to mutualistic (both microbe and animal benefit) (Klasing 2005). Most wild animals support small numbers of parasites that increase the cost of maintaining the digestive tract and the body but do not impair survival or reproduction until the infection rises. Nematodes that infect the gastric mucosa can increase the cost of repairing the mucosa, and reduce both food intake and body mass in Svalbard reindeer, which ultimately decreases the likelihood of surviving winter and giving birth to a calf in spring, especially when infections are high and winter weather is severe (Irvine et al. 2001).

The specificity of the relationship between microbe and host is difficult to define because diet and digestive function influence the number and diversity of microbes. The population of commensal microbes is a part of the community that is readily recognized by the immune system. This commensal microflora dominates the microbial community of small carnivores and young animals because their diets are highly digestible and retention times are short, which provides few substrates and little time for microbial populations to be sustained in the digesta. Commensal bacteria provide protective benefits to the animal by excluding more virulent pathogens; this close association between bacteria and mucosa is fostered by secretions of mucins and lectins by the animal (Gaskins 1997; Cash et al. 2006). The balance between mutualism and parasitism is often delicate; some benign microbes such as *Escherichia coli* can become pathogenic when energy supplies are limited (Klasing 2005). The lymphatic system of the digestive tract responds to microbes that invade the mucosa and maintains a record of these antigens for accelerating the antibody response to subsequent infections. The risk of pathogenic outbreak is reduced by these specific immune responses and by maintaining conditions in the digestive tract that favor a stable and benign community that utilizes the food rather than the tissues of the host.

A large number of transient organisms can join the commensal microflora as diet breadth expands, and as digestive capacity and retention times increase from carnivores to herbivores (Chapter 5). The diversity of microbes in the digestive system is enormous; all three domains of life are represented (Bacteria, Archaea and Eukarya) in every community which can include eubacteria, archeabacteria, protozoa, fungi and viruses (Mackie et al. 2000) (Table 6.1). The functional diversity within groups is also large because many of these organisms can alter their genes and their expression of enzymes quickly through rapid replication, small genomes and the lateral transfer of genetic material as plasmids between closely related taxa. Consequently many groups overlap in function across the wide range of animals and their diets. Protozoa can degrade cellulose in termites but mainly digest starches in the vertebrate digestive system (Table 6.1). The best way to track microbes and their function is to identify groups with genes for common machinery (e.g., the 16sRNA fragment in ribosomes) and to relate those identities to the activity of enzymes (Forsberg et al. 2000; Mackie et al. 2000). Genetic identities indicate that many bacteria can switch their enzymes and fermentation products as diets and digestive function are altered. Consequently, the yield of energy and nutrients from fermentation changes with the microbial community. Microbial populations in sheep are altered when animals switch from grasses that contain cellulose to seaweeds containing mixed polymers; seaweed consumption reduces the population of cellulolytic species but increases the numbers of species with broader substrate specificities such as *Butyrivibrio fibrosolvens* (Orpin et al. 1985).

Anaerobic conditions in the digestive tract do not allow the microbes to extract all the energy from the food. Fermentation oxidizes the carbohydrates to a series of short-chain fatty acids (SCFA), which can be absorbed by the animal. Acetate, propionate and butyrate are the principal acids produced by fermentation of carbohydrates (Fig. 6.12). The total concentration of SCFA and the proportions of each acid in the digesta depend on the composition of the diet, the predominant metabolic pathways of the microbial community, and the rate at which the acids are removed from the chamber by absorption or digesta flow. Acetate predominates in fermentations of animals that consume grasses and sedges because acetate is a common product of bacteria, especially those that utilize cellulose (*Fibrobacter succinogenes*, *Ruminococcus albus* and *R. flavefaciens*). Propionate and butyrate production is associated with fermentations of less fibrous foods that contain more soluble carbohydrates such as pectin and starch, which are used by *Butyrivibrio fibrisolvens*, *Selenomonas ruminantium* and *Prevotella* sp. (Russell 2002; Dehority 2003).

Fermentation acids are absorbed by facilitated diffusion into the mucosal cells, followed by export to the blood in both the foregut and the hindgut (Hume et al. 1993; Rechkemmer et al. 1995; Gabel and Aschenbach 2000). Acetate, propionate

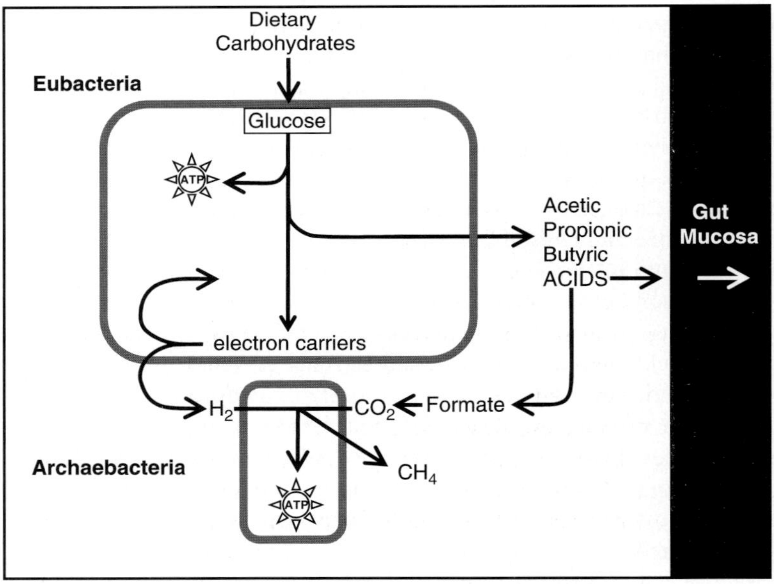

Fig. 6.12 Metabolic pathways for fermentation of carbohydrates in the digestive tract. Eubacteria produce ATP by converting carbohydrates to acids that can be absorbed by the animal. Hydrogen (H_2) is produced as a waste product because electrons cannot be discharged from carriers without oxygen (O_2) in this anaerobic system. Archaebacteria can use the H_2 to produce ATP and methane (CH_4) from acids such as formate that are converted to CO_2. Production of H_2, CO_2 and CH_4 provide the highest yield of ATP for the microbial community. These gases are part of the cost of fermentation because their loss reduces the efficiency of converting dietary energy to metabolizable energy for the animal

and butyrate can enter the TCA cycle either to be oxidized for current energy demands of the animal, or to be deposited in fat as an energy store (Fig. 6.8). Propionate also can be converted to glucose and then to glycogen for storage. Thus propionate can be an important source of glucose when animals ferment fibrous foods that contain little starch or sugar at the end of winter. Some energy from the fermentation is lost as methane (CH_4), whose carbon has been oxidized to only a small fraction of its biochemical potential. Oxidation of carbon from sugars results in the progressive transfer of electrons onto carriers such as NADH and $FADH_2$. In the absence of oxygen, those carriers cannot be regenerated for further production of ATP. Instead, the electrons can be removed from the carriers in the conversion of H^+ ions to H_2 gas. Archaebacteria such as *Methanobrevibacter ruminantium* use the energy from H_2 to produce ATP by reducing formate ($HCOO^-$) or CO_2 to CH_4. Methane and H_2 are not metabolized by the animal but are expelled at the lungs and at either end of the digestive tract.

The efficiency of retaining energy released from the digestion of carbohydrates is reduced by the myriad processes of sustaining the microbial community. The metabolizability of energy from soluble carbohydrate is therefore reduced when microbes convert the carbon to SCFA. Thus carbohydrates have a fuel value less than 16.7 $kJ \cdot g^{-1}$ when they are digested by fermentation (Chapter 1). However, cellulose and mixed polymers are indigestible without fermentation, so the unavoidable loss of energy to microbes is more than offset by the large amount of energy released from the abundant fiber.

In contrast, the loss of energy from microbial fermentation of starches and soluble sugars, which can be easily digested by enzymes secreted by the animal, is normally wasteful. Much of these losses can be avoided by fermentation in the hindgut because most of the soluble carbohydrate is digested in the foregut and the small intestine, which leaves only unresorbed secretions and undigested fiber to be fermented in the hindgut. Fermentation of soluble carbohydrate in the foregut of adult ruminants and kangaroos (Chapter 5) is largely unavoidable and, consequently, fermentations in the foregut typically produce more methane than those in the hindgut (Hume 1997). In young ruminants fermentation of milk in the forestomach can be almost completely prevented by closure of a groove (the esophageal or reticulo-omasal groove) that connects the esophagus directly to the omasum and the acid region of the stomach (abomasum) (Langer 1994). The reticulo-omasal groove is reduced as the rumen expands during growth of the young ruminant; ruminal development is stimulated by SCFA released from the fermentation of plant material (Van Soest 1994). However, even in adult ruminants some soluble carbohydrates may continue to escape ruminal fermentation. The expression of glucose transporters (SGLT) in the small intestine of adult roe deer, fallow deer and moose suggests that some soluble carbohydrates escape ruminal fermentation when the diet is rich in starches from seeds, fruits and new leaves (Wood et al. 2000), probably via rapid passage in the fluid phase of ingesta.

The SCFA produced in the foregut of sheep and cattle can satisfy 30–80% of the animal's maintenance energy requirement (Stevens and Hume 1995). Similar values (23–67%) have been reported for wild ungulates, with slightly lower values (21–42%)

for wallabies (Stevens and Hume 1995). Hindgut fermentations generally make smaller contributions to maintenance energy requirements: 20–40% in larger hindgut fermenters such as pigs and wombats, and 8–17% in smaller hindgut fermenters such as the koala, beaver and guinea pig (see Stevens and Hume 1995 for a detailed list). In birds, fermentation in the ceca has been found to contribute 5–10% of the maintenance energy requirement. In reptiles, the contribution from fermentation to energy requirements varies widely from 15% to virtually 100% because of the low maintenance requirements for ectothermy.

Energy and nutrients from the food consumed by the animal are ultimately lost from the fermentation chamber as microbes are washed out. Ruminants, kangaroos and other foregut fermenters can use the microbes from fermentation in the foregut because the cells can be digested and absorbed in the acid stomach and small intestine (Chapter 5). However, most of the microbes are lost after hindgut fermentation in large herbivores such as elephants and horses, and in ectotherms such as fish and reptiles. Some small herbivores including rabbits, voles, ptarmigan and marsupial possums can recover the contents of the hindgut fermentation by ingestion of feces formed from cecal contents, a practice called cecotrophy. The role of microbial fermentation as a source of protein, energy and vitamins is further discussed in Chapters 8, 9 and 11.

6.8 Summary: Carbohydrates

Most of the carbon that flows from plants to animals is in the form of carbohydrate. Carbohydrates are formed as simple sugars, and as structural and non-structural polysaccharides. Non-structural polysaccharides such as starch are important forms of carbon storage in plant cells that are readily digested by the endogenous enzymes of animals. Structural polysaccharides in plant cell walls (dietary fiber) are degraded mainly by enzymes produced by microbes inhabiting the foregut and hindgut. The SCFA produced in these fermentations can make substantial contributions to the maintenance energy requirement of animals: up to 80% of the energy required by mammalian foregut fermenters and up to 40% of the energy required by hindgut fermenters. Contributions are smaller in birds (up to 10%), but ectothermic reptiles can derive virtually their entire energy requirement from fermentation.

Chapter 7
Lipids: Fatty Acids and Adipose Tissue

All tissues contain small amounts of lipid because phospholipids are part of the membranes of both animals and plants. Lipid fractions in the skins of fruit act to waterproof tissues with coats of wax, and attract animals with colored carotenoids or aromatic essential oils. Lipid fractions in animals include cholesterol, which stiffens membranes and forms the steroid hormones that integrate development and reproduction. Cells contain droplets of lipid as energy depots in the muscle, liver and adipose tissue of animals and in the seeds and fruits of many plants. Foods that are rich in lipid are often selected by animals because lipids contain more gross energy (kilojoules per gram, $kJ \cdot g^{-1}$) than either carbohydrates or proteins (Chapter 1). Omnivores such as chickadees select oily seeds for their winter caches and rodents are routinely trapped with baits based on oily peanut butter. Carnivores also favor the fatty parts of their prey. Polar bears choose blubber over the muscles of seals (Stirling and McEwan 1975) and carnivorous marsupials choose the thorax and fat bodies over the legs of insects (Chen et al. 2004). Lipids are related to survival and reproduction of many animals because the energy from fat is used during winter fasts, cold spells, migration, incubation and lactation. Lipid stores in animals also contain contaminants that bio-accumulate at each trophic level, resulting in peak concentrations in apex predators such as eagles and polar bears (Moriarty 1999).

7.1 Functional Chemistry of Fatty Acids

Fatty acids are the building blocks for all lipids, especially those that constitute fat depots (triglycerides) and membranes (phospholipids). The smallest lipids are the water-soluble SCFA produced by fermentation of carbohydrates (Chapter 6). Fatty acids consist of a hydrocarbon chain that is poorly soluble in water. Increasing the length of the hydrocarbon chain decreases water solubility but increases the affinity for other fatty acids in membranes and fat (Fig. 7.1). Long chains of fatty acids readily associate to form oily droplets or micelles in the watery matrix of the digesta and within the adipose cells of fat depots. Total lipid in food and tissue is determined by the Soxhlet method, which extracts lipids with solvents such as chloroform, methanol and petroleum ether. Solvent-extracted lipids may be further

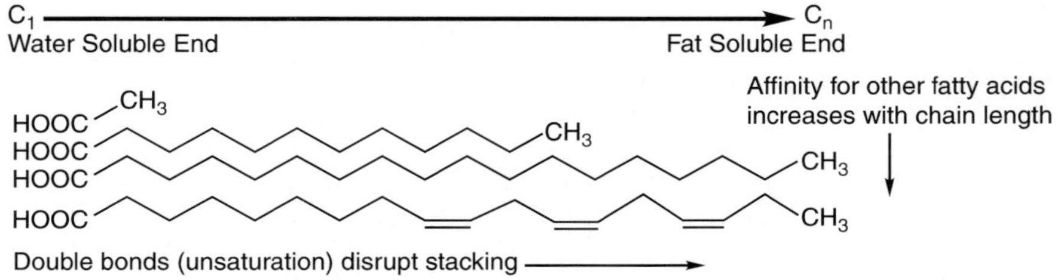

Fig. 7.1 General structure of fatty acids. Numbering of carbon atoms begins at the water-soluble end. Short fatty acids are water soluble. Increasing the length of the hydrocarbon chain increases the ability to stack adjacent molecules and dissolve the acid in membranes. Double bonds in the chain alter the associations between fatty acids

separated by thin-layer chromatography to quantify lipid classes (triglycerides vs. phospholipids). Gas chromatography (GC) is used to quantify individual fatty acids as dietary markers (e.g., QFASA or alkane; Chapter 4) or essential nutrients (see below).

Fatty acids without double bonds between carbons are called saturated. Saturated fatty acids with 16 (palmitic acid) and 18 (stearic acid) carbons are synthesized by all animals. Saturated fatty acids stack tightly to form a compact store of energy in fat depots. Monounsaturated fatty acids contain a single double bond between carbons, while acids with two or more double bonds in the carbon chain are called polyunsaturated fatty acids (PUFA) Double bonds can disrupt the orderly stacking of the hydrocarbon chains (Fig. 7.1). Unsaturation increases the fluidity of lipids, especially at low temperatures; fatty acids in plants and ectothermic animals contain greater proportions of unsaturated fatty acids than those of birds and mammals (Fig. 7.2). Lipids in the legs of Arctic caribou contain more monounsaturated fatty acids than those at the core of the animal; tissue temperatures decline from the core (38°C) of the body to the hooves (0°C) (Irving 1972; Blix 2005). Decreasing ambient temperature induces an enzyme for producing monounsaturated fatty acids in carp, which allows the fish to maintain the fluidity of its membranes in cold water (Tiku et al. 1996).

Fatty acids are identified by both common and systematic names as well as numbers that describe their structure. For example, the systematic name for oleic acid is *cis*-9-octadecenoic acid, which is abbreviated to 18:1 $c\Delta9$, which denotes an 18-carbon fatty acid with a single double bond starting at carbon 9 with both hydrogen atoms bonded on the same (cis) side of the double bond. Oleic acid is part of a series of 18-carbon fatty acids (Fig. 7.3) that are found in a wide variety of organisms, from the endosperm of seeds (e.g., rice and wheat) and the flesh of fruits (e.g., olive) to muscle, fat, eggs and mammalian milk (Sargent et al. 2002) (Fig. 7.2). Double bonds in each fatty acid are introduced by desaturase enzymes that are specific to the position on the hydrocarbon chain. Animals express the gene for $\Delta9$ desaturases but do not express the $\Delta12$ and $\Delta15$ desaturases. Consequently, animals cannot produce the double bonds at C_{12} of linoleic acid (six carbons from the terminus = n-6)

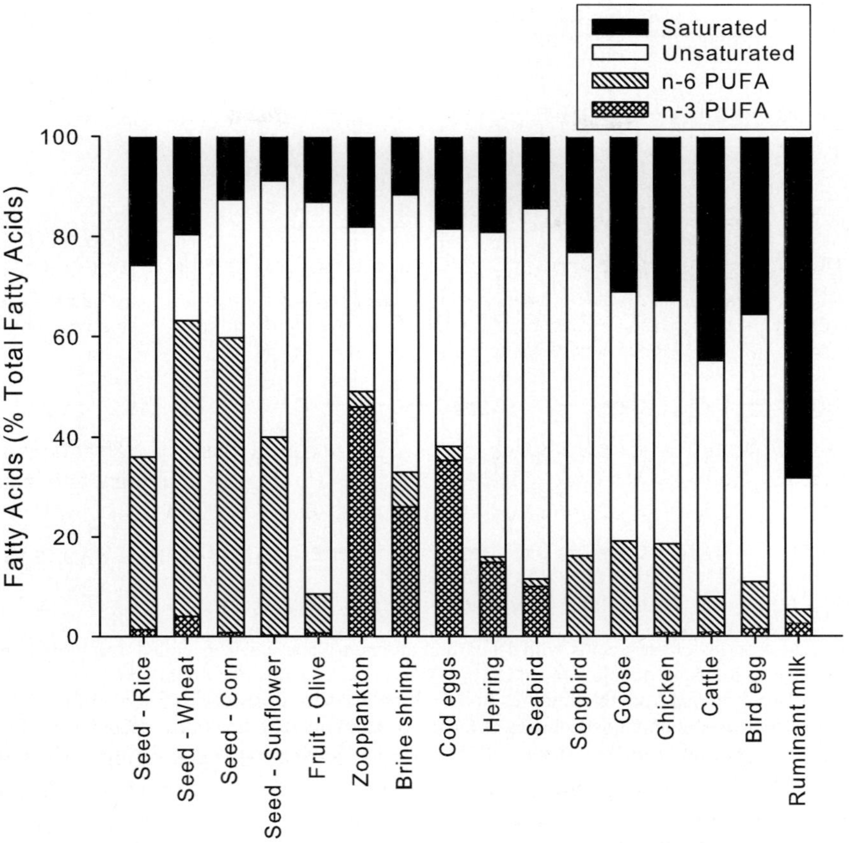

Fig. 7.2 Percentage of saturated, unsaturated and polyunsaturated (PUFA) fatty acids in the tissues of selected plants and animals. Plants: whole seeds of rice and wheat (National Research Council 2006); oil from corn, sunflowers and olives (National Research Council 2006). Animals: marine zooplankton, brine shrimp and eggs of the marine cod (Sargent et al. 2002); whole herring (Cooper et al. 2005); subcutaneous fat from northern fulmar (Wang et al. 2007), red eyed vireo (Pierce and McWilliams 2005) and Canada goose (Thomas and George 1975); muscle and bones from domestic chicken and cattle (National Research Council 2006); average of eggs from chickens, ducks, turkeys and Japanese quail; average of milks from domestic cattle, sheep, goats and water buffalos (Agricultural Research Service 2007)

or at C_{15} of linolenic acid (three carbons from the terminus = n-3) (Fig. 7.3). Therefore fatty acids with double bonds at these positions are required in the diet and are classed as 'essential fatty acids'. Linoleic and linolenic acids are modified with elongases that extend the hydrocarbon chain, and with desaturases that introduce double bonds at Δ5 and Δ6 in the elongated chains. The modified acids retain the last double bond at the same distance from the terminus, that is either at n-6 or at n-3. Linoleic (18:2) and arachadonic (AA; 20:4) acids are the principal essential fatty acids among the n-6 series, whereas linolenic (18:3), eicosapentaenoic (EPA; 20:5) and docosahexanoic (DHA; 22:6) acids are the principal essential fatty acids

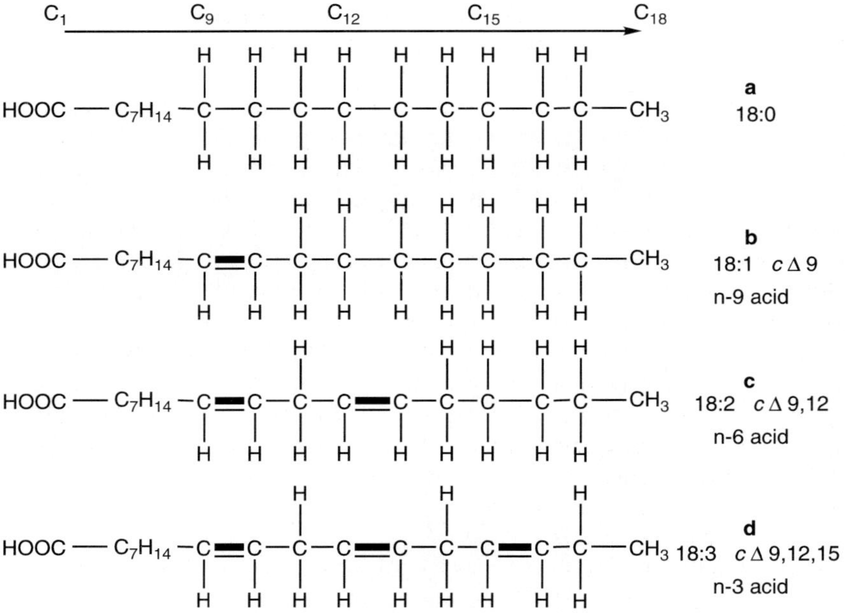

Fig. 7.3 A series of fatty acids with 18 carbon atoms. **a** Stearic acid is called a saturated fatty acid because it has no double bonds (18:0). **b** Oleic acid contains one double bond (18:1) at C_9 that is also 9 carbons from the terminus (n-9). **c** Linoleic acid is part of a series of essential fatty acids with the last double bond starting at 6 carbons from the terminus (n-6). **d** Linolenic acid is part of another family of fatty acids with the last double bond starting at 3 carbons from the terminus (n-3)

in the n-3 series. Carnivores such as Atlantic salmon, rainbow trout and domestic cats may require dietary sources of AA, EPA and DHA because the activities of enzymes for modifying linoleic and linolenic acids are not sufficient to meet their requirements (Sargent et al. 2002; National Research Council 2006).

Fatty acids found in animal tissues are incorporated from the diet or synthesized de novo by the animal. Although large depots of body fat can include a wide variety of dietary acids, membrane fatty acids are less variable because the composition is closely associated with functions of plasma membranes and mitochondria such as energy metabolism and homeostasis. Essential PUFA participate in many functions of membranes in blood cells, blood vessels, skin, nerves and muscles. PUFA are particularly important during the development of brain and eyes among vertebrates; n-3 acids can comprise half the fatty acids of neural membranes (McKenzie 2005). Small changes in PUFA concentrations may alter the function of tissues. For instance, increasing proportions of n-6 acids in muscle are correlated with greater running speeds among species of mammals and with greater swimming speeds among individual Atlantic salmon (McKenzie 2005; Ruf et al. 2006). Changes in stores of fatty acids can affect their use as a fuel for flight in birds and for thermoregulation in mammals (Frank et al. 2004; Pierce et al. 2005). Dietary changes that alter the fatty acid composition of tissues may therefore affect patterns of migration and hibernation in wildlife (Geiser et al. 1994; Harlow and Frank 2001; Pierce and McWilliams 2005).

Deficiencies of essential fatty acids are rare in wild animals because plants synthesize both the n-6 and n-3 PUFA that are used by herbivores and their predators. Requirements for essential fatty acids are less than 1% of the dietary DM in domestic dogs and cats. Requirements for fish are also less than 1% of DM for adult fish but are much higher during larval stages of development (1–6% of DM) (Sargent et al. 2002; National Research Council 2006). Marine algae and plankton are rich in n-3 acids whereas terrestrial plants contain proportionately more n-6 acids (Fig. 7.2). Subcutaneous fat depots of seabirds and songbirds reflect these differences in PUFA between marine and terrestrial food webs (Fig. 7.2). Essential fatty acids can be used as digestible markers in carnivores by comparing the signatures of fat depots in seals or seabirds with the signatures of prey items (Iverson et al. 2004). Hunters may also perceive these shifts in fatty acid composition of the animal as taste and smell; waterfowl hunters on the Chesapeake Bay first reported a change in the flavor of canvasback ducks that was later associated with a shift in the diet of those ducks from tubers to clams (Jorde et al. 1995).

Comparisons of fatty acids between diet and the tissues of herbivores are affected by digestive and metabolic transformations. Bio-hydrogenation by microbes saturates many of the PUFA in the foregut of ruminants before they are absorbed in the small intestine (Fig. 7.4). Fat depots of cattle, sheep, red deer and camels contain proportionately more saturated fatty acids than those of hindgut fermenters such as geese,

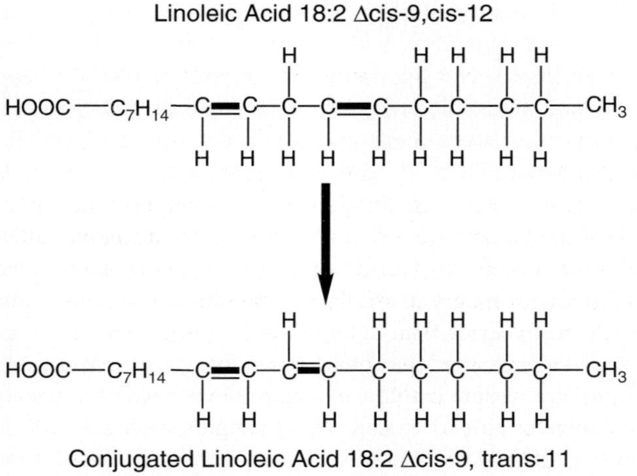

Fig. 7.4 Bio-hydrogenation of plant acids alters the position of the double bond and the relative orientation of hydrogen on the double bond. Double bonds with 'cis' orientation place hydrogen on the same side of the bond whereas 'trans' bonds place the hydrogen in opposition. Ruminal bacteria can bio-hydrogenate plant acids to conjugates. These conjugates may be further modified by the animal once they are absorbed

horses and rabbits (Fig. 7.2) (Van Soest 1994). Ruminal microbes produce SCFA by fermentation and synthesize novel fatty acids that have odd-numbered hydrocarbons and branched chains (Russell 2002). These novel fatty acids are used for microbial membranes. SCFA and branched chain acids are deposited in the fat depots of the animal and in its milk (Van Soest 1994; Agricultural Research Service 2007). Ruminal bacteria such as *B. fibriosolvens* also modify (conjugate) linoleic acid by altering the position and the orientation of the double bond (Fig. 7.3). The concentration of conjugated linoleic acid (CLA) in the muscle, fat and milk of ruminants reflects both absorption of the conjugate produced in the rumen as a well as bio-hydrogenation of linoleic acid in the tissues of the animal (Russell 2002; Bauman et al. 2006). The proportions of fatty acids such as CLA in the muscles of game animals may have several beneficial effects on human consumers, which include a reduced risk of cardio-vascular disease and cancer (Hudson and Jeon 2003).

7.2 Classes of Lipids

Fatty acids are combined in four general classes of lipids: wax esters, triglycerides, phospholipids and sterols (Fig. 7.5). The water-soluble acid end on the hydrocarbon chain of fatty acids is used to form an ester linkage that can be broken during digestion and reformed for deposition in tissue. Two fatty acids are joined at the acid end to form a wax ester. Wax esters are secreted to exclude water from the surface of plant cell walls, the cuticle of invertebrates and the feathers of birds. Plant waxes that are comprised of long odd-numbered hydrocarbons (more than 20 C) may be used as dietary markers in herbivores because they are relatively indigestible, and easily distinguished from the fatty acids produced by animals (Mayes and Dove 2000). Not all waxes are indigestible; wax esters are a major source of PUFA and digestible energy for fish, seabirds and whales that consume marine invertebrates because copepods and krill deposit wax esters in their tissues as an energy store (Sargent et al. 2002). Vertebrate animals only store lipids as triglycerides, which are the combination of a carbohydrate (glycerol) and three fatty acids (Fig. 7.5). Triglycerides do not require additional organelles or water for storage because the lipid readily forms a single compact droplet within the cells of adipose tissue (fat depots), muscle and liver. Consequently, animals store much more carbon in lipid as triglyceride than in carbohydrate as glycogen (Chapter 6).

The substitution of a charged phosphate group for one fatty acid in the triglyceride changes the lipid to a molecule that can associate with water at the phosphate end and with lipids at the hydrocarbon end. Phospholipids spontaneously form micelles that contain fatty acids and triglycerides for transport in the blood. Opposing layers of phospholipid are the basis for membranes in all cells. Membranes also contain sterols, which are lipids with a planar form that enhance the interactions between fatty acid chains and so reinforce the membrane (Fig. 7.6). Sterols are also used as hormonal messengers between tissues because they readily dissolve in membranes. Plants contain a wide variety of sterols (e.g., phytoestrogens), but cholesterol is

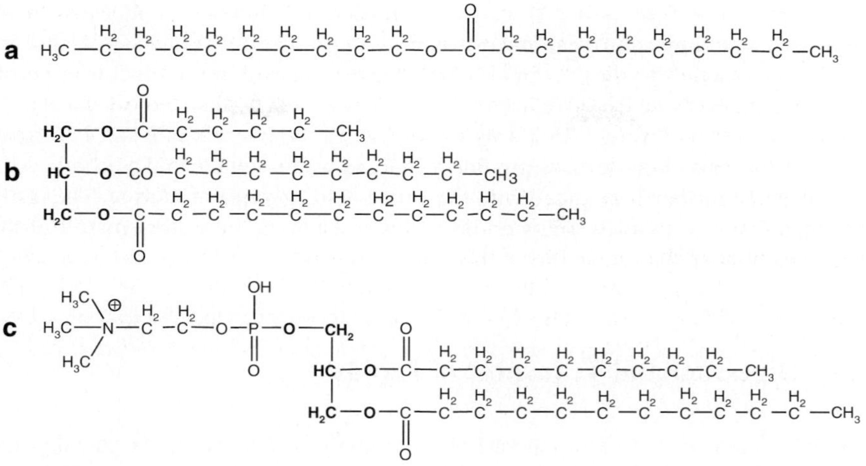

Fig. 7.5 The water-soluble ends of fatty acids form ester bonds with other acids and with the carbohydrate glycerol (*bold*). **a** Ester linkages between two fatty acids form wax esters on cell surfaces of plants and animals. **b** Triglycerides are large molecules that can be stored in body fat. **c** Substitution of one fatty acid with a water-soluble molecule such as phosphatidyl-choline produces the phospholipids that constitute cell membranes

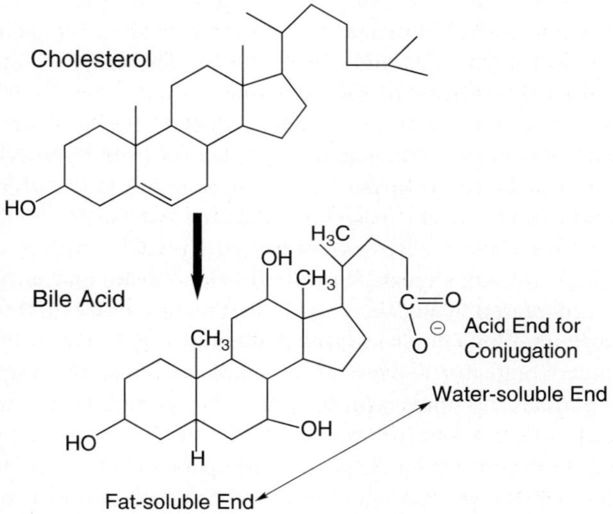

Fig. 7.6 Cholesterol forms several compounds in animals but most cholesterol is used for synthesis of bile, which is used to help digest dietary fat. Cholic acid is the simplest form of a bile acid that can be conjugated with amino acids such as taurine and glycine to form bile salts (see Chapter 8). Bile salts dissolve fats in water by providing an interface with both fat-soluble and water-soluble ends

only found in animal tissues (Fig. 7.6). Cholesterol is used to synthesize many hormones including androgens and estrogens for reproduction. Plant sterols that are similar in structure to these animal hormones may disrupt reproduction in herbivores that ingest these PSMs (Chapter 3) (Harborne 1993). Most cholesterol is used for lipid digestion through the formation of bile in the liver. A bile salt is formed by modifying the cholesterol to provide a charged water-soluble end as cholic acid, which can be further conjugated with the amino acids glycine or taurine. The result is a soap that can dissolve triglycerides and wax esters in the watery matrix of the digesta so that enzymes can digest the lipid.

7.3 Digestion and Transport of Lipids

Lipids spontaneously form a separate phase from the watery mixture of the digesta. Birds of the Family Procellaridae (petrels, shearwaters, fulmars) are called 'oil birds' because ingested lipids separate from water and particles in a specialized crop. Wax esters from the diet of invertebrates are retained as an oily phase at the top of the crop whereas free fatty acids and watery solutes drain from the bottom (Place et al. 1989; Wang et al. 2007). The oil in the crop provides an energy store for chicks while their parents are foraging at sea (Roby et al. 1997).

Lipids are digested by breaking the ester linkages to release fatty acids, cholesterol and glycerol that are absorbed by the mucosa. Lipid digestion can start in the foregut with lipases that are secreted in the mouth and the stomach of infant mammals consuming milk. Lipid digestion is completed in the midgut by pancreatic lipases and esterases (Stevens and Hume 1995). The entry of lipids into the duodenum causes the release of the hormone cholecystokinin (CCK), which stimulates secretions of bile from the liver and the release of both lipase and co-lipase from the pancreas. Three steps are used for lipid hydrolysis: bile salts stabilize the lipid in water, co-lipase forms a protein interface with the stabilized lipid, and lipase binds to co-lipase to break the esters (Fig. 7.7). The time required to complete these steps is controlled by CCK, which slows gastric emptying and the flow of digesta into the duodenum. Retention in the midgut may be further increased by an 'ileal brake' that reverses digesta flow when lipid enters the lower small intestine (Karasov and Hume 1997). Birds that digest wax esters from invertebrates or berries reflux small intestinal contents back to the gizzard, which mixes the lipids with the intestinal secretions and provides more time for digestion (Place 1992).

Long retention times also favor the passive absorption of fatty acids and cholesterol that diffuse through the plasma membrane and coalesce into droplets within the mucosal cell. The intestinal mucosa of fish may be packed with a temporary store of lipid droplets after a meal; snakes and caimans also increase the volume of their mucosal cells as lipid is absorbed and processed following a large meal (Corraze 2001; Starck 2005). Although some fatty acids may flow directly into the blood, most fatty acids and cholesterol are repackaged into lipoproteins for transport and

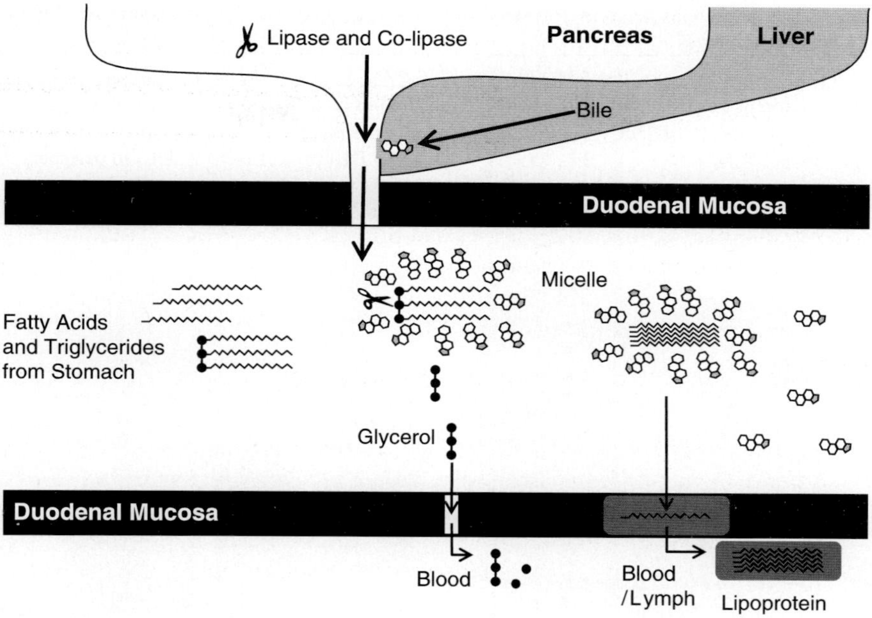

Fig. 7.7 Digestion of triglycerides and free fatty acids is achieved by dissolving the lipids in bile salts, which provides an interface for pancreatic co-lipase to facilitate digestion by lipase. Glycerol is absorbed and transported through the blood stream. Micelles of lipid are absorbed by the mucosal cells and repackaged as lipoproteins for transport

delivery to specific tissues. Lipoproteins are phospholipid envelopes embedded with an address of surface proteins (apo-lipoproteins) that activate cell receptors of target tissues. Mucosal cells re-form fatty acids into compact triglycerides before they are incorporated into lipoproteins called chylomicrons or protomicrons. Mammals secrete large chylomicrons of dietary lipids into the lymph, while birds secrete smaller protomicrons into the blood (Table 7.1) (Stevens 1996). Fish produce lipoproteins similar to the chylomicrons of mammals, but these may be distributed in either lymph or blood, depending on the species (Corraze 2001). A lymphatic distribution of chylomicrons in mammals results in the deposition of dietary acids in body fat without modification by the liver. Thus the fatty acid compositions of body fat can be used to estimate diets of seals because the fatty acids in food correlate with those in chylomicrons and blubber (Cooper et al. 2005) (Fig. 7.8).

Lipoprotein lipase is an enzyme bound to the surface of cells lining the vascular system. Apo-lipoproteins on the surface of the chylomicrons activate cellular lipoprotein lipase to break the esters in the triglycerides and release the fatty acids for absorption by the surrounding muscle and adipose cells. Lipoprotein lipase activity can reflect gains of body fat; activity of the enzyme increases during autumn as black-capped chickadees fatten, and it remains elevated during winter when birds must feed each day to rebuild fat lost during the night (Sharbaugh 2001).

Table 7.1 Lipoproteins that transport fatty acids (FA), phospholipids (PL) and cholesterol (CL) in mammals and birds

Lipoprotein	Content	Mammal	Bird
Chylomicron or Protomicron	FA, PL, CL	Intestine to lymph	Intestine to blood
Very Low Density (VLDL)	FA, PL, CL	Liver to periphery	Liver to periphery and ovary
Low Density (LDL)	PL, CL	Liver (VLDL) to periphery	Liver (VLDL) to periphery
High Density (HDL)	PL, CL	Adipose to liver	Adipose to liver

Fig. 7.8 Lipid transfer from mother to pup in gray seals. The blood serum of pups is often opaque with lipoproteins after a large meal of milk, which contains as much as 54% lipid, in contrast to the serum of the mother which is clear when she is fasting on the shore

Remnants of the chylomicrons return to the liver where the remaining lipid and protein is repackaged. The liver integrates lipid metabolism because it synthesizes lipids and produces the bile required to digest lipid. Lipids accumulate in the liver as snakes and birds process large meals (Starck and Beese 2002; Battley and Piersma 2005); lipid storage in marine cod and sharks can increase the size of the liver to 25% of body mass (Rust 2002). The liver exports lipid to peripheral tissues in Very Low Density Lipoprotein (VLDL). Triglycerides, cholesterol and phospholipids in VLDL are exported to adipose and muscle. Lipoprotein lipase allows triglycerides to be removed from the VLDL, which increases the density of the lipoprotein to Low Density Lipoprotein (LDL; Table 7.1). Cholesterol is released separately from triglycerides under the control of LDL and apo-lipoproteins for cell synthesis. High Density Lipoprotein (HDL; Table 7.1) returns cholesterol from adipose to liver and other tissues between meals. Aberrant delivery of cholesterol to cells lining the blood vessels is associated with cardiovascular problems (e.g., atherosclerosis) that are found in humans and domestic animals. These problems are rare in wild carnivores even though they consume diets high in cholesterol and triglycerides (Pond 1984).

In birds, VLDL is the principal lipoprotein for distributing dietary lipids because protomicrons are repackaged in the liver (Stevens 1996). Birds also use VLDL to provision yolk of the developing egg with triglycerides, cholesterol and phospholipids. The liver exports yolk VLDL with vitellogenin, a complex of protein and phospholipid (Stevens 1996). The concentrations of vitellogenin and VLDL in the blood of birds and fish are indicative of egg formation; these yolk precursors provide a non-lethal method for determining the reproductive state of seabirds at rookeries and waterfowl at nesting grounds (Vanderkist et al. 2000).

7.4 Fat Synthesis and Mobilization

The triglycerides that comprise body fat can be formed directly from dietary fatty acids or by de novo synthesis from carbohydrate. Animals fatten fastest on fat because liver and adipose can readily form the ester link between glycerol and dietary fatty acids. Bears fatten quickly on spawning salmon because the brain and eggs they select from the fish are rich in lipids that are easily absorbed and quickly incorporated into triglycerides in the body. Omnivores such as turkeys and waterfowl that consume starch from seeds also gain fat rapidly because all the carbon in glucose can be converted to triglyceride (Fig. 7.9). Glycolysis and the pentose phosphate pathway (ribose synthesis) produce glycerol, acetyl-CoA (2-Carbon fragments) and the reducing equivalents (NADPH) required for the synthesis of triglyceride. The acetyl-CoA from glucose can also be used to synthesize cholesterol, which allows animals to maintain the production of bile without dietary cholesterol (Fig. 7.9). Herbivores use SCFA from fermentation of carbohydrate to build long-chain fatty acids; ruminants use acetate (2:0) to build palmitate (16:0) for adipose tissue as well as milk (Van Soest 1994).

Fat metabolism is integrated with glucose homeostasis through the hormones insulin and glucagon. The rise in insulin after a meal stimulates the synthesis of fat as glucose is removed from the blood and deposited in fat. Conversely, glucagon stimulates mobilization of fat, which conserves blood glucose. Fat breakdown (lipolysis) is promoted by glucagon during the normal daily pattern of feeding and by epinephrine during an emergency demand for energy such as the 'flight or fight' response (Chapter 11). The activation of lipolysis in adipose cells is tightly controlled to avoid wasting energy stores; binding of hormone to the cell surface receptor is followed by an intracellular cascade of four enzymes to activate hormone-sensitive lipase. The lipase breaks the ester linkages to release glycerol and fatty acids into the bloodstream. Some of the long-chain acids are carried on albumin (a protein), but most fatty acids circulate in the non-esterified form and are readily absorbed by most tissues. Fatty acids can be completely oxidized in the TCA cycle to produce CO_2 and ATP (Figs 6.1 and 7.9).

Migrating shorebirds, songbirds and waterfowl use fatty acids as their primary fuel for flight (McWilliams et al. 2004). Plasma concentrations of fatty acids and ketone bodies increase during migratory and sedentary fasts of birds and mammals

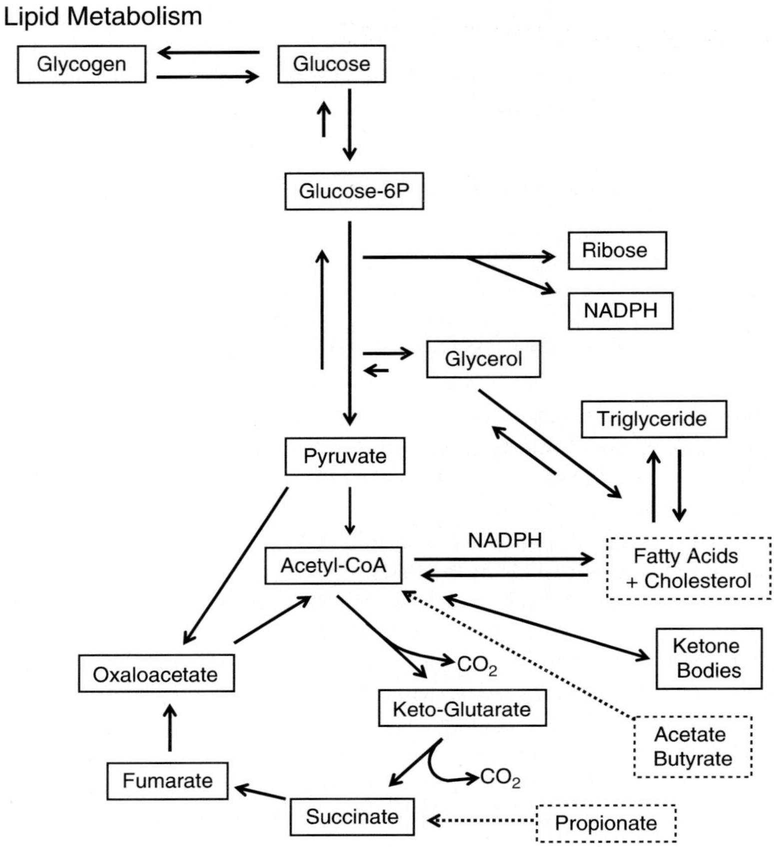

Fig. 7.9 Metabolic pathways for utilizing lipids for energy or for storing carbohydrate as triglyceride (fat)

(Le Maho et al. 1981; Gannes 2001). Ketone bodies are formed by a spillover of carbon from the oxidation of fatty acids in the TCA cycle in the liver. Acetyl-CoA is converted to ketones (acetoacetate and β-OH-butyrate) when the amount of oxaloacetate is insufficient to carry the 2-carbon fragments through the TCA cycle (Fig. 7.9). Ketosis is associated with glucose conservation because the cycle can be supplied with more oxaloacetate from glucose (Fig. 7.9). Ketones produced in the liver can be subsequently oxidized by muscles and other tissues where oxaloacetate is more available.

Fat mobilization spares critical stores of glucose in glycogen as well as the amino acids from body protein that would otherwise be needed to replenish glycogen stores (Chapter 6). Glycerol released from mobilized fat can be used to produce glucose (gluconeogenesis) in the liver. Fat mobilization provides enough glycerol to meet more than 70% of the demand for glucose and over 90% of the energy expended by bears during dormancy (Barboza et al. 1997). The use of fat for energy sustains body functions by maintaining body protein and can even support the additional demands of lactation in fasting seals and bears. The energy demands and body composition of animals during fasting and feeding are discussed further in Chapter 10.

7.5 Summary: Lipids

Fats are the most energy-dense substrates, and foods high in fat are often selected by wildlife. Fat is the principal form of energy storage in animals. Body fat stores are mobilized in times of energy deficits such as the migratory flights of birds and the torpor bouts of mammals. Some polyunsaturated fatty acids cannot be synthesized in the body and are therefore required in the diet. Essential fatty acids play important roles in a number of tissues such as the developing eyes and brain. Dietary changes that alter the fatty acid composition of tissues can affect annual patterns of migration and hibernation in wildlife.

Chapter 8
Nitrogenous Substrates: Nucleic Acids to Amino Excretion

Nitrogen (N) may be the most limiting nutrient for many populations of wildlife because nitrogenous compounds in crude protein are the basis for the structure and function of animals (Fig. 1.4) (White 1993). The organic N in plants and animals includes amino acids and the wide array of proteins they form, the genetic program in DNA and its transcripts in RNA, as well as numerous intermediary metabolites and vitamins. Protein is the largest fraction of N in the animal body, with contractile functions in all muscle fibers, structural functions in connective tissue, skin, tendons and arteries, and specialized functions such as hormones, enzymes, antibodies, transport molecules and clotting factors. Animals may use stored protein for energy during migration and hibernation, and in the production of milk and eggs. Both herbivores and carnivores use dietary proteins for new tissue and for energy, but herbivores also may need to recycle N to conserve the seasonally low supply of protein from their diets.

8.1 Amino Acids and Essentiality

Proteins are a sequence of amino acids linked by peptide bonds. The side chains of 20 amino acids give proteins their structural and functional properties (Fig. 8.1). Structural proteins have repeating sequences that polymerize into fibers (e.g., collagen). The sequences for enzymes (e.g., pancreatic lipase) may be subdivided into sections that fold into globules containing binding sites for cofactors (e.g., co-lipase) and catalytic pockets for substrates (e.g., triglycerides). In a globular protease, the charged side chain on lysine helps to dissolve the protein in water; the small side chain of glycine allows tight folds between helical sections; and the hydrocarbon side chains of valine and leucine line the interior of the protein where the large weakly charged side chain of histidine is used for catalytic reactions (Fig. 8.1) (Mathews and Van Holde 1996). Protein synthesis requires the full complement of the 20 common amino acids in varying proportions that depend on the amino acid sequence of the proteins being formed.

Animals cannot synthesize all the amino acids they require and must therefore rely on a dietary supply from prey, plants or microbes to sustain protein synthesis.

P.S. Barboza et al., *Integrative Wildlife Nutrition*,

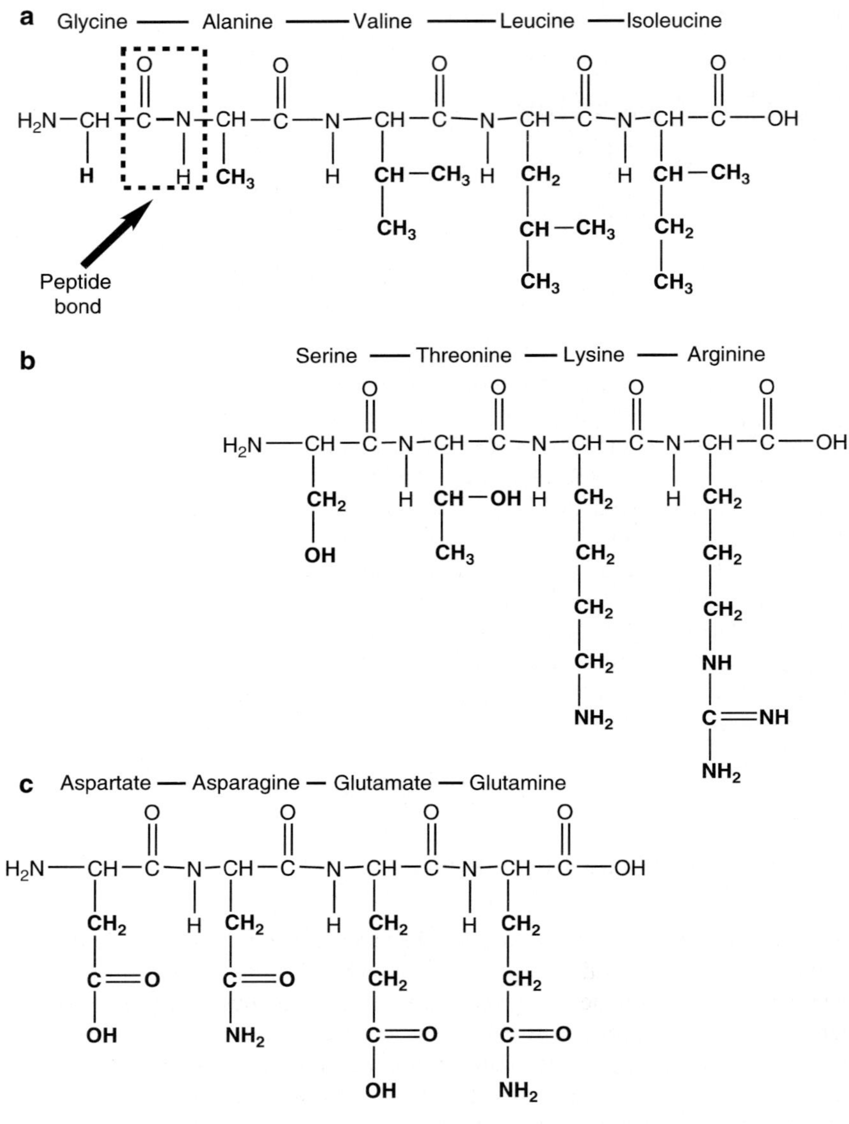

Fig. 8.1 Amino acids used by animals. Amino acids are grouped according to the structure of the side chain (*bold*) and are presented as a peptide. **a** Hydrocarbon side chains. **b** Side chains with hydroxyl (OH) and amine (NH_2) groups. **c** Amide (CONH) and acid (COOH) side chains. **d** Cyclic side chains. **e** Sulfur amino acids. Cystine consists of two cysteine side chains connected through a sulfhydryl bridge. Taurine is not found in proteins but is used as a conjugate for bile salts

Amino acids can be considered in two parts: the α-amino group used in the peptide bond, and a C skeleton that includes the side chain (Fig. 8.1). Non-essential or dispensable amino acids can be synthesized completely from a pool of C and

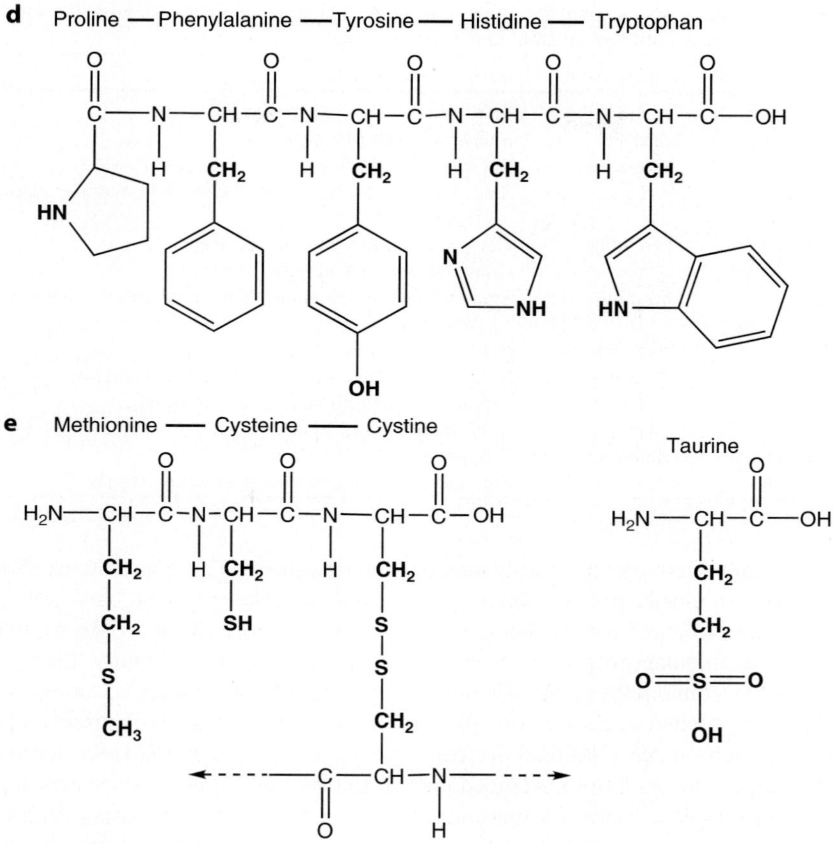

Fig. 8.1 (continued)

amino-N in the diet. A dietary supply of amino acid is required at all life stages when both parts of the amino acid cannot be synthesized (complete essentiality) or when the C skeleton cannot be formed (side-chain essential). Completely essential amino acids such as threonine and lysine are replaced quickly by the body because they cannot be reformed if the amino group is removed or the carbon skeleton is altered. These acids can be used as dietary markers because their isotopic composition tracks those of dietary proteins in growing animals (Fantle et al. 1999). Dietary amino acids also may be required when the synthetic pathways are inadequate for the demand (conditional essentiality). The rate of histidine synthesis in mammals, for example, is sufficient for maintenance of adults but not for growth (Table 8.1).

The quality of a dietary protein is evaluated by comparing the proportions of essential amino acids in the food protein with those in the animal. Muscle proteins are consistent in composition among vertebrates and are generally similar to those of eggs and milk (Fig. 8.2). Consequently, carnivores can easily meet their requirements for essential amino acids because the diet closely matches the proteins synthesized

Table 8.1 Amino acids essential for protein synthesis in animals that lack a significant source of microbial protein from fermentation in the digestive tract

Side chain group	Amino Acid	Basis for essentiality
Hydrocarbon	Valine	Side chain – all life stages
Hydrocarbon	Leucine	Side chain – all life stages
Hydrocarbon	Isoleucine	Side chain – all life stages
Hydroxyl	Serine	Side chain – conditional on glycine and threonine supply in birds
Hydroxyl	Threonine	Complete amino acid – all life stages
Amine	Lysine	Complete amino acid – all life stages
Amine	Arginine	Side chain – conditional on N load for urea synthesis in mammals
Cyclic	Phenylalanine	Side chain – all life stages
Cyclic	Tyrosine	Side chain -conditional on phenylalanine supply
Cyclic	Histidine	Side chain – conditional on rate of growth
Cyclic	Tryptophan	Side chain – all life stages
Sulfhydryl	Methionine	Side chain – all life stages
Sulfhydryl	Cysteine	Side chain – conditional on methionine supply

by the animal. Plant proteins rarely match the composition of animal proteins. Plant leaves contain mostly photosynthetic proteins that are relatively well balanced, but seeds contain storage proteins that are often very imbalanced (Moir 1994). Animals may use an imbalanced protein from one food with a complementary source of amino acids from another item. Thus many primates obtain a balanced amino acid intake by supplementing a diet of fruits and leaves with prey such as insects, eggs and other vertebrates (National Research Council 2003). Alternatively, animals need to ingest more of an imbalanced protein to meet their daily requirement for a limiting amino acid. Low concentrations of lysine and methionine often limit the use of protein from a plant diet (Fig. 8.2). In herbivores, microbial protein may complement plant proteins in essential amino acid composition. Ruminal microbes synthesize proteins with higher proportions of lysine and methionine than many plant proteins (Fig. 8.2) by utilizing both protein and non-protein forms of N and sulfur (S) in plants (Nolan 1993). Grazing bison and kangaroos can therefore use the imbalanced proteins in senescent grasses because they are supplemented with the more balanced microbial protein synthesized in their foregut.

Dietary requirements for essential amino acids are also affected by interactions with other amino acids. Phenylalanine requirements are increased by low supplies of tyrosine because the cyclic side chain for phenylalanine is hydroxylated to form tyrosine (Fig. 8.1D). Conversely, high supplies of tyrosine save phenylalanine and thus lower the requirement for the latter amino acid. The high concentrations of the

Fig. 8.2 (continued) Research Council 2006); skin and wool of domestic sheep (National Research Council 2007b); feathers of domestic chicken (National Research Council 2003); whole eggs from mixed species of fish (Mambrini and Guillaume 2001); average of eggs from chickens, mallard ducks, turkeys and Japanese quail; average of milks from domestic cattle, sheep, goats and water buffalos (Agricultural Research Service 2007)

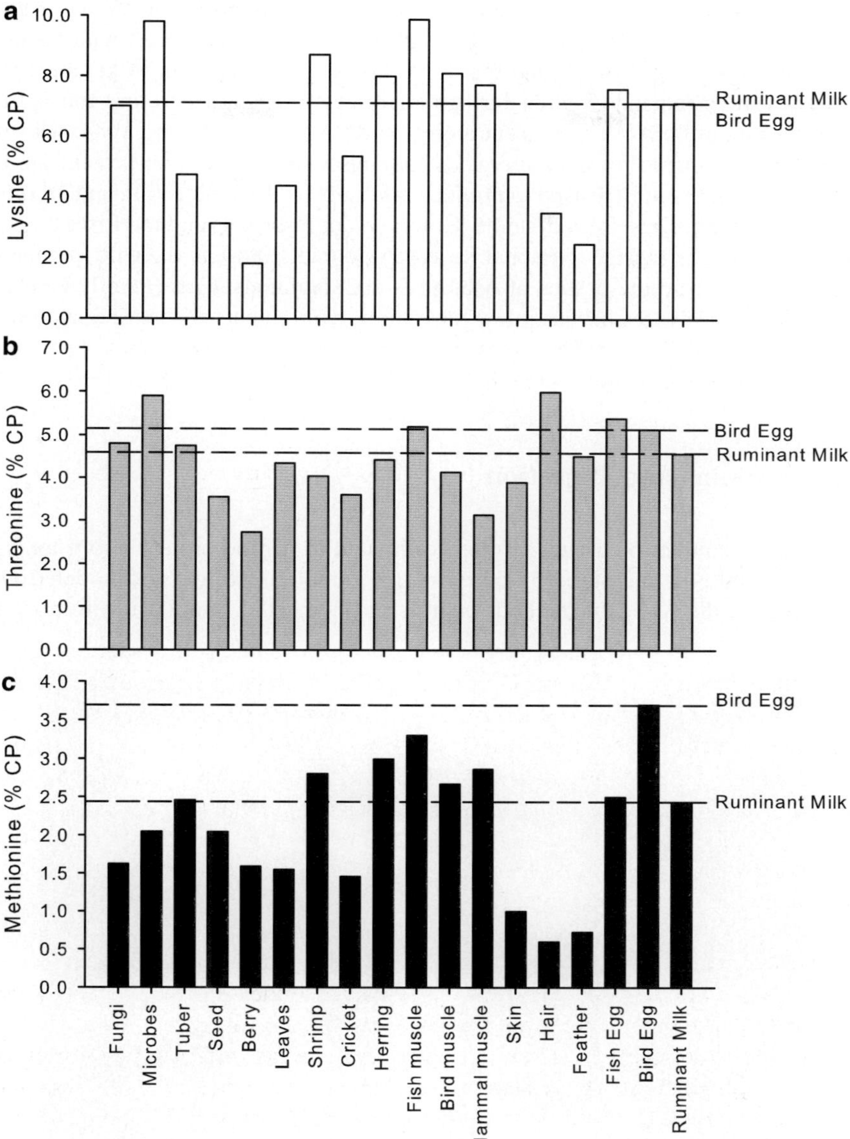

Fig. 8.2 Contribution of three essential amino acids to crude protein (CP) in the tissues of selected microbes, plants and animals. **a** Lysine. **b** Threonine. **c** Methionine. *Dashed horizontal lines* provide a reference for comparison with bird eggs or milk from ruminants. Microbes: yeast (National Research Council 2003) and mixed ruminal organisms from domestic sheep (National Research Council 2007b). Plants: tubers of sweet potato, seeds of corn, blueberries and leafy alfalfa (National Research Council 2003). Animal: muscle from mixed species of shrimp (National Research Council 2006); whole cricket and herring (National Research Council 2003); muscle from mixed species of fish (Mambrini and Guillaume 2001), domestic chicken and cattle (National

S-containing methionine in bird eggs reflect the requirements of the chick for both methionine and cysteine. Feathers, skin and hair are rich in cysteine, which forms the sulfhydryl bridges of cystine that crosslink the protein fibers of keratin (Fig. 8.1E). The requirements of birds during molt or egg laying are therefore considered to be the sum of both of these sulfur amino acids (Murphy and King 1984; Klasing 1998). Requirements for S-amino acids may also include the non-protein amino acid taurine that is mostly used to produce bile salts for lipid digestion and, in most animals, is synthesized from cysteine. However, domestic cats and red foxes require a dietary source of taurine (National Research Council 2006). In the wild, predators obtain taurine from the tissues of fish, birds and mammals. Cats (Family Felidae) are obligate carnivores because they have high requirements for taurine and other amino acids, both dispensable and essential; those requirements are most easily met by consuming other animals (Fig. 8.3).

8.2 Proteins and Digestion

The protein content of tissues is often estimated as 'crude protein' from total N content. Total N is measured by digesting the sample in strong acid to determine ammonia by the Kjeldahl method, or by combusting the sample in a furnace to

Fig. 8.3 Carnivores derive most of their dietary energy from the oxidation of amino acids in prey. **a** Felids such as the African lion may be considered obligate carnivores because of their requirements for specific amino acids such as arginine and taurine, and their high N requirement for maintenance. **b** Indian cobras use special salivary glands to inject neurotoxins (venom) and proteolytic enzymes to immobilize their prey

determine N_2 gas released in an elemental analyzer. The average N content of a wide variety of proteins is 16 g N·100 g^{-1} protein, and so crude protein is usually calculated as 6.25 g crude protein·g^{-1} N (Robbins 1993). However, the N content of a protein is affected by its amino acid composition; some legume and seed proteins, for example, are 18.9% N (5.29 g crude protein·g^{-1} N). Also, not all the N in the diet is associated with protein, but may be in the form of inorganic nitrates that are not available to animals.

Dietary proteins have a wide diversity of structures and amino acid sequences that must be degraded to peptides and amino acids for absorption. The simple peptide bond between two amino acids has 400 possible sequences (20 × 20 amino acids). Protein digestion (proteolysis) has broad specificity for amino acid sequences, which allows the same system to digest many proteins. Proteolysis is carefully controlled to avoid self-digestion; loss of these controls at death results in continued proteolysis that degrades the tissues of the digestive tract. Consequently, the digestive tract is quickly removed from harvested fish and game to avoid changes in flavor or microbial contamination of the meat as the viscera degrades. Proteolytic enzymes are secreted in the stomach and duodenum as zymogens, which are inactive enzyme precursors that are activated by acid or another enzyme in the digestive tract. Salivary glands of snakes also secrete proteases that are active only in venom and serve to increase the penetration of neurotoxins by degrading skin and muscle proteins (Zug 1993) (Fig. 8.3).

Hydrochloric acid secreted into the stomach denatures proteins by disrupting the weak bonds that fold and hold proteins in tissues, opens membranes by releasing embedded proteins, disaggregates fibrous proteins and precipitates soluble proteins. Acid secretion is stimulated by the presence of protein in the stomach and stopped by negative feedback from the mucosa when digesta pH falls below 1.5 (Fig. 8.4). Denaturation increases the time and surface area for enzymatic digestion by precipitating soluble proteins into a slower-moving solid phase and by opening the structure of all proteins to enzymes. Enzyme secretion is combined with acid secretion in the gastric glands. In mammals, inactive pepsinogen is produced by chief cells at the base of the gland and activated to pepsin with acid produced by parietal cells at the top of the gland (Fig. 8.4). Gastric glands in fish are simpler than those of mammals; carnivores such as pike (Family Esocidae) and sculpin (Family Cottidae) produce pepsinogen and acid together from a group of oxyntopeptic cells in the stomach (Rust 2002). The combination of acid and pepsin breaks the internal peptide bonds of the protein and prevents the molecule from reforming in the higher pH of the midgut.

Gastric proteolysis does not completely digest protein, but it increases the rate of digestion of denatured proteins and peptides in the midgut. Gastric proteolysis is most advantageous to carnivores such as snakes and lizards that consume large meals of protein and frogs that swallow whole prey. Acid secretion and the subsequent regulation of blood pH contribute to large increases in energy expended after a meal, and ultimately reduce the net gain of energy from the diet of these predators (Wang et al. 2005). Animals that consume smaller amounts of proteins that are more easily degraded require less gastric proteolysis and thus have lower costs of digestion. Most fish do not secrete acid or pepsin in the early larval stages, and some species

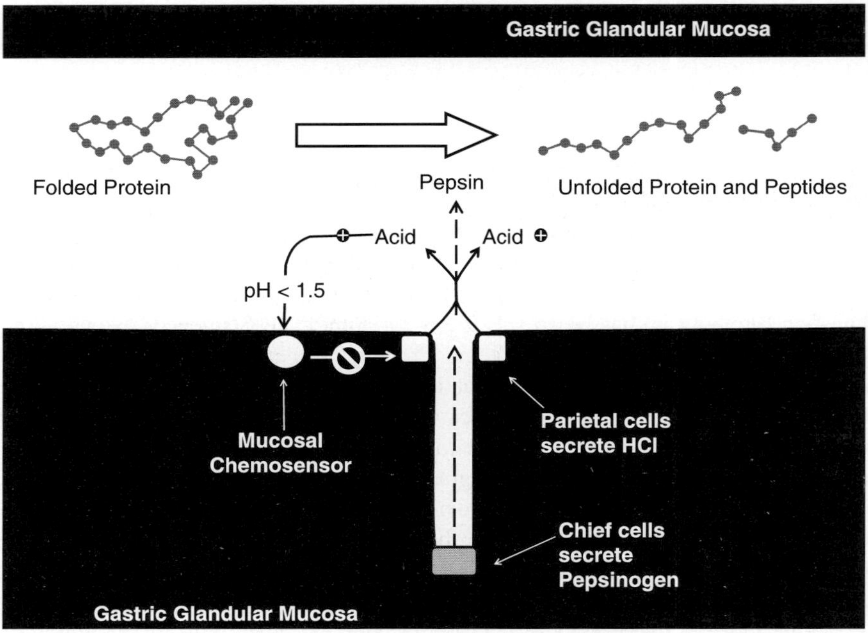

Fig. 8.4 Gastric secretion and its control by feedback inhibition in mammals. Pepsinogen is activated to pepsin by acid hydrolysis soon after the enzyme leaves the gastric gland. Low pH in the digesta inhibits further acid secretion. Proteins are unfolded and partly degraded to peptides by the combined effect of pepsin and acid

are agastric as adults (Chapter 5) (Guillaume and Choubert 2001). Larval fish may rely on easily digested proteins as well as free amino acids from zooplankton (Rønnestad and Conceição 2005). Young mammals also rely on readily digestible milk proteins (caseins) that can be denatured with less acid. Suckling mammals produce a form of pepsin (chymosin or rennett) that is active at a higher pH (pH 4–5) than adult pepsin (pH 2–3). Young animals may therefore expend less energy on protein digestion and direct those savings to deposition of dietary protein and growth. The costs of digestion and growth are discussed further in Chapter 10.

The flow of acid digesta into the duodenum stimulates the release of the hormone secretin into the blood and the secretion of mucus that protects the mucosa. Secretin stimulates the pancreas to release alkaline buffers that raise the digesta to pH 7–8. Pancreatic secretions also contain a series of zymogens that form an activation cascade. The key to the activation cascade is enterokinase, which is released from the duodenum when acid and protein enter the midgut (Fig. 8.5). Enterokinase activates trypsin, which subsequently activates other enzymes such as chymotrypsin and carboxypeptidase (Fig. 8.5). The cascade amplifies and accelerates proteolysis because enterokinase can activate multiple trypsin molecules. Each enzyme has a different specificity for amino acid side chains that complements the rest of the cascade, which results in the rapid degradation of a long chain of peptides from the

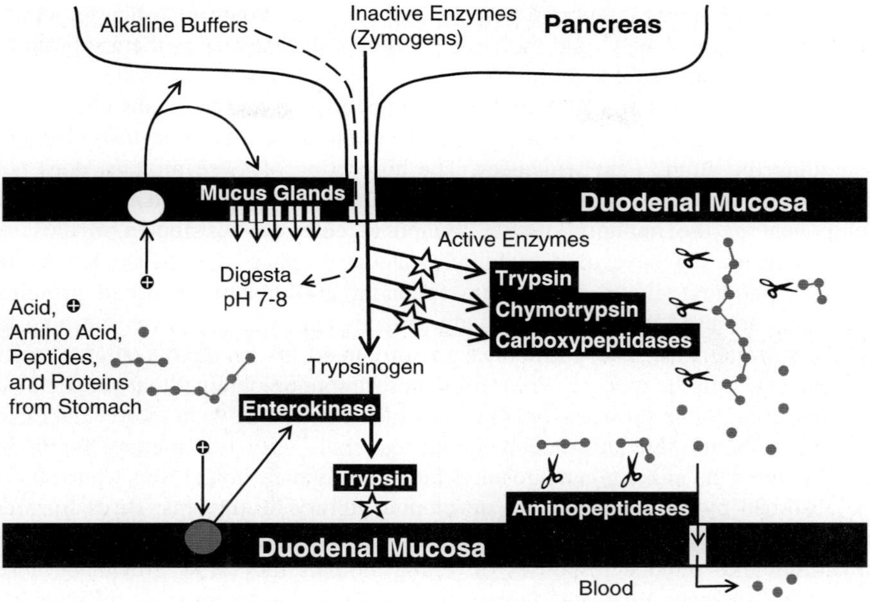

Fig. 8.5 Digestion of protein in the small intestine. Acidic digesta stimulate the release of buffers from the mucosa and from the pancreas. Pancreatic enzymes are activated by a cascade in the digestive tract. Enteropeptidase from the mucosa activates trypsinogen to trypsin. Trypsin (*star*) activates other trypsinogen molecules as well as chymotrypsinogen, pro-carboxypeptidases and other pancreatic zymogens. Amino peptidases at the surface of the mucosa degrade small peptides. Amino acids and small peptides are transported into the cells of the mucosa. Peptides are further cleaved in the cell before amino acids are released into the blood

center and both ends. Peptides are further cleaved by amino-peptidases that are bound to the mucosal surface.

Mucosal cells absorb both individual amino acids and short peptides of two or three amino acid residues (dipeptides and tripeptides) (Breves and Wolffram 2006). Amino acids are absorbed through the mucosal cell and secreted into the blood by an array of transporters with different side-chain specificities. Amino acid transporters use ATP and can therefore oppose concentration gradients when an acid has a higher concentration in the blood than in the digesta; herbivores therefore can absorb amino acids even when the concentrations in the diet and the digesta are low. Over half the amino acids may be absorbed as short peptides by PEPT1 transporters (Breves and Wolffram 2006). Most of the short peptides are hydrolyzed to amino acids within the mucosal cell before being secreted into the blood. The number of amino acid and peptide transporters declines precipitously from the ileum to the hindgut, so that amino acids are poorly absorbed from the cecum and colon (Hume et al. 1993). Most microbial protein produced by fermentation in the hindgut is therefore lost in the feces. Small hindgut fermenters such as marsupial possums, hares, lemmings and grouse may recover some of this protein by practicing cecotrophy (Chapter 5); the ingestion of cecal contents allows microbial protein to be digested in the foregut and the midgut in the same manner as dietary proteins (Stevens and Hume 1995).

A very small fraction of the proteins and peptides in digesta is absorbed when cell membranes invaginate and bring substances into the cells (pinocytosis) without degrading the amino acid sequences. The absorption of these proteins does not contribute significantly to the amino acid uptake of the animal but is an integral component of the immune system. Lymphoid cells are distributed in clusters throughout the digestive tract as follicles at the esophagus (tonsils), duodenum (Peyer's patches), ileum and cecum (Klasing 2005). The absorbed proteins (antigens) are used to produce antibody proteins that will recognize and bind to invading organisms. The antibodies are produced by immune response cells (lymphocytes) in the mucosa. Free antibodies (immunoglobulins) that are secreted in mucus bind to the surface proteins of organisms and make them more vulnerable to proteolysis and thus destruction. Infant mammals receive immunoglobulins in milk. Some of the milk immunoglobulins bind the proteins of organisms while others are absorbed by the infant for the developing library of antigens. Birds transfer immunoglobulins to their eggs for immune function in the chick (Klasing 2005). Infections and immune responses of young animals may slow growth because energy and nutrients are diverted to repairing tissue and to responses at the intestine that can impair absorption when the mucosa is infiltrated with lymphocytes. The cost of maintaining immune function and its relationship with trace minerals and vitamins are discussed in Chapters 9 and 10.

8.3 Intermediary Metabolism of Amino Acids

The organs of the body have different demands for amino acids because they synthesize proteins with different amino acid sequences. The C and N in amino acids are therefore exchanged within a cell and between organs. Transaminases catalyze reversible reactions that transfer amino groups between the C skeletons of all 20 common amino acids. For example, alanine aminotransferase (ALT) uses the amino acid alanine and the C skeleton keto-glutarate to produce a C skeleton pyruvate and another amino acid, glutamate. Hibernating bears use ALT to make glucose in the liver from alanine in muscle; alanine is exported from muscle to the liver where the amino acid is converted to pyruvate for gluconeogenesis (Koebel et al. 1991). The transaminases are intracellular enzymes that are elevated in blood serum when cells are damaged; serum ALT increases as liver cells are damaged in dogs but decreases during hibernation in bears, which indicates that the liver and other tissues are not damaged by the long fast (Blood and Studdert 1988; Barboza et al. 1997).

Amino acids can be oxidized for energy (ATP) or stored as lipid or glycogen once the amino group is removed by a transaminase. All 20 common amino acids can be converted to C skeletons that can enter the TCA cycle (Fig. 8.6). Consequently, the C from all amino acids can be used as a fuel during fasting or exercise. All amino acid C can be stored as fatty acids (ketogenic) in triglycerides (fat). Glucogenic amino acids

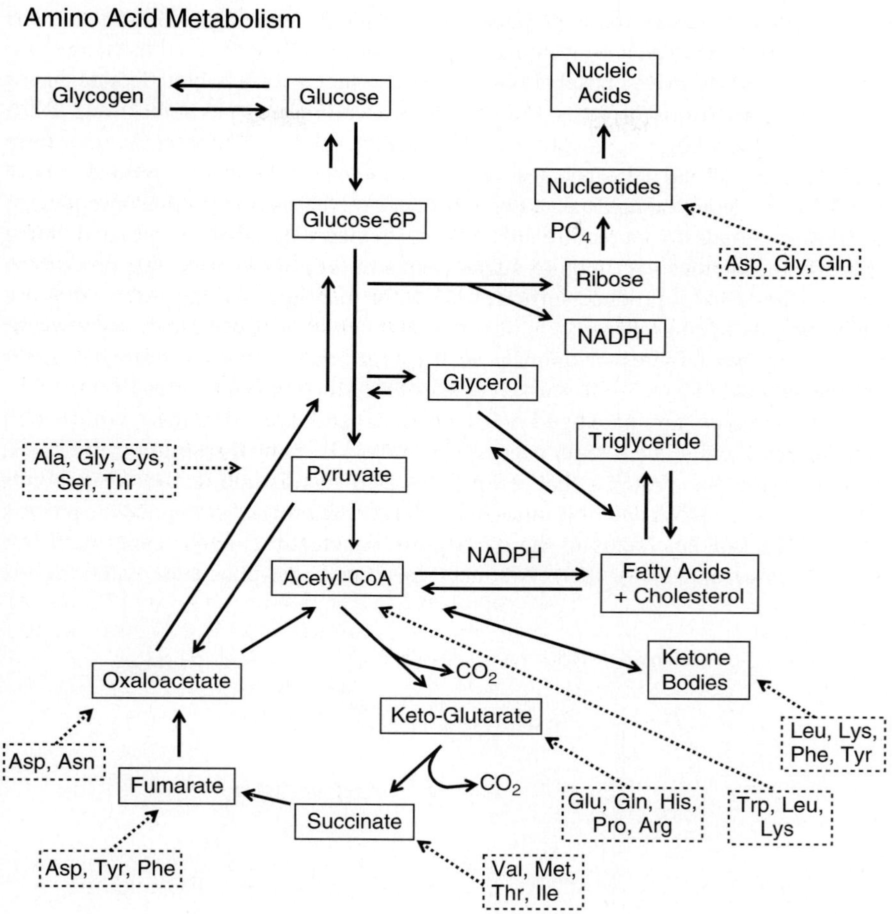

Fig. 8.6 Metabolic pathways for utilizing amino acids for energy or for storing the C as fat (triglyceride) or glycogen

can be used to produce glucose, glycerol and glycogen (Fig. 8.6). Body proteins are used to restore glucose reserves during prolonged fasting (e.g., hibernation in bears) or exercise (e.g., migratory flights of birds) because all but two of the amino acids are glucogenic. The use of body protein for energy is discussed in Chapter 10.

8.4 Nucleic Acids and Digestion

The bases of DNA and RNA contain N; bases with double rings are purines (adenine, guanine) whereas those with a single ring are pyrimidines (thymine, cytosine, uracil; Fig. 8.7). A sugar (ribose or deoxy-ribose) added to a base forms a nucleoside that then becomes a nucleotide when phosphate is added to the sugar. Phosphate esters on each nucleotide (e.g., adenosine-mono-phosphate or AMP) are used as energy currencies in ADP and ATP. Phosphate esters between sugars form polynucleotide chains for the various forms of RNA (messenger, transcriptional, ribosomal) and DNA (nuclear, mitochondrial, cytosolic). Nucleic acids can be synthesized de novo from phosphate and amino acids because glucogenic acids can be used to produce ribose and dispensable amino acids are used to produce the bases (Fig. 8.6). Consequently, vertebrate animals do not require nucleic acids in their diet. On the other hand, biting flies and mosquitoes may require nucleic acids in their diet during egg production because they cannot produce sufficient DNA for their eggs. Biting insects that use the N from blood, muscle and skin of vertebrates subsequently affect the movements and energy expenditures of animals such as caribou and moose during summer (Renecker and Hudson 1990; Russell et al. 1993; Mörschel and Klein 1997).

Nucleic acids are packaged with proteins and are contained within cell membranes. Acid in the foregut exposes the nucleic acids by disrupting membranes, by denaturing the proteins that wrap DNA and by disrupting the weak bonds between bases in DNA and ribosomal RNA. Pancreatic proteases continue the process of unraveling nucleic acids in the duodenum. Pancreatic endonucleases for RNA (ribonuclease) and DNA (deoxyribonuclease) cut sugar-phosphate esters at the

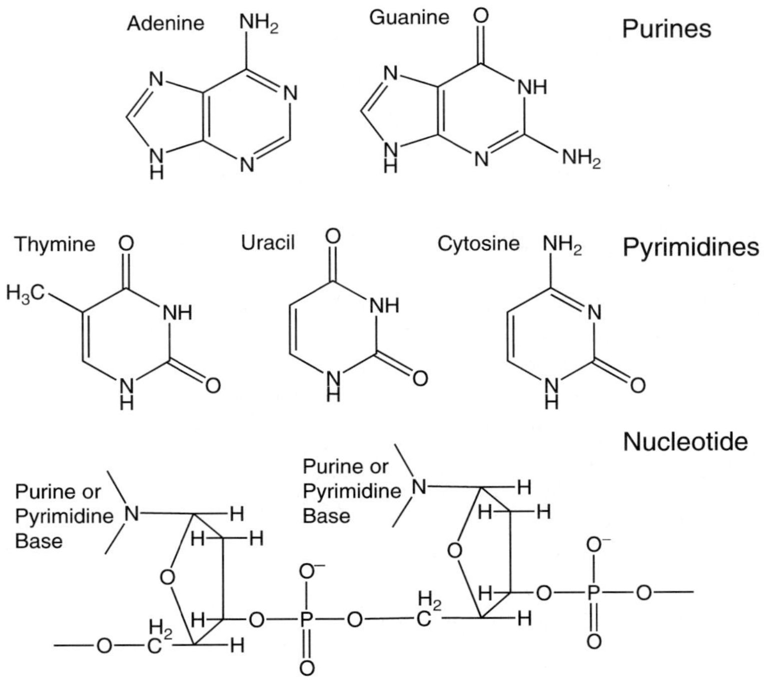

Fig. 8.7 Pyrimidine and purine bases used to form nucleotide chains in RNA and DNA with phosphates and either ribose (RNA) or deoxy-ribose (DNA)

center of the chains to produce polynucleotides. Intestinal phosphodiesterases cleave mononucleotides from the ends of the chains before nucleotidases remove the phosphate to produce nucleosides (base + sugar) (Mathews and Van Holde 1996). Nucleosides and inorganic phosphate are absorbed by active transport into the mucosa (Breves and Wolffram 2006).

Nucleic acids may account for 20% of the microbial N flowing into the duodenum of ruminants because every microbial cell produced by fermentation has a genome (Stevens and Hume 1995). Elk, bison, sheep, goats, cattle and kangaroos have high concentrations of endonuclease in the pancreas because nucleic acids are an important source of nucleosides as well as phosphate for foregut fermenters. Absorbed nucleosides are readily phosphorylated by tissues for incorporation into nucleotides. The sugar from excess nucleosides can be oxidized for energy or used for synthesis of glucose or triglycerides (Fig. 8.6). Excess purines and pyrimidines are oxidized and then excreted in the urine.

8.5 Nitrogen Metabolism

Proteins and nucleic acids are both forms of organic N that are linked by their functions and by the metabolism of the N they contain. Organic N flows through several interconnected metabolic pools in the body. Body protein is the largest pool of N, and is constantly turned over as cells maintain their function. Proteins are degraded and synthesized continuously; the difference between the two rates results in net synthesis or net loss (Fig. 8.8). Total protein turnover is similar between summer activity and winter hibernation of bears; body protein is gained in summer when the rate of synthesis exceeds degradation, but is lost during hibernation when the rate of degradation exceeds synthesis (Barboza et al. 1997). Protein turnover releases and consumes all 20 common amino acids that exchange C and N by transamination in a combined pool of amino acids. Cellular RNA and DNA are also turned over as proteins are synthesized and degraded throughout the life of cells, from growth and maintenance to their eventual death. Nucleic acid turnover therefore results in turnover of the subsidiary pools of purine and pyrimidine N in the body (Fig. 8.8).

Waste N is formed whenever the C in either protein or nucleic acids is oxidized for energy. Body N is excreted during normal cell turnover for maintenance of tissues and when cellular constituents are used during fasting or exercise. Dietary N is excreted when N intake exceeds the demand for cell replacement and growth, that is, when dietary protein is oxidized for energy. Waste N is routed to pathways that are already available for oxidizing the C from amino acids or purines (Fig. 8.8).

8.5.1 Ammonia

Amino acids are the most abundant form of organic N and the principal source of excretory N from the diet or from body tissue. The simplest route for using C from

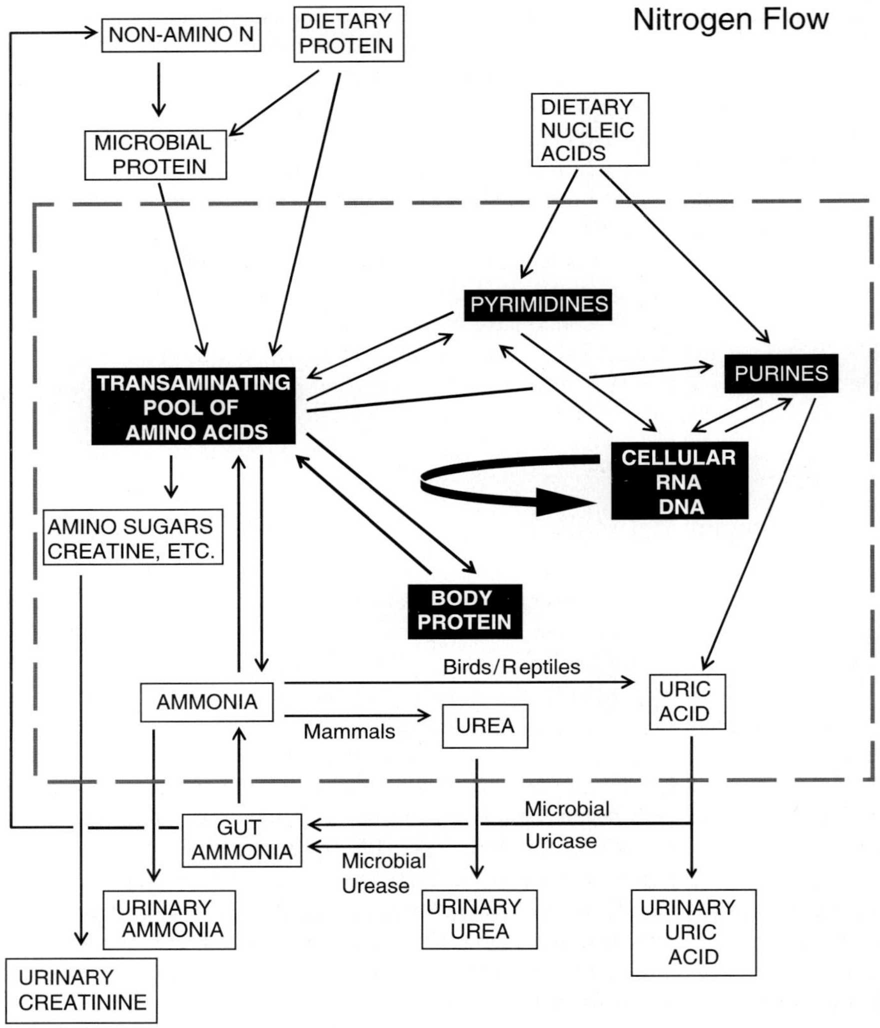

Fig. 8.8 Exchanges of N within the animal body (*broken line*) and their relationships to dietary intake and excretion of N metabolites

amino acids is to remove the amino groups as ammonia (NH_3). Amino groups are routed to a common amino acid such as glutamate by transamination; for example, the amino group from excess alanine is transferred to α keto-glutarate with ALT to produce glutamate (Figs. 8.1c, 8.6 and 8.8). The amino group is removed from glutamate by glutamate dehydrogenase to release ammonia (Fig. 8.9). Glutamine also carries an amino group on the side chain; an amino group is added to the side chain of glutamate to produce glutamine and subsequently removed to produce ammonia (Fig. 8.9). Ammonia is produced at the liver and the kidney from glutamate

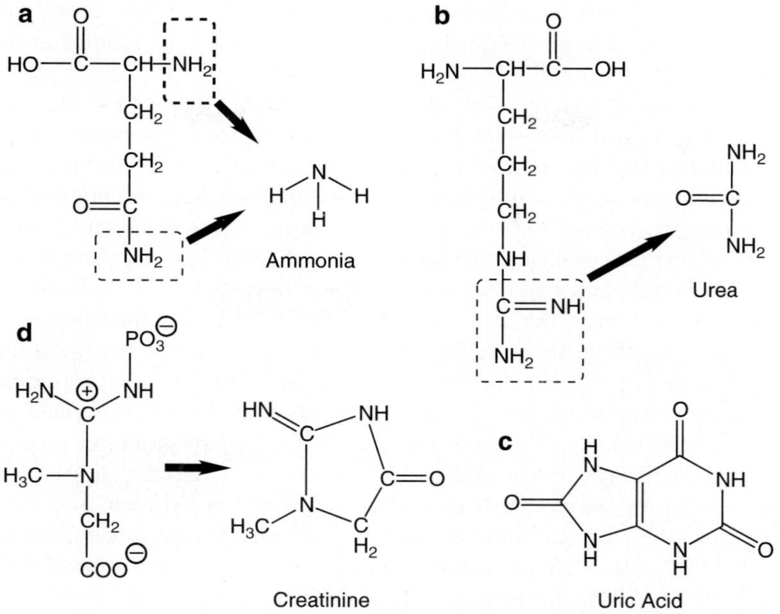

Fig. 8.9 Excretory metabolites of N. **a** Ammonia is formed from the amino groups of amino acids such as glutamine. **b** Urea is produced by hydrolysis of the arginine side chain. **c** Uric acid is produced from purines such as adenine, guanine and their amino acid precursors. **d** Creatinine is spontaneously formed from creatine phosphate in muscle

and glutamine. However, ammonia is toxic because it spontaneously forms ammonium in water (NH_4^+) and acts as a potent alkali that alters the pH of body fluids. Consequently, animals cannot tolerate accumulation of ammonia in the body. Most of the waste N from fish is excreted as ammonia at the gills where ammonium diffuses readily into the surrounding water; this is an inexpensive route for N excretion that allows larval fish to rely on amino acids as their primary source of C and energy (Mambrini and Guillaume 2001).

8.5.2 Urea

Mammals only use ammonium excretion to control urinary pH for acid–base balance. Mammals excrete most of their waste N as urinary urea, which is non-toxic and readily diffuses across membranes. Transaminated amino groups and ammonia are used to produce urea in the liver (Waterlow 1999). Urea synthesis is a modification of the pathway for synthesizing the amino acid arginine (Fig. 8.1B); the addition of one enzyme (arginase) hydrolyzes the side chain to produce urea (Fig. 8.9). Dietary arginine serves as a carrier for routing excess N from the diet to urea; obligate carnivores such as cats have high demands for arginine and thus urea synthesis (National Research Council 2006) because they oxidize large amounts of protein for energy.

Urea is a primary osmolyte for marine sharks; urea is excreted at the gills and the kidney but accumulates in the body when filtered urea is returned to the blood as the salinity of water changes (Withers 1992; Pillans et al. 2005). Terrestrial mammals concentrate urea in the kidneys, especially as water availability declines; desert mammals increase the concentration of urea in both urine and blood. The ability to retain urea conserves water in camels and desert kangaroos because these animals can excrete less water in urine (Schmidt-Nielsen et al. 1957; Hume 1999). Urea also diffuses from the blood into the saliva and across the mucosa of the foregut and hindgut through urea transporters (Martin et al. 1996; Marini et al. 2006). Microbes in the gut degrade the urea with urease to produce ammonia, which can be used for microbial protein synthesis. Waste N that is recycled as urea in ruminants and kangaroos sustains the microbial community and its fermentation of low-protein fibrous forage (Kennedy et al. 1992). Urea recycling is also an important route for conserving N in winter when diets are both low in N and high in fiber. In reindeer and caribou, for example, 71% of the urea produced in winter is recycled to the gut when animals consume low-N diets such as lichens (Barboza and Parker 2006, 2008). In muskoxen, 87% of the urea produced on a fibrous diet of grass in winter is recycled to the gut, and 45% of this recycled N is returned to the amino acid pool in the body (Barboza and Peltier, unpublished data). Recycled urea-N can return to the pool of amino acids in two ways: when the microbial protein synthesized from the recycled N is digested and the amino acids are absorbed in the small intestine, or when the microbial amino acids are deaminated in the gut and the ammonia is absorbed and attached to C-skeletons in the liver (Fig. 8.8). In dormant bears, 97% of the urea-N produced in the body is recycled to the gut, and 100% of the recycled N is returned to the amino acid pool, which conserves body protein and allows young animals to survive their first winter (Barboza et al. 1997).

8.5.3 *Uric Acid*

Uric acid is produced from a pathway for degrading purine bases (adenine and guanine) in most tissues, especially the liver (Figs 8.8 and 8.9). All animals produce uric acid, but the compound is processed further to more soluble products such as urea in fish and allantoin in most mammals other than humans. The urinary excretion of purine metabolites such as allantoin can be used to indicate the digestion of nucleic acids, which is mainly associated with digestion of microbial cells in herbivorous mammals (Balcells et al. 1998). Purine excretion can be used to indicate the microbial yield and indirectly the relative metabolizable energy intake by domestic ruminants and free-ranging elk from urine samples collected in snow (Garrott et al. 1996).

In reptiles and birds, most waste N is excreted as uric acid. Excess N from amino acids is transaminated to aspartate, glycine and glutamine which are incorporated into purines (Fig. 8.6). Uric acid is less soluble than urea and can therefore be safely stored in a solid form within the developing egg of birds and excreted with minimal water loss by desert birds and reptiles (Campbell 1994). Uric acid is secreted by active transport at the renal tubules rather than filtered like urea. Consequently, uric acid excretion allows carnivorous birds and reptiles to excrete large amounts of N in a semi-solid form even though their kidneys can only produce a dilute urine (Kirschner 1991). Uric acid is maintained in a colloidal form in the kidneys by forming microspheres coated with small amounts of protein; the acid would otherwise precipitate and block the tubules (a condition called renal gout) (Braun 1999). In reptiles, uric acid may also facilitate excretion of excess minerals such as Na or K that can be complexed with urate salts for safe storage in the urinary bladder (Minnich 1972). In ducks and grouse, urine from the cloaca is refluxed into the colon and ceca where water and ions are absorbed and where urates are degraded by microbial uricases (Braun 1999; Hughes et al. 1999). The ammonia so produced is subsequently absorbed into the blood and returned to the liver. Recycling of uric acid in birds is not as well researched as urea recycling in mammals; recycled urate-N could conserve body protein in birds such as grouse that consume low-N winter browse, but the proportions of urate-N recycled and returned to amino acids have not been measured (Laverty and Skadhauge 1999).

8.5.4 Creatinine

Creatinine is derived from the high-energy store of phosphate in muscle (creatine phosphate; Fig. 8.9). Creatine phosphate spontaneously forms creatinine, which cannot be re-used and is therefore excreted in the urine. Creatinine in blood and urine is derived from muscles in the diet and the body of carnivores. In herbivores, daily urinary excretion of creatinine (milligrams per day, $mg{\cdot}d^{-1}$) is proportional to muscle mass and is not affected by muscle activity. The concentration of creatinine (milligrams per milliliter, $mg{\cdot}mL^{-1}$) can therefore be used as a reference for concentrations (milligrams per milliliter, $mg{\cdot}mL^{-1}$) of other metabolites by calculating a ratio to creatinine (milligrams of metabolite per milligram of creatine, mg metabolite·mg $creatinine^{-1}$).

The urinary ratio method has been used to evaluate the relative condition of wild ruminants in winter where urine can be collected from snow (DelGiudice et al. 1989). Creatinine ratios have been applied to deer (*Odocoileus* sp.) for urea from amino acids, allantoin from purines, 3-methylhistidne from muscle, hydroxyproline from connective tissue, and cortisol and corticosterone from stress hormones (DelGiudice et al. 1988, 1996, 1998; Saltz and White 1991; Parker et al. 1993a; Vagnoni et al. 1996). Increases in the urinary ratio of urea to creatinine concentrations have been used to indicate the oxidation of amino acids from body protein in winter for white-tailed deer, elk and bison (Moen and DelGiudice 1997; DelGiudice et al. 2001). Creatinine concentrations do not remain constant in all species because clearance of creatinine at the kidneys changes with season in reindeer, caribou and muskoxen (Peltier et al. 2003; Parker et al. 2005; Barboza and Parker 2006). Also, urea to creatinine concentrations may not accurately reflect body protein loss when renal function is altered such as by water availability or solute load, or when dietary N is excreted as urea (Barboza et al. 2004). One alternative to using urea to creatinine concentration ratios as a measure of body condition is to compare the N isotopes in urinary urea with those from dietary N and muscle creatinine or blood cells; in reindeer and caribou the isotopic similarity between urinary urea and body proteins increases as body N is oxidized to urea (Parker et al. 2005; Barboza and Parker 2006).

8.6 Nitrogen Balance and the Requirement for N

N balance is the net gain or loss of N from the body of an animal, which is usually measured by mass balance in a metabolism cage or tank (Chapter 4). The minimum requirement for dietary N is the intake that supports zero N balance, that is, when the rates of protein synthesis and degradation are equal (Fig. 8.10). Positive N balance reflects deposition of dietary N in tissues for either short-term stores in the liver or for growth and seasonal gain of tissue such as muscle. N balance does not continue to increase with intake of N because mass gain is controlled by the genetic program for

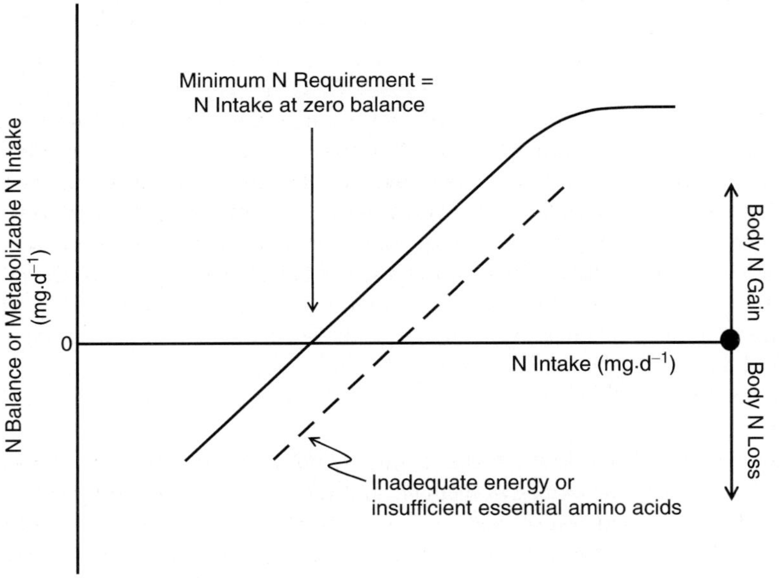

Fig. 8.10 Estimating the minimum requirement for N from the relationship between N balance and N intake (*solid line*). Animals retain less N when supplies of energy or essential amino acids are inadequate (*broken line*). Inadequate intakes of energy increase N loss because dietary and body proteins are used for energy. Inadequate supplies of essential amino acids decrease N gain by limiting the synthesis of body protein. N balance reaches an asymptote when intakes of N exceed the maximum for protein deposition in the body

tissue turnover and synthesis. Protein synthesis requires a complete supply of amino acids; diets that are low in one or more essential amino acids will increase the apparent N requirement because the animal must consume more N to meet the demand for the limiting amino acids (Fig. 8.10). Animals in negative N balance use stores of body protein to sustain the amino acid pool for protein synthesis in critical tissues. Labile proteins such as albumins from the liver are used to sustain the amino acid pool between meals, whereas proteins in muscle may be used during prolonged fasting or exercise (Wannemacher and Cooper 1970; Waterlow 1999; Bordel and Haase 2000).

The minimum N requirement for maintenance of an animal can be estimated from endogenous losses in urine and feces. Daily and seasonal requirements for N are minimized by reducing losses of N in both the urine and the feces (Table 8.2). Those losses are lowest when energy intakes are adequate for protein turnover, that is, when digestible energy intakes are sufficient to maintain body mass. The interactions between energy demands and body protein of fasting, growing and reproducing animals are discussed further in Chapter 10.

8.6.1 Endogenous Urinary N

The minimum urinary N loss that is predicted at zero N intake is called endogenous urinary N (EUN). The principal source of N in EUN is urea or uric acid from amino acids that were released by normal protein turnover (Fig. 8.8). Amino N is excreted when the C skeleton is used for energy (ATP) or for glucose synthesis (Fig. 8.6). Carnivores have higher EUN than herbivores and consequently require more dietary N at zero balance. High EUN in cats and other obligate carnivores reflects their inability to down-regulate amino acid oxidation when protein intakes decline

Table 8.2 Calculating N balance in a ruminant (e.g., caribou)

Parameter	Calculation	Measure
Ingested N ($g \cdot d^{-1}$)	A	56
Fecal N ($g \cdot d^{-1}$)	B	12
Digestible N intake ($g \cdot d^{-1}$)	C = A – B	44
N digestibility ($g \cdot g^{-1}$)	D = C ÷ A	0.79
Metabolic fecal N (MFN) loss ($g \cdot d^{-1}$)	E	9
Truly digestible N intake ($g \cdot d^{-1}$)	F = A – (B – E) = C + E	53
True N digestibility ($g \cdot g^{-1}$)	G = F ÷ A	0.95
Urinary N loss ($g \cdot d^{-1}$)	H	8
Metabolizable N retained ($g \cdot d^{-1}$) (N balance)	I = C – H	36
N metabolizability ($g \cdot g^{-1}$)	J = I ÷ A	0.64
Endogenous urinary N (EUN) loss ($g \cdot d^{-1}$)	K	4
Truly metabolizable N intake ($g \cdot d^{-1}$)	L = C – (H – K) = I + K	40
True N metabolizability ($g \cdot g^{-1}$)	M = L ÷ A	0.71
Biological value ($g \cdot g^{-1}$)	N = L ÷ F	0.75

(Hendriks et al. 1997). High energy demands may contribute to increasing EUN if amino acids are used for energy. Eutherian mammals such as deer have higher EUN than marsupials (kangaroos), partly because their basal rates of energy metabolism are higher than those of marsupials (Hume 1999). High glucose availability reduces EUN because amino acids are rarely needed for gluconeogenesis. Nectarivorous marsupials (e.g., honey possums) and birds (e.g., hummingbirds) have very low EUN losses and thus low N requirements (Brice and Grau 1991; Bradshaw and Bradshaw 2001). Recycling of waste N further reduces EUN and N requirements. Herbivores with extensive fermentation systems such as grazing kangaroos, wombats and camels have low EUN because they can re-use a large proportion of the urea they produce (Hume 1999).

8.6.2 Fecal N Losses

Losses of N in feces depend on the N content and structure of the diet and the function of the digestive tract (Chapter 5). Fecal N losses are lowest for nectarivores because their liquid diet is highly digestible. Diets of fibrous plants result in more feces and a greater loss of indigestible N and unresorbed N. Indigestible N is mainly associated with proteins and non-amino N bound to the cell walls of plants. Some soluble dietary proteins can also escape digestion if they are bound to tannins, a group of PSMs found in the leaves of woody plants (Chapter 3). A standard approach to measure protein binding of forage tannins is to measure the ability of the food to precipitate protein from cattle blood (bovine serum albumen or BSA). This quantification of tannin-binding capacity is used to more accurately calculate the reduction in digestible protein and digestible dry matter consumed by ruminants (Hanley et al. 1992). Tannins may also bind to proteins in the secretions of some animals such as saliva (Robbins et al. 1987) which reduces the impacts of tannins as digestion-reducing agents.

The unresorbed fraction of fecal N that is not of immediate food origin is called metabolic fecal N (MFN) because it is comprised of mucus secretions and cells. MFN increases with food intake and with abrasive dietary components such as plant fiber consumed by herbivores and the particles of sand ingested by omnivores that feed on underground fungi, insects and plant tubers (Young and Hume 2005). MFN also includes microbes that are associated with the solute phase as well as those bound to the fiber matrix in the particulate phase of digesta (Mason 1969). Hindgut fermenters such as horses lose most of the microbial N produced from fermentation in their feces, whereas cecotrophy in ringtail possums recovers enough microbial N from the feces to halve their dietary N requirement (Chilcott and Hume 1984). The fecal loss of microbial N is influenced by the structure and composition of the diet, and the composition of the microbial community (Chapter 6). Modified amino acids can serve as bacterial markers in the feces (e.g., 2,6-diaminopimelic acid or DAPA) of herbivores. DAPA can be used as an index of the dietary energy available to the foregut fermentation in ruminants such as white-tailed deer (Brown

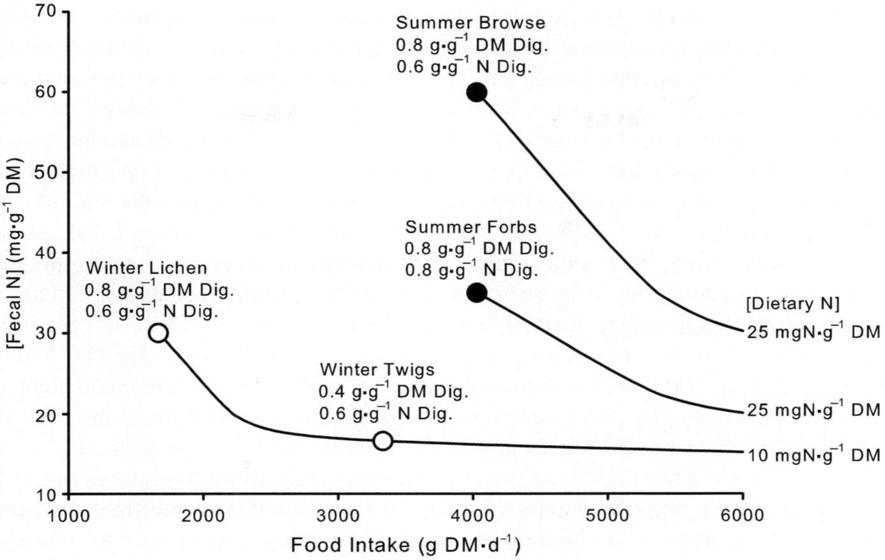

Fig. 8.11 A seasonal model of fecal N concentration (mg N·g^{-1} DM) in relation to dietary N for caribou. Food intakes are based on meeting energy demands of 24 and 58 MJ·d^{-1} in winter and summer respectively. Food intakes vary as the energy digestibility of the diet changes from 0.8 to 0.2 kJ·kJ^{-1} when the gross energy content is 18 kJ·g^{-1} DM (digestible energy contents of 3.6–14.4 kJ·g^{-1} DM). Sources of fecal N include metabolic fecal N (MFN) as well as the undigested residues from the food; MFN is calculated as 1 g N·100 g^{-1} fecal DM. This calculation may underestimate fecal N because abrasion can increase MFN at low digestibilities and high intakes of DM

et al. 1995); increasing energy intake provides more energy for microbial growth and the production of more DAPA in the rumen (Van Soest 1994).

Fecal concentrations of N have been used to indicate the N content of food for herbivores, especially ruminants such as deer and sheep. However, the concentration of any metabolite in feces is affected by the DM digestibility of the diet. Less digestible diets result in more feces that are mainly associated with plant fiber or ash which lower the concentration of all N fractions. Fecal N concentration declines as an animal eats more food to compensate for the decreasing digestibility of energy in the diet, that is, indigestible DM dilutes fecal N as food intakes increase. The model in Fig. 8.11 predicts a decline in fecal N when caribou switch from a highly digestible food such as lichen to less digestible forage such as twigs that contain the same concentration of N. Increases in the N content of the diet will increase fecal N concentration because more N is lost with the indigestible fraction. Fecal N concentrations typically increase when herbivores switch from senescent winter twigs to new summer forbs because DM digestibility and dietary N content increase together (Fig. 8.11) (McKinney et al. 2006). Tannins, which often tend to be higher in summer browse, may enhance the indigestible N and thus the concentration of N in feces (Osborn and Ginnett 2001). Variation in the energy and N digestibility of foods as well as changes in food intake and digestive function (Chapter 5) alter the relationship between fecal N and dietary N concentrations (Fig. 8.11). The concentration of fecal N from white-tailed deer does not increase with increasing dietary protein when diets are highly digestible but does increase with dietary protein on low-energy diets (Brown et al. 1995). Similarly, the model in Fig. 8.11 predicts that fecal N concentration would change very little when caribou switch from winter lichen to summer forbs because both diets are high in digestible energy.

Thus fecal N may be more useful as an indicator of dietary N for some species and habitats than others. Long-term data sets on fecal N reflected annual variation in forage quality for growth of bighorn sheep but patterns of fecal N were less indicative of forage growth and body condition of mule deer (Kucera 1997; Blanchard et al. 2003). The relationship between fecal N concentration and dietary N varies because digestive responses may enhance or diminish the rate at which fecal N increases with dietary N. Fecal N is just one of several indicators that may be used to relate food quality and abundance to the body condition of herbivores; these complex interactions are best evaluated with multiple markers and an appreciation of the underlying physiology. Body composition of animals is further discussed in Chapter 10.

8.6.3 Protein Quality

The quality of dietary N for an animal is related to the efficiency of digesting and retaining the ingested N. Fecal and urinary losses of N include dietary N as well as MFN and EUN. The apparent or net uptake of N from the diet is therefore the digestible N intake, and the apparent or net retention of that dietary N is the metabolizable N intake or N balance (Table 8.2). The gross uptake of N from the diet is the truly digestible N intake, that is, the apparent digestible N intake plus MFN (Table 8.2). Similarly, the addition of EUN to apparent metabolizable N intake is the true retention of N from the diet (Table 8.2).

The proportion of MFN in total fecal N is highest when foods are low in N. Consequently, apparent digestibilities of N decline with dietary N content and therefore underestimate the quality of low-N foods (Robbins 1993). True N digestibilities are a better indicator of the availability of protein for digestion, which is estimated as $1.00\,g \cdot g^{-1}$ for prey consumed by bears (Pritchard and Robbins 1990) and $0.93\,g \cdot g^{-1}$ for grasses and legumes consumed by mule and white-tailed deer (Robbins et al. 1987). High true digestibilities reflect the wide specificity and speed of proteolysis in the small intestine (Fig. 8.5). True N digestibilities may be reduced by PSMs that form complexes with dietary proteins (e.g., tannins in woody browse) or with gut proteases (e.g., protease inhibitors in seeds) (Harborne 1993).

The proportion of the absorbed N that is retained by the animal is the biological value, which is calculated as the ratio of truly metabolizable to truly digestible N intakes (Table 8.2). Nitrogen metabolizabilities and biological values are highest

Table 8.3 The effect of dietary protein quality on a growing carnivore (e.g., the American mink)

Parameter	Calculation	Low-quality protein	High-quality protein
Biological value ($mg \cdot mg^{-1}$)	A	0.75	0.95
Ingested N ($mg \cdot d^{-1}$)	B	1,200	1,200
N digestibility ($mg \cdot mg^{-1}$)	C	0.9	0.9
Digestible N intake ($mg \cdot d^{-1}$)	D = B × C	1,080	1,080
Metabolic fecal N (MFN) loss ($mg \cdot d^{-1}$)	E	6	6
Truly digestible N intake ($mg \cdot d^{-1}$)	F = D + E	1,086	1,086
Truly metabolizable N intake ($mg \cdot d^{-1}$)	G = F × A	815	1,032
Endogenous urinary N (EUN) loss ($mg \cdot d^{-1}$)	H	200	200
Metabolizable N retained ($mg \cdot d^{-1}$) (N balance)	I = G – H	615	832
Maintenance requirement for N ($mg \cdot d^{-1}$)	J	560	560
Body protein gain ($mg \cdot d^{-1}$)	K = (I – J) × 6.25	341	1,698

when the supply of amino acids from the diet closely matches the requirement for tissue synthesis. Large differences between the composition of animal tissues and plant proteins may result in biological values as low as 0.58 for pollen (Bradshaw and Bradshaw 2001). Egg and milk proteins have high biological values at or near 1, which indicates that very little amino-N is lost by oxidation of these proteins. High biological values for egg or milk protein in growing animals reflect natural selection for maximizing the transfer of N from mother to offspring. In the example in Table 8.3, a growing mink is projected to gain body protein four times faster on high-quality egg protein than on the same amount of lower-quality plant protein (i.e., with a biological value of 0.95 vs. 0.75). Growth and reproduction may be limited by the biological value of proteins available to small animals with limited home ranges. Habitat disturbances that reduce the proportions of forbs with high N content and DM digestibility lower the biological value of available proteins for growth of cottontail rabbits (Lochmiller et al. 1995). Similarly, granivorous birds may rely upon the high biological value of some seed proteins to meet their demands for reproduction (Allen and Hume 1997; Valera et al. 2005). The energy and protein requirements of growing and reproducing animals are further discussed in Chapter 10.

8.7 Summary: Nitrogen

Nitrogen is often the most limiting nutrient during growth and reproduction because nitrogenous compounds (proteins, nucleic acids) are the basis for the structure and function of animals. Proteins consist of 20 common amino acids, of which approximately half cannot be synthesized by the animal and so must be supplied by the diet or by microbial synthesis in the digestive tract. The quality of food protein is therefore determined by its amino acid composition; high-quality food proteins (high biological value) closely match those synthesized by the animal. Amino acids are deaminated before the carbon is used for energy or glucose production. The N released from amino acid catabolism is potentially toxic. Ammonium is highly toxic but easily excreted by fish into the surrounding water. Urea and uric acid are non-toxic forms of waste N that are produced by terrestrial animals. Herbivores that consume low-N diets can recycle waste N by degrading urea or uric acid with microbes in the digestive tract. Nitrogen recycling minimizes the loss of fecal and urinary N, which reduces the requirement for N.

Chapter 9
Metabolic Constituents: Water, Minerals and Vitamins

Water is the reactive medium for metabolic pathways in plants and animals. Although water loss is potentially lethal to cells, some organisms (e.g., bacterial and fungal spores, plant seeds and brine shrimp eggs) undergo arrested metabolism and tolerate desiccation. The availability of water affects primary production in terrestrial habitats; rainfall stimulates the growth and reproduction of plants and animals in hot deserts, and snow melt provides water for annual production in boreal forests (Chapter 1).

Water holds a wide range of solutes, from simple mineral ions to sugars, fatty acids, amino acids and complex proteins and vitamins. Enzymes are affected by mineral ions in the same way that hydrogen ion concentration (pH) affects protein structure and digestion (Chapter 8). High concentrations of sodium (Na) and chlorine (Cl) ions cause proteins to unfold and denature in a phenomenon called 'salting out'. Interactions with ions also facilitate enzyme activity; muscle contraction and blood clotting both involve activation by calcium (Ca) ions. Moreover, mineral ions form the catalytic core or active site of several enzymes. For example, zinc (Zn) is essential for protein digestion by carboxypeptidase A, and iron (Fe) in hemoglobin is required to bind oxygen. Complexes of minerals and carbon (C) comprise a wide variety of structures found in vitamins, which then participate in the transfer of electrons or functional groups. For example, cobalt (Co) is the catalytic center of vitamin B_{12}, which participates in the transfer of methyl groups ($—CH_3$) during the formation of fatty acids and nucleic acids in the production of blood cells. The metabolism of animals therefore depends on the reactive mix of ions and trace nutrients in body fluids.

9.1 Water and Electrolytes

Water availability influences the distribution of wildlife because all animals require water to maintain the volume and composition of their body fluids. Although animals can use a variety of mechanisms to conserve the water in their bodies, impaired regulation of water and solutes results in rapid death. Oriental-white backed vultures die quickly after consuming carcasses of livestock that were treated with the anti-inflammatory drug diclofenac. Vultures die from impaired regulation of body fluids

P.S. Barboza et al., *Integrative Wildlife Nutrition*,

because the drug is highly toxic to the kidney of this bird (Meteyer et al. 2005). Water functions as a solvent for salts, ions and gasses, and as a transport medium for metabolic products in the body. It is important for digestion, excretion and temperature regulation, as well as lubrication of joints and as the medium to conduct sound and light.

Water is a small molecule that readily diffuses across membranes in the body. A single oral dose of water labeled with isotopes of hydrogen (^2H_2O; ^3H_2O) or oxygen ($H_2{}^{18}O$) diffuses from the digestive tract and into the blood to equilibrate with tissues throughout the body within 90 minutes in ducks and within 180 minutes in larger animals such as reindeer and seals. The dilution space of labeled water is used to measure the total body water pool and is related to the amount of lean tissue and lipid in the animal (see Chapter 10). Most water in the body is distributed in the intracellular space, which is closely regulated to maintain the activity of metabolic pathways within cells (Fig. 9.1). The extracellular space includes interstitial fluids between cells in the blood (plasma or serum), connective tissues between joints and muscles, and peritoneal fluids between organs. Extracellular spaces are more variable in volume and composition than the intracellular space because they are a buffer between the cells of the animal and its environment. The most variable of these extracellular spaces are those in the digestive tract and urinary bladder (Fig. 9.1). Fluids in the rumen of desert goats and the urinary bladder of desert tortoises vary in composition and volume with feeding and excretion; those volumes may be more than 10% of body mass and serve as stores of water between drinking bouts (Shkolnik et al. 1980; Nagy and Medica 1986). Dehydrated goats can drink a volume of water equivalent to 30% of their body mass; the excess water dilutes the ruminal fluid by 90% but only

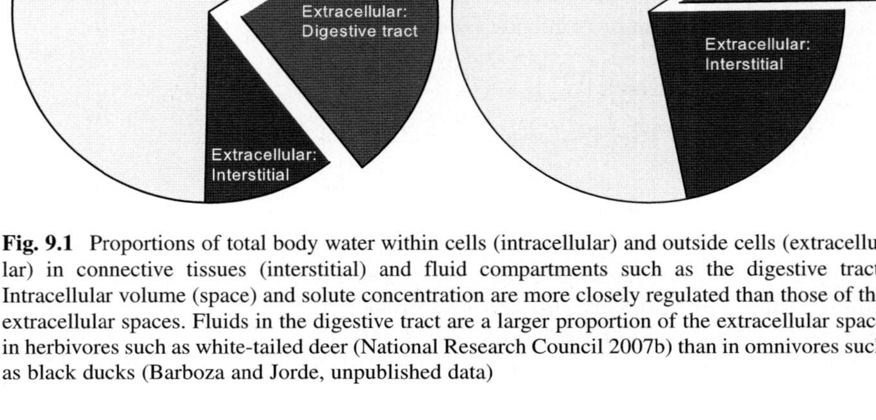

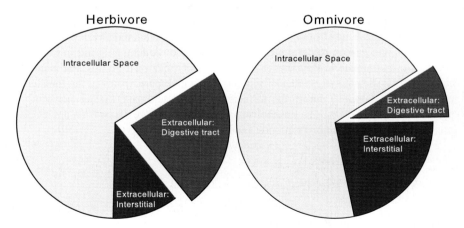

Fig. 9.1 Proportions of total body water within cells (intracellular) and outside cells (extracellular) in connective tissues (interstitial) and fluid compartments such as the digestive tract. Intracellular volume (space) and solute concentration are more closely regulated than those of the extracellular spaces. Fluids in the digestive tract are a larger proportion of the extracellular space in herbivores such as white-tailed deer (National Research Council 2007b) than in omnivores such as black ducks (Barboza and Jorde, unpublished data)

dilutes the concentration of the blood by 10% because water is only slowly absorbed from the digestive tract (Brosh et al. 1983; Choshniak et al. 1984).

A molecule is dissolved in water when the solute is surrounded by a solvation shell of water molecules. The solute and its solvation shell are usually too large to move across membranes without specific channels or transporters. Unbound water molecules that are not associated with solvation shells can continue to diffuse through membranes by osmosis. Water moves down concentration gradients from high to low concentrations of unbound water, that is, from low to high concentrations of solutes. Osmotic potential of a fluid increases with the concentration of solutes, which is measured as osmolarity (milliosmols per liter, mOsmol•L^{-1}); sea water is 1,000 mOsmol•L^{-1} whereas pure water is free of solutes and is 0 mOsmol•L^{-1} (Fig. 9.2). The concentration of charged solutes such as mineral ions may also be measured as the electrical conductivity or resistivity of the solution. Electro-fishing uses the electrical conductivity of body fluids to temporarily immobilize fish for capture by disrupting nerve and muscle function; captures require less voltage in fresh water than sea water because the fluids of the fish are more conductive than the surrounding fresh water. Solutes are also measured as total dissolved solids (TDS); TDS includes large uncharged molecules as well as the smaller charged ions. Water sources that are below 1,000 ppm TDS are suitable for consumption by most animals. Some animals can ingest solute concentrations similar to that of sea

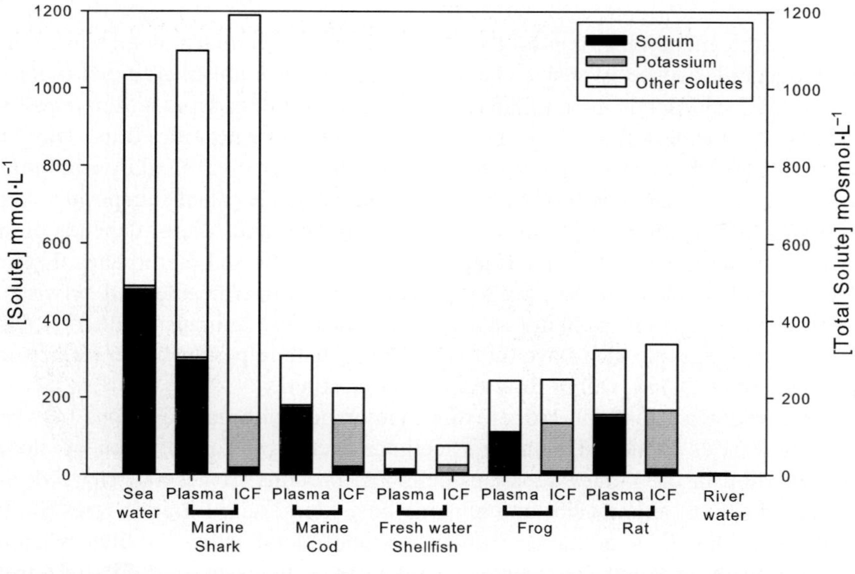

Fig. 9.2 Concentrations of sodium, potassium and other solutes (mmol•L^{-1}, mOsmol•L^{-1}) in sea water, fresh water from a river, and the body fluids of animals (Schmidt-Nielsen 1997). Solute concentrations are also compared between extracellular (ECF; blood plasma) and intracellular (ICF; muscle) fluids of dogfish shark, marine cod, freshwater shellfish, edible frogs and brown rats (Kirschner 1991)

water (approximately 3,000 ppm TDS), but more concentrated solutions draw water from the body into the digestive tract, causing osmotic diarrhea in a manner similar to the malabsorption of lactose (Chapter 6) (National Research Council 1974; National Research Council 2005).

Most of the osmotic potential of sea water is attributed to Na^+ and Cl^-, with smaller contributions from magnesium (Mg^{2+}), sulfate (SO_4^{2-}) and Ca^{2+} (Fig. 9.2). Blood plasma and interstitial fluids are also high in Na^+ and Cl^- but are low in potassium (K^+) ions. Conversely, intracellular osmotic potential is mainly attributed to K^+, phosphate (PO_4^{3-}) and proteins, with smaller contributions from Mg^{2+} and SO_4^{2-} (Linder 1991). The differences in solute concentrations produce gradients that are maintained in the animal by hydrolysis of ATP, which powers membrane transporters or pumps. The imbalance in Na^+ and K^+ is produced by active transport through the Na/K pump in the cell membrane. This gradient also produces an imbalance of charge between the cell contents and the plasma that is used for other cellular exchanges and for the action of nerves and muscles. The activity of the Na/K pump accounts for 31% of the energy expended by domestic animals at rest, which contributes to the energy required for maintenance of the whole animal (Chapter 10) (Underwood and Suttle 2001).

9.1.1 Transport Mechanisms

Exchangeable membranes control the composition of an animal's blood plasma and the extracellular space. Water is channeled while solutes are filtered and pumped across gills, salt glands, rectal glands, kidneys, urinary bladders and the digestive tract. Water channels allow frogs and tortoises to resorb water from urine stored in the bladder. Such water exchanges across membranes are controlled by combining the osmotic potential of solutes with transcellular channels called aquaporins (Agre et al. 2002). The permeability of a membrane to water and solutes depends on its lipid composition and structure (Chapter 7). Thick shells and membranes that are invested with cholesterol and waxes are relatively impermeable and provide a strong barrier to water exchange at the surface of desert animals (Withers 1992). Conversely, lungs and gills have thin membranes that are permeable to respiratory gases (O_2 and CO_2) as well as to water and some solutes.

In the glomerulus of the kidney, thin membranes with gap junctions between cells allow both paracellular and transcellular exchanges for filtration of blood. Ultrafiltration in the kidney separates cells and proteins from small charged and uncharged sugars, amino acids and mineral ions. Glomerular filtration rates (GFR) are increased by changing the pressure and flow of blood across the filter when an excess of water or solutes is cleared from the extracellular space. GFRs of terrestrial animals are typically greater for species from water-abundant (mesic) areas than those from arid (xeric) areas (Yokota et al. 1985). Selective pumps return glucose and amino acids to the blood, but leave mineral ions (e.g., Na, Cl) and nitrogenous wastes in the urine for excretion (Chapter 8). Fish, amphibians, reptiles and birds produce dilute urines with an osmotic pressure similar to their blood. These animals can easily clear water loads through the kidneys but risk losing solutes in the large volume of filtrate. Palestine sunbirds, for example, consume up to six times their body mass as watery nectar each day and excrete most of that water by increasing GFR to produce dilute urine. The ability of their kidneys to recover 98% of the glucose from the nectar may be partly due to a reduction in water absorption from the digestive tract, which reduces the load of water on the kidneys (McWhorter et al. 2004). Conversely, dehydrated desert goats reduce GFR and urine excretion after drinking a large volume of water, which regulates the loss of Na^+ and Cl^- as water is slowly absorbed from the digestive tract (Choshniak et al. 1984).

In mammals, the kidney can produce urine with a higher concentration of solutes than the blood by combining solute pumps with a counter-current flow that can return more water to the blood. Kidneys with long concentrating tubules have proportionately large central medullary sections relative to the volume of glomerular tissue in the peripheral cortex. The relative medullary thicknesses (RMT) of desert mammals are greater than those from mesic areas among both placental mammals and marsupials (Calder and Braun 1983; Brooker and Withers 1994). RMT correlates with urine concentrating ability in rodents (Al-kahtani et al. 2004); for example, the large RMT of the hopping mouse is associated with a maximum urine concentration of 9,400 mOsmol•L^{-1}, which is 9.4 times the concentration of sea water and 25 times the concentration of blood plasma (Schmidt-Nielsen 1997). In marsupials, the urine concentrating ability of young animals is low at birth but increases with the development of the kidney medulla and with increases in RMT (Wilkes and Janssens 1988). Kidney mass and GFR may vary among adults within a population as water and solute intakes change during the year, especially in species with large seasonal changes in activity and reproduction (McAllan et al. 1996, 1998).

The removal of salt loads from the blood at chloride cells in the gills of fish and salt glands of birds, reptiles and sharks is accomplished via specific transporters such as the Na/K pump. The ability to secrete a fluid with high concentrations of Na and Cl allows gulls and waterfowl to gain water from drinking in brackish estuaries or the ocean because the solute excretion requires little water (Fig. 9.3) (Schmidt-Nielsen 1997). Secretions from salt glands account for 40% of Na excretion and only 11% of the total water loss of mallard ducks drinking brackish water (300 mM NaCl, 67% sea water) at the maximum concentrating ability of their salt gland (Bennett et al. 2003). The combined effect of Na/K pumps and water resorption in the mammalian kidney also allows harp seals and desert wallabies to drink sea water (Kinnear et al. 1968; How and Nordoy 2007).

9.1.2 Aquatic Exchanges of Water

Fish readily exchange water with their surroundings, especially at the thin vascularized surfaces of the gills. The net loss or net gain of water at the gills depends on

Fig. 9.3 Water for some mammals and birds can be gained by drinking sea water if Na^+ and Cl^- ions can be removed and secreted at a higher concentration than the blood. **a** Seabirds such as the herring gull eliminate excess salt in secretions from the nasal salt glands. **b** Marine mammals such as the South American sea lion eliminate excess salt in urine from the kidney

the osmotic gradient between the blood and the medium. Salty surroundings may be osmotically opposed by maintaining a lower osmolarity (e.g., bony fishes, Family Teleostei) or complemented with alternative solutes to Na^+ and Cl^-. Many bacteria and invertebrates use sugars, amino acids and their derivatives to maintain an osmotic balance with marine and hyper-saline environments such as salt lakes (Hochachka and Somero 2002). Some of those solutes also are used for cellular exchanges within vertebrate tissues such as the kidneys (Yancey 1988). The elasmobranch fishes (e.g., sharks and rays) use urea as an uncharged internal solute to maintain plasma osmolarity near that of the surrounding ocean by retaining urea at the kidney and at the gills (Fig. 9.2; Chapter 8) (Moyle and Cech 2004). Plasma osmolarities of bony fish are lower than sea water and greater than fresh water (Figs 9.2 and 9.4), which results in a net efflux of water in the ocean and a net influx of water in a freshwater stream. Water is replaced by drinking in the ocean; the excess Na^+ and Cl^- are excreted by α-chloride cells in the gills whereas loads of Mg^{2+} and SO_4^{2-} are excreted by the kidneys (Moyle and Cech 2004). Freshwater teleosts do not drink but urinate to excrete the continuous influx of water (Fig. 9.4), resulting in higher GFRs for fresh water than for marine species of fish (Yokota et al. 1985). Solutes lost in the urine of freshwater fish are replaced by solutes in the food and by absorbing Na^+ and Cl^- at the gills by β-chloride cells. Fish that move between the ocean and rivers can adjust to the changing demands for ion excretion by changing the number and activity of the α-chloride and β-chloride cells to favor salt excretion and absorption respectively. Survival of young Atlantic salmon is related

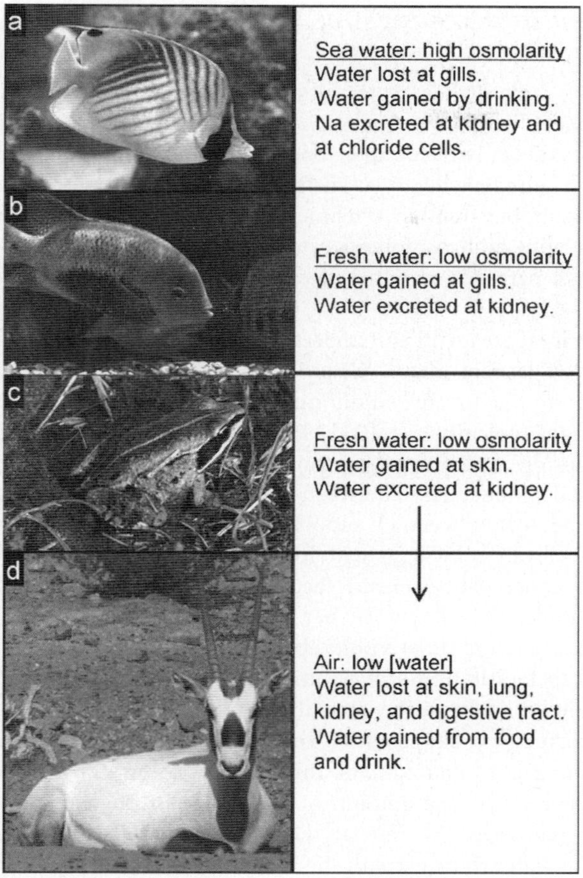

Fig. 9.4 Exchanges of water depend on the difference in the concentration of solutes (osmolarity) and unbound water between the animal and its environment. **a** Marine teleosts such as butterfly fish lose water at the gills because the medium has a higher osmolarity than the body. Salt gained from feeding and drinking is excreted at the kidneys and at the salt glands in the gills. **b** Freshwater teleosts such as spotted tilapia gain water at the gills because the concentration of solutes in the medium is lower than the body; excess water is excreted at the kidneys. **c** Terrestrial animals continuously lose water to air at the skin and lungs. Wood frogs can gain water through the skin in summer, but minimize evaporative losses by hibernating in the leaf-litter on the forest floor during winter. **d** Desert animals such as the Arabian oryx minimize excretory losses of water by producing dry feces and concentrated urine

to the activity of chloride cells in the gills as they move from their natal streams to the ocean (McCormick et al. 1999). The energy required to regulate the water and solute balance in fish is less than 5% of the energy used at rest, but that cost may increase and detract from growth as the osmotic difference between the animal and the medium increases and diverges from the optimal range for the species (Withers 1992; Moyle and Cech 2004).

9.1.3 Terrestrial Exchanges of Water

Fish can always replace their water losses from the medium as long as they can resolve imbalances in solutes. Terrestrial vertebrates must resolve imbalances of both water and solutes between their body and the environment (Fig. 9.4). Water evaporates from animals when their surfaces are exposed to air. Vertebrates cannot gain water from air, but small invertebrates such as ticks and mites can use specialized structures with high osmolarities to condense water vapor (Withers 1992), which allows the parasites to remain hydrated while waiting for their next vertebrate host.

Evaporation increases with air temperature as more energy is available for liquid water to enter the gaseous phase. Water vapor equilibrates with the liquid phase in still air; moving air disrupts the equilibrium between phases and increases evaporative loss. Bird eggs typically lose 10–12% of their laid mass as water is lost at warm temperatures during incubation even though air flow in the protected nest is low. Pores in the shell conduct gases between the air and the growing embryo. Gas conductance is a trade-off between O_2 supply and the loss of water from the egg. Birds that nest at low elevation lay eggs with low conductances that minimize water loss, whereas higher conductances ensure the diffusion of O_2 to eggs laid by birds that nest at high altitudes (Vleck and Bucher 1998).

Respiratory tracts are most vulnerable to evaporation because their surfaces are moistened to facilitate the dissolution of O_2 and CO_2 and because ventilation removes moist air from the animal each time it exhales. Long air passages and counter-current exchanges with the blood reduce the temperature of the exhaled air in ostriches and camels; this reduces the water-holding capacity of the air and thus reduces the amount of water lost through respiration (Withers 1992). Flying increases the rate at which birds breathe and thus the rate at which they lose water during respiration (Adams et al. 1997). Respiratory water loss may limit the flight range of migratory shorebirds; predicted flight ranges are longest at altitudes where cool temperatures minimize evaporative losses (Klaassen 1995; Landys et al. 2000).

Skin surfaces are most resistant to evaporation because water permeabilities are reduced by multiple layers of skin cells and waxy coats. However, amphibians such as frogs that can absorb O_2 and water through the skin are also vulnerable to losing water by evaporation when they emerge from ponds (Fig. 9.4). Frogs that estivate during droughts or hibernate during winter reduce evaporative losses by secreting layers of skin or mucus and by using burrows or leaf-litter where the air is humid and still (Fig. 9.4). Surface evaporation also may be used for thermoregulation; the energy used by water molecules to enter the gaseous phase removes heat from the surface, and this produces evaporative cooling. Heat transfer is most efficient when the evaporative surface is close to an area that produces heat (e.g., muscle) or can conduct the heat (e.g., blood vessels). Most animals can use high-frequency breathing or panting to evaporate air from the tongue and mouth where the thin skin covers a dense network of blood vessels. Although large ungulates such as elk and eland commonly rely on cutaneous evaporation and the smaller ungulates such as mule deer and hartebeest use respiratory evaporation for thermoregulation in summer (Parker and Robbins 1984), moose take advantage of free water in their environment for evaporative cooling. Saliva also can be used to transfer water to the body surface. For instance, red kangaroos lick their forelimbs to cool the blood flowing through a capillary bed close to the skin (Dawson 1983), and heat-stressed wood storks and black vultures urinate on their legs (Withers 1992). In horses and humans, sweating secretes water from glands on the skin surface, especially near the major muscle groups at the shoulder and rump where most heat is generated during running.

Animals including small ground squirrels and African gazelles that need to conserve water at high temperatures may reduce evaporation by allowing body temperature to rise; this reduces the difference in temperature between the animal and the air (Withers 1992; Maloney and Dawson 1998). Birds that are routinely exposed to high temperatures may reduce evaporative water loss further by changing the ultrastructure of their skin and by reducing their metabolic rate and heat production (McKechnie and Wolf 2004). The effects of temperature on energy metabolism are further discussed in Chapter 10.

9.1.4 Water Turnover and Balance

The water requirements of terrestrial animals are determined by the rate at which water enters (influx) or leaves (efflux) the body water pool (Fig. 9.5). Water efflux is the combined loss of water in the excreta (urine and feces) and the evaporative or insensible losses of water from the skin (cutaneous) and respiratory tract. Intracellular and extracellular spaces of water are maintained when rates of influx and efflux of water are balanced (Fig. 9.5). The body water pool can be maintained over a wide range of complementary influxes and effluxes. Red knots balance water influx and efflux at daily rates equal to eight times the body water space when the shorebirds feed on clams; daily water efflux declines to 0.3 times the body water pool when the same birds stop feeding (Visser et al. 2000). Water turnover (also referred to as water flux) generally increases with energy requirements because animals consume more food, excrete more solutes and lose more water to respiration as activity increases (Macfarlane and Howard 1972). Conversely, low water turnovers are associated with low energy requirements in spinifex pigeons from deserts and in sloths from tropical rainforests (Nagy and Montgomery 1980; Williams et al. 1995). Water requirements can be estimated on the basis of energy expenditures for domestic dogs and cats at 2.93–4.18 g $H_2O \bullet kJ^{-1}$ (National Research Council 2006).

Water turnovers increase with the energy demands of lactation and with the secretion of water in milk, especially for species that produce large volumes of dilute milk such as horses (Maltz and Shkolnik 1984; National Research Council 2007a). Water requirements may be increased further by enhanced insensible losses at high ambient temperatures (Fig. 9.5). Water turnovers of desert tortoises, kanga-

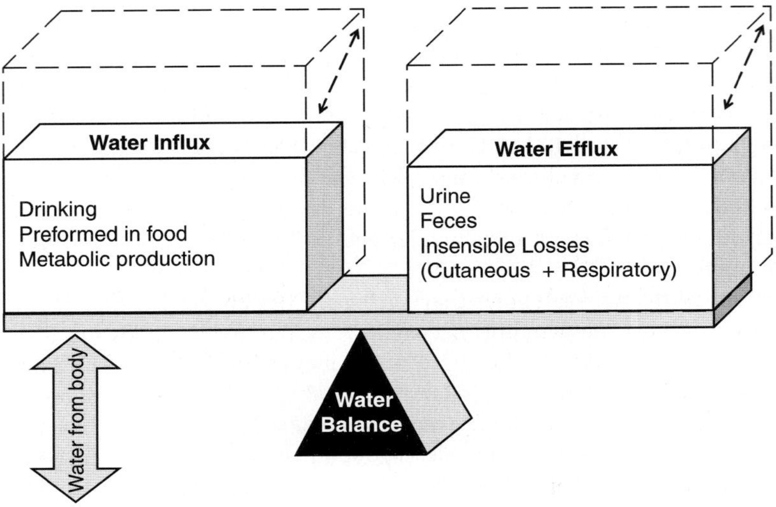

Fig. 9.5 Water enters and leaves the body water pool by several routes in terrestrial animals. Water also is produced within the body during metabolism and when water is absorbed from the digestive tract and urinary bladder. The balance between total influx and total efflux of water is zero when animals maintain total body water. Dehydration reduces the body water pool and results in negative water balance. Accumulation of water in extracellular spaces results in positive water balance. Water turnover or flux (ml•d^{-1}) is the rate at which water enters the body water pool. The turnover changes with both influx and efflux of water (*broken lines*) when food intakes, solute loads and respiration rates are altered

roos and koalas increase with activity, air temperatures and radiant heat loads from winter to summer (Ellis et al. 1995; Henen et al. 1998; Bradshaw 2003). Water turnovers and insensible water losses of nestling birds and young mammals are usually high because growing animals have high energy demands and proportionately large surface areas in relation to their small body masses (Chapter 1). Young red kangaroos are more susceptible to the heat and water restrictions of droughts than adults because their insensible water losses are as high as those of adult females three times their body mass (Munn and Dawson 2001).

Birds and mammals may rely on sources of surface water for drinking in hot, dry habitats such as deserts and savannahs. The distribution of watering holes influences the foraging areas used by animals; in semi-arid rangelands, domestic cattle and sheep heavily graze the areas around watering holes (Landsberg et al. 1999). In Australia, water from wells allows feral cats and goats to persist in dry habitats often to the detriment of native marsupials such as bilbies that do not require surface water for drinking (Hume 1987; Hume et al. 2004). Drinking behaviors may be influenced by other factors such as the presence of predators (e.g., crocodiles) and the quality of the water. Ephemeral ponds and seasonal wetlands become increasingly salty as the water evaporates and they become contaminated by decomposing plants and animals. In cold climates, drinking water and eating snow incur an added energy cost because ingested water must be warmed to body temperature; the energy required to warm ingested water is equivalent to 3–14% of the digestible energy intake of muskoxen and 3–30% of the total energy expended by smaller mammals (Berteaux 2000; Crater and Barboza 2007).

Many animals rely on water in food because drinking water may be scarce and drinking may incur other risks to survival such as predation, thermal or solute loads and microbial pathogens. Preformed water in foods varies from less than 30% of the mass of seeds and winter twigs to more than 80% of the mass of algae and fruit (Fig. 1.2) or lush green forage in early spring. The water content of animal foods is 73% for lean tissues such as muscle, but declines with increasing concentrations of fat. However, the dry portion of foods also can yield metabolic water when reducing equivalents (e.g., NADH and $FADH_2$) are oxidized by the following reaction (Fig. 6.1):

$$O_2 + 4H^+ \rightarrow 2H_2O$$

Metabolic water production is therefore related to the fuel value of each substrate: oxidation produces 1.07g $H_2O{\bullet}g^{-1}$ of fat but only 0.56 g $H_2O{\bullet}g^{-1}$ of carbohydrate. Protein oxidation yields a similar amount of water to that of carbohydrate, but some metabolic water must be used to excrete the amino-N. Protein oxidation therefore produces 0.50 g $H_2O{\bullet}g^{-1}$ when N is excreted as uric acid and only 0.40 g $H_2O{\bullet}g^{-1}$ when N is excreted as urea (Chapter 8). Carnivores such as desert kit foxes may derive most of their water requirements from their prey because lean tissues are rich in preformed water, and fat provides oxidative water. Omnivores such as bilbies and Merriam's kangaroo rats survive in deserts without drinking water by using preformed and metabolic water from plants and insects (Tracy and Walsberg 2002). Metabolic water also is produced by oxidation of the animal's own tissues; oxidation of fat provides most of the water required by fasting northern elephant seal pups, which maintain the osmolarity of their blood plasma after they are weaned by their mothers (Adams and Costa 1993; Ortiz et al. 2002). During hibernation, bears, ground squirrels and prairie dogs also rely on water produced from tissue breakdown. Birds that catabolize lean body tissues during long migratory flights produce metabolic water from oxidation of protein and also release intracellular and extracellular water from cells and connective tissues (Lindström and Piersma 1993).

Protein catabolism has the potential to increase urinary water loss because urea and uric acid are solutes that must be removed from the extracellular space. High N intakes therefore may increase urinary volumes when animals can increase their water intakes from food or drink (National Research Council 2007a). Restrictions on water influx can alter N excretion as animals reduce urinary water losses; recycling of N-catabolites decreases excretory N loads. The proportion of urea-N recycled increases when drinking water is restricted for desert sheep, goats, camels, kangaroos and wombats (Mousa et al. 1983; Barboza 1993b; Freudenberger and Hume 1993). In birds, reflux of urine from the cloaca to the hindgut allows recycling of urate-N by microbes and resorption of water and ions in the ceca (Hughes et al. 1999; Laverty and Skadhauge 1999). Although high solute loads can increase urinary water losses, the concentrating abilities of the kidney and/or salt glands may minimize changes in water flux. Urine flow rates of emus were similar among diets with different concentrations of NaCl because kidney function changed with solute load and water availability (Dawson et al. 1991). Similarly, water fluxes only increased by 2–3% in mallard ducks when fresh drinking water was replaced with brackish water (Bennett et al. 2003).

The amount of water lost in feces varies with the digestive function of the animal and the structure of the diet. Coarse residues from fibrous diets of leaves and stems result in more feces with a greater water-holding capacity than the dense residues from more digestible items such as seeds or animal prey. Desert carnivores such as cats and foxes lose little water in the feces because their diet is highly digestible and the feces are mainly comprised of the skin and bones of prey. Water is removed from the digesta by muscular contractions that extrude fluid from the digesta toward the intestinal wall and by osmotic exchanges across the wall between intestinal contents and blood. Fecal water losses by desert species are usually lower than those of closely related species from mesic regions, even on the same diet. For example, common wombats from mesic regions produce feces with large particles that are similar to feces produced by horses, whereas hairy-nosed wombats from the desert produce feces with fine particles in small, dry pellets (Barboza 1993a, 1993b). Particles that are fine and dense have large surface areas and small volumes that facilitate water resorption across the intestinal mucosa. Also, the longer distal colon of hairy-nosed wombats increases the time and surface area for water resorption from feces as they form pellets (Barboza and Hume 1992). Several ruminants such as caribou, white-tailed deer, sheep and goats can produce small fecal pellets from a long distal colon that is arranged in concentric loops called the spiral colon (Hofmann 1989). Hindgut morphology may reflect evolutionary selection for water conservation in some groups of ungulates; the desert-adapted dromedary camel also has a spiral colon whereas the aquatic hippo has only a short hindgut (Stevens and Hume 1995).

Food intake is dependent on hydration, and mammals and birds often reduce their intake of DM when water availability declines (Langhans et al. 1995; Forbes 2007). Consequently, water deficits may cause secondary deficits of energy and nutrients; white-tailed deer decreased food intake and lost body mass when drinking water was restricted to 33% of ad libitum levels (Lautier et al. 1988). The effects of limited water availability on animals can be measured by comparing food intakes and balances between periods when drinking water is available ad libitum and when water is either restricted (Table 9.1) or completely withdrawn. Kangaroos and goats responded to moderate restrictions of drinking water (43–68% of ad libitum) by reducing water turnover without changing body mass or total body water (Freudenberger and Hume 1993).

Water turnovers decline with reductions in influxes of drinking, preformed and metabolic water, and with commensurate reductions in both fecal and urinary effluxes of water (Table 9.1). Fecal water losses are minimized by decreases in the moisture content of feces whereas urinary losses are minimized by increases in urine osmolarity. Water resorption at the distal colon and the kidney is facilitated

Table 9.1 Water balance of a ruminant when drinking water that is either freely available (ad libitum) or restricted to 50% of ad libitum. Values are derived from responses of white-tailed deer and goats to water restriction (Lautier et al. 1988; Freudenberger and Hume 1993; National Research Council 2007b)

Parameter	Units	Calculation	Ad libitum	Restricted	Change (%)[1]
Body mass	kg	A	60	60	0
Drinking water	$g \cdot d^{-1}$	B	1,200	600	–50
Food intake	$g \cdot d^{-1}$	C	2,000	1,600	–20
Food moisture	$g \cdot g^{-1}$	D	0.48	0.48	0
Preformed water	$g \cdot d^{-1}$	$E = C \times D$	960	768	–20
DM intake	$g \cdot d^{-1}$	$F = C - E$	1,040	832	–20
Metabolic water content	$g \cdot g^{-1}$	G	0.35	0.35	0
Metabolic water gain	$g \cdot d^{-1}$	$H = F \times G$	364	291	–20
Total water influx (turnover)	$g \cdot d^{-1}$	$I = B + E + H$	2,524	1,659	–34
Fecal output	$g \cdot d^{-1}$	J	1,100	704	–36
Fecal moisture	$g \cdot g^{-1}$	K	0.6	0.5	–17
Fecal water loss	$g \cdot d^{-1}$	$L = J \times K$	660	352	–47
Fecal DM loss	$g \cdot d^{-1}$	$M = J - L$	440	352	–20
DM digestibility	$g \cdot g^{-1}$	$N = (F - M)/F$	0.58	0.58	0
Urinary water loss	$g \cdot d^{-1}$	O	947	574	–39
Excretory water loss	$g \cdot d^{-1}$	$P = L + O$	1,607	926	–42
Insensible water loss	$g \cdot d^{-1}$	$Q = I - P$	917	733	–20

[1]100 × (Restricted – Ad libitum) ÷ Ad libitum.

by the anti-diuretic hormones (ADH) arginine vasotocin (AVT) and arginine vasopressin (AVP). AVP is found in mammals whereas AVT is found in fish, reptiles and birds (Norris 1997). These hormones open water channels in the mucosa of the colon and in the renal tubules to return water to the blood, and secondarily act on the brain to stimulate drinking behavior. Responses to plasma ADH and water restriction are more pronounced in desert animals than similar species from mesic habitats (Macfarlane and Howard 1972). Concentrations of plasma ADH were inversely related to water turnovers of sympatric wallabies during a drought; the most mesic bettongs had water turnovers that were 2.8 times higher than those of an arid-zone species of hare-wallaby, but plasma ADH concentrations of the bettong were only 2% of those measured in the hare-wallaby (Bradshaw 2003).

Loss of body mass during water restriction may be associated with changes in both extracellular and intracellular water spaces (Fig. 9.1). Most of the reduction in the extracellular water space is often in the digestive tract, especially in large herbivores such as sheep, kangaroos and donkeys (Macfarlane et al. 1961; Maloiy et al. 1978). Although increases in rumen osmolarity may serve as a post-ingestive feedback that reduces food intake, desert ruminants are able to maintain fermentation and digestion during bouts of dehydration and rehydration (Brosh et al. 1983; Langhans et al. 1995). Prolonged periods of low availability of both food and water during drought may be associated with a loss of body tissue, which reduces the intracellular water space but may also reduce energy demand. Sand gazelles lost dry mass from the digestive tract, kidney, heart and muscle during 4 months of food and water restriction, which resulted in reductions in the rates of energy metabolism and insensible water loss (Ostrowski et al. 2006).

Animals that survive hot dry environments and droughts use both physiological and behavioral mechanisms to minimize water losses and maximize water gains. Antelope ground squirrels tolerate elevated body temperature but also evade radiant heat loads in burrows, where higher humidity also increases the water content of stored seeds (Robbins 1993; Schmidt-Nielsen 1997). Springbok antelope feed at night on moist flowers and leaves to maximize water intake at lower air temperatures and avoid high radiant heat loads (Nagy 1994). Populations of some desert animals may depend on refuges where supplies of water and food are adequate for a minimal viable number of animals to survive drought. Populations of black-tailed jack rabbits fluctuate widely because the species is vulnerable to drought, but high fecundity restores the population from a few individuals that survive in refugia during severe drought (Nagy 1994). The energy and nutrient requirements of animals for survival and reproduction are further discussed in Chapter 10.

9.2 Minerals

Water conveys dissolved minerals from rocks and soils to plants and up the trophic chain of animals. Minerals released from underwater hydrothermal vents sustain unique communities of invertebrates and fish in the immediate vicinity while currents distribute those minerals throughout the ocean. Minerals are conveyed by lava flows and volcanic eruptions to the surface of the earth and are weathered into water courses that flow below and above ground through riparian systems. Geology and hydrology affect the availability of minerals for plants and animals over large geographic areas. Sodium (Na) and iodine (I) are relatively abundant in marine systems and are thus more available to plants and animals in coastal habitats than at the center of continents (Underwood and Suttle 2001). Conversely, selenium (Se) concentrations of plants are low on the Pacific and Atlantic coasts of the United States of America in high rainfall areas where leaching occurs, but some plants (e.g., milk vetch) can accumulate Se to toxic levels in the center of the country. Soil minerals therefore may limit animal populations in some regions; low Se in soils is associated with poor reproduction of black-tailed deer in northern California (Flueck 1994). The diversity among plants and foraging areas, however, can ameliorate some of these local affects on mineral supplies. Herbivores can generally avoid Se toxicity from milk vetch by consuming other plants as long as forage abundance is high (Van Soest 1994). Similarly, inland populations of herbivores may supplement their diet with salts of Na and Mg from natural deposits and springs at mineral licks (Tankersley and Gasaway 1983; Ayotte et al. 2006).

The metabolizability of minerals depends on their form. Complexes and compounds that contain minerals are indigestible if the element cannot dissolve into an absorbable form in the digestive tract. Iron (Fe) is abundant in soils as rust or ferric oxide (Fe_2O_3), but the compound is insoluble and therefore unavailable to animals. Conversely, ferrous sulfate ($FeSO_4$) can dissolve to form Fe^{2+} ions that can be absorbed. Organic complexes of minerals are often more metabolizable than inorganic forms because the mineral is either absorbed with the complex or more easily released; Fe in blood and muscle is readily absorbed as heme and retained in hemoglobin and myoglobin. Dietary minerals are likely to be more available for carnivores than herbivores because the tissues of prey contain minerals in readily digestible forms. Minerals in plants may be contained within indigestible cell walls or complexed with a wide diversity of PSMs and other minerals that can interfere with absorption and retention.

Mineral concentrations vary widely among the tissues of plants and animals because each element can serve several functions. Silicon (Si) is deposited as silica in the cell walls of grasses that are readily consumed by bison, and in the spicules of sponges that are consumed by hawksbill turtles. Silicon has only a minor biochemical role in the tissues of vertebrates and is therefore present in low concentrations in bison and turtles despite the abundance of the element in their diet. Elements that are required for structural compounds or osmotic balance are accumulated to high concentrations in the organism and are considered macrominerals (typical units are grams per hundred grams or grams per kilogram, $g \cdot 100\,g^{-1}$ or $g \cdot kg^{-1}$). The common macrominerals of vertebrate animals include Na, Cl, K, Ca, phosphorus (P), magnesium (Mg) and sulfur (S) (Table 9.2). Each macromineral can serve more than one function in the same organism, as in birds that use Ca for bones, enzyme activation cascades and egg shells (Table 9.2). Trace minerals usually serve catalytic roles in enzymes or hormones that may be involved in many biochemical pathways and tissues, but nonetheless only require small amounts of the element (typical units are miligrams per kilogram or micrograms per kilogram, $mg \cdot kg^{-1}$ or $\mu g \cdot kg^{-1}$). Trace minerals commonly required by vertebrate animals include manganese (Mn), Fe, copper (Cu), Zn, I and Se (Table 9.3).

Mineral dynamics are similar to those of water (Fig. 9.5) because influxes and effluxes are equal when the body pool is maintained. Positive balances indicate the accumulation of mineral in tissues whereas negative balances indicate depletion of body pools. Body pools of macrominerals are large and widespread in contrast to pools of trace minerals that are small and distributed among intracellular spaces. Bone is a large pool of Ca, whereas Zn is associated with numerous enzymes in both the cytosol and nucleus of cells in several organs. Mineral fluxes reflect the turnover of their principal pools. Na turnovers are high and commensurate with the turnover of the extracellular space, whereas Ca turnovers may be slow, especially in adults with their low rates of bone deposition. Mineral status is usually evaluated as the amount of mineral in an indicator tissue or the activity of an associated enzyme in the body. For example, net changes in Ca flux are often reflected in mass, density and Ca content of bone, and the plasma concentration of vitamin D (a hormone that stimulates Ca absorption). Homeostatic

Table 9.2 Common forms and functions of macrominerals in animals

Name	Chemical form	Biologically active forms	Chemical function	Body function
Sodium	Na^+	Free ion outside cells	Gradient	Nerves, muscles, body fluid volumes
Potassium	K^+	Free ion within cells	Gradient	Nerves, muscles, acid-base balance
Chlorine	Cl^-	Free ion	Gradient	Gastric acidity, acid-base balance
Calcium	Ca^{2+}	Free ion	Activation cascades	Nerves, muscles
	$Ca_{10}(PO_4)_6(OH)_2$	Hydroxyapatite	Structural rigidity	Bone, teeth
	$CaCO_3$	Calcium carbonate	Structural rigidity	Egg shell
Phosphorus	HPO_4 esters	Phospholipids	Linkage	Membranes
		ATP, ADP, AMP	Energy store	Metabolism, RNA, DNA
	$Ca_{10}(PO_4)_6(OH)_2$	Hydroxyapatite	Structural rigidity	Bone, teeth
Magnesium	Mg^{2+}	Free ion	Activation cascades	Energy metabolism
Sulfur	SO_4^{2-}, —SH	Methionine Glutathione Vitamins B_1 and B_8	Proteins, electron (e^-) transfer	Feathers, hair Oxidative defense, TCA cycle

Table 9.3 Common forms and functions of selected trace minerals in animals

Name	Chemical form	Biologically active forms	Chemical function	Body function
Manganese	Mn^{2+}	Superoxide dismutase	e^- transfer	Oxidative defense
Iron	Fe^{2+}	Hemoglobin, transferrin	e^- transfer	O_2 transport, respiration
Copper	Cu^{2+}	Superoxide dismutase, ceruloplasmin	e^- transfer	Oxidative defense, immunity
Zinc	Zn^{2+}	Metallothionein, superoxide dismutase	e^- transfer	Gene expression, oxidation
Iodine	I^-	Thyroid hormones	Signal transduction	Energy metabolism
Selenium	SeO_3^{2-} Se-Cysteine Se-Methionine	Glutathione peroxidase	e^- transfer	Oxidative defense, thyroid function
Cobalt	Multi-ring complex with Co at center	Cobalamines, vitamin B_{12}	C transfer	Synthesis of amino acids, bases and SCFA

controls regulate the availability of minerals to tissues by modulating excretion from the kidney, absorption from the digestive tract and release from tissues. Consequently, the concentrations of mineral ions in blood may not indicate body stores. Plasma concentrations of Ca^{2+} are maintained by the large store of Ca in bone and do not decline until the exchangeable pool of Ca is depleted. Apparent digestibilities of minerals reflect both the demand for a mineral and the availability of that mineral in a diet. An animal that is replete in Ca will not absorb as much Ca as an animal with depleted bones and high concentrations of vitamin D, even though both animals consume the same diet.

Minerals in plants and animals are measured by removing organic material and by dissolving the residue in acid. Combustion in a furnace at 450–600°C removes all the C, N, H and O to leave a residue of minerals or ash, which is dissolved in acid for further analysis. Unfortunately, some minerals are volatile (e.g., Se and I) and escape from the ash during combustion. 'Wet digestion' in a boiling mixture of strong oxidants such as nitric (HNO_3), hydrochloric (HCl), perchloric ($HClO_4$) and sulfuric (H_2SO_4) acids retains the volatile elements but removes the organic material. Some macrominerals can be assayed by electrochemistry (e.g., Na^+, Cl^-) or by reactions that produce a compound that absorbs wavelengths of light for spectrophotometry (e.g., P). Spectral methods are also used for direct analysis of the minerals in atomic absorption spectrophotometers or plasma spectrometers (e.g., inductively coupled plasma, or direct current plasma spectrometers). These devices measure the absorption or emission of light when the minerals enter a flame or an electrical arc. Mineral analysis is often expensive because the equipment is costly and the procedures use specialized gases and generate hazardous wastes. Samples that are intended for analysis of trace minerals should be handled carefully because they are vulnerable to contamination with dust from other samples and soils. The quality of analysis should be monitored by including certified standards in each batch of samples and by comparing results between laboratories.

9.2.1 Sodium, Chlorine and Potassium

Concentrations of Na^+, Cl^- and K^+ contribute largely to acid–base balance in the animal; they account for most of the balance in osmotic pressure and charge between intracellular and extracellular fluids (Fig. 9.2). Exchanges of these ions allow nerves to transmit impulses, muscles to contract, glands to secrete acids and membranes to transport solutes. Large changes in transcellular gradients are potentially fatal because they disrupt the ratio of $Na^+:K^+$. A rapid rise in plasma K^+ can cause heart failure, whereas rapid rises in plasma Na^+ cause osmotic changes in the spinal cord and brain that produce convulsions (National Research Council 2005). Na, K and Cl homeostasis is mainly controlled by excretion at the kidney. The hormone aldosterone from the adrenal gland stimulates resorption of Na at the kidney as plasma Na^+ declines. Concentrations of aldosterone in plasma of common wombats and gray kangaroos are low on the coast and high in the mountains, reflecting the high concentrations of Na in coastal environments and low concentrations of Na in plants and soils of montane habitats (Hume 1999).

Excretion of excess K^+ also is stimulated by aldosterone because it restores the high Na^+: K^+ ratio in plasma and other extracellular fluids such as saliva (Underwood and Suttle 2001; Wang 2004). Increases in plasma Na^+ cause the release of atrial naturetic hormone (ANH) from the heart, which opens Na channels in the kidney and allows excess Na^+ to be removed in the urine. Concentrations of plasma Na and K are affected by water balance and dietary loads of these electrolytes. Ornate dragons are lizards that maintain transcellular gradients during a drought by allowing concentrations of both Na and K to rise in plasma and also tissue. These lizards continue to acquire Na from their diet of ants, but allow Na to accumulate in the extracellular space until rain provides sufficient drinking water for excretion (Bradshaw 1992).

Marine fish mainly absorb Na and Cl at the gills from the surrounding water in the sea but rely on dietary sources of K (Fig. 9.2) (Kaushik 2001). Terrestrial animals and freshwater fish rely on their diet for all three electrolytes. Dissolved Na^+ and K^+ are absorbed across the mucosa of the digestive tract by a variety of exchanges with H^+, Cl^-, Mg^{2+} and other solutes that are linked to ATP and the Na/K pump. Chlorine is an abundant mineral in both plant and animal tissues and is therefore rarely limiting for wildlife (Underwood and Suttle 2001). Some plants in deserts and island habitats contain potentially toxic concentrations of Na, but animals such as kangaroos and rodents usually avoid the Na loads that would accompany consumption of halophytic shrubs such as salt bush and blue bush (Hume 1999). Concentrations of Na in plants consumed by animals are usually lower than those of animal tissues (Fig. 9.6). Conversely, the leaves of legumes, grasses and forbs are rich in K (Fig. 9.6), which may impose large excesses of K on herbivores. Desert reptiles can eliminate K loads with a minimum loss of water by using salt glands and by storing K as stable urate salts in the urinary bladder (Minnich 1972; Bradshaw 1992). High intakes of K, however, may impair net absorption of low concentrations of Na in leafy plants. Ratios of Na^+:K^+ in plasma and saliva decline from winter to spring as reindeer switch from lichens that are low in K to forbs and browse that are 5–10 times richer in K but contain the same levels of Na (Staaland et al. 1980). Mineral licks are an important source of Na for ungulates such as elk, moose, mountain goats, muskoxen and Stone's sheep in habitats far from the sea (Risenhoover and Peterson 1986; Klein and Thing 1989; Ayotte et al. 2006) (Fig. 9.7). Tree bark also may provide a source of Na for rodents such as red-backed voles and porcupines, which can result in considerable damage to trees planted for commercial forestry (Roze 1989; Hansson 1991).

Demands for Na and K increase with the growth of body tissues and thus gains in both intracellular and extracellular spaces. Requirements for these electrolytes therefore increase with water turnover and energy expenditure in ungulates such as white-tailed deer (Hellgren and Pitts 1997). Incremental demands for Na during reproduction are more likely to limit herbivores, especially fecund species with rapid growth rates such as California voles because of low Na concentrations in plants compared to prey tissues (Batzli 1986). Because Na is not accumulated in high concentrations in plants, deficiencies in herbivores are most likely to occur in

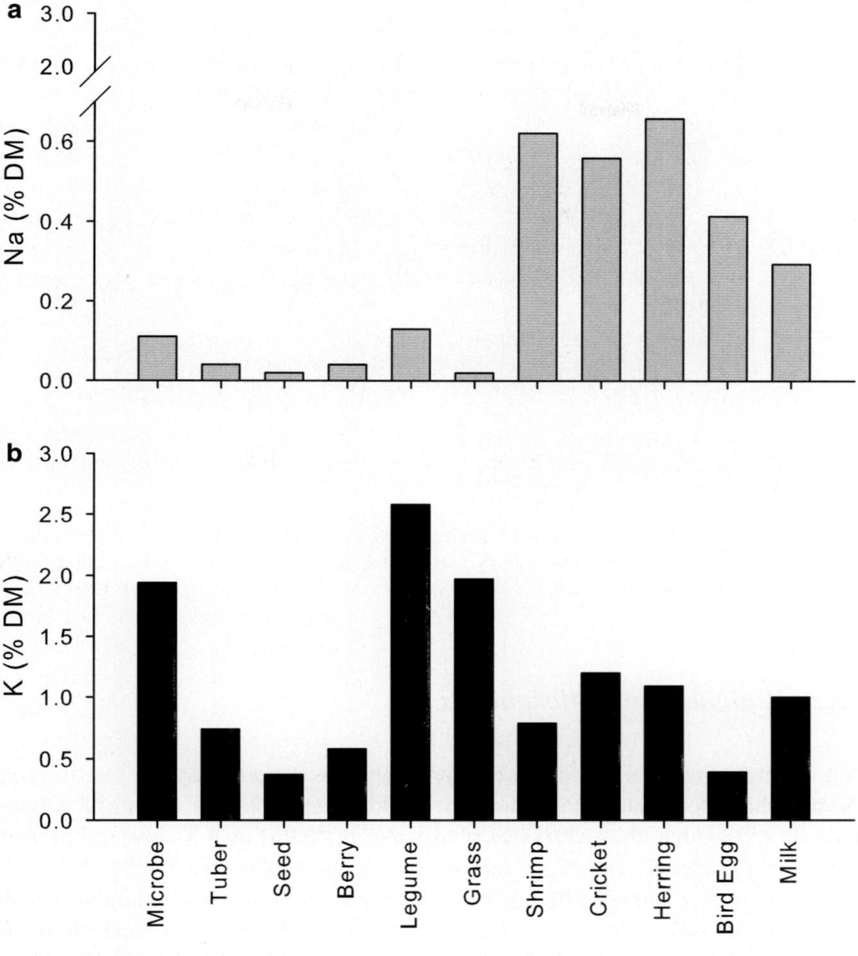

Fig. 9.6 Concentrations of two electrolytes in the dry tissues of selected microbes, plants and animals. **a** Sodium (Na). **b** Potassium (K). Microbes: yeast (National Research Council 2003). Plants: tubers of sweet potato, seeds of corn, blueberries, leafy alfalfa (National Research Council 2003) and hay from mixed species of temperate grasses (National Research Council 2007a). Animals: muscle from mixed species of shrimp (National Research Council 2006); whole cricket and herring (National Research Council 2003); chicken eggs with shells (Johnson 2000; National Research Council 2003); average of milks from domestic cattle, sheep, goats and water buffalos (Agricultural Research Service 2007)

areas where alpine and mountain soils are leached by rain and snow. Increases in activity also may increase the Na demands of animals that secrete high concentrations of Na in sweat, such as horses (National Research Council 2007a).

Fig. 9.7 Inland populations of herbivores such as moose may be limited by Na, especially during summer when high forage intakes for body mass gains, antler growth and reproduction result in the ingestion of large amounts of K

9.2.2 Calcium and Phosphorus

Ca and P influence skeletal formation, and contribute to nerve impulse transmission, muscle contractions and metabolism. The predominant biological form of P is phosphate (PO_4^{3-}), which links with other phosphates in ATP (Fig. 9.8), glycerol in phospholipids (Chapter 7; Fig. 7.5) and sugars in DNA (Chapter 8; Fig. 8.7). The biologically active forms of Ca are ionic; Ca^{2+} activates enzymes by binding to carboxyl groups (—COO^{2-}) on proteins and to both carbonate (CO_3^{2-}) and phosphate (Fig. 9.8). Calcium carbonate is the mineral component of a protein matrix that forms the shells of bird eggs and mollusks. Bone is a matrix of proteins called collagens that are integrated with a complex of Ca and phosphate called hydroxyapatite (Fig. 9.8). The matrix of bone is not static but continuously exchanging Ca and P with the blood (Fig. 9.9) as bone cells mineralize and demineralize the protein matrix in response to both blood Ca^{2+} and the physical stimuli of gravity and activity. Bears can maintain the concentrations of Ca and P in serum as well as the mineral content of their bones through the long inactivity of their winter dormancy (Floyd et al. 1990). The exchangeable nature of the bone matrix allows the gradual incorporation of other ions. The trace mineral fluoride (F^-) substitutes for hydroxyl groups in teeth to produce a less porous matrix; excess exposure to F^- from ground waters may alter bone structure and tooth hardness, especially in young animals (National Research Council 2005), and is associated with tooth breakage and reduced life span of elk wintering near hot springs in Yellowstone National Park (Garrott et al. 2002).

Homeostatic controls of Ca and P in vertebrates are cued by changes in plasma Ca^{2+} and the large reserve of both Ca and P in bone. Declines in blood Ca^{2+} are opposed by the action of parathyroid hormone (PTH), which stimulates release of Ca from bone, resorption of Ca at the kidney and absorption of Ca from the digestive tract mediated by activated vitamin D. Bone mobilization releases both Ca and P, but the relative excess of P usually is cleared at the kidney with the stimulus of PTH. Rises in blood Ca are opposed by calcitonin, which suppresses Ca release from bone. Calcitonin suppresses the effect of PTH on the kidney to allow P to be retained in blood and Ca to be lost in urine. Calcium homeostasis in fish differs

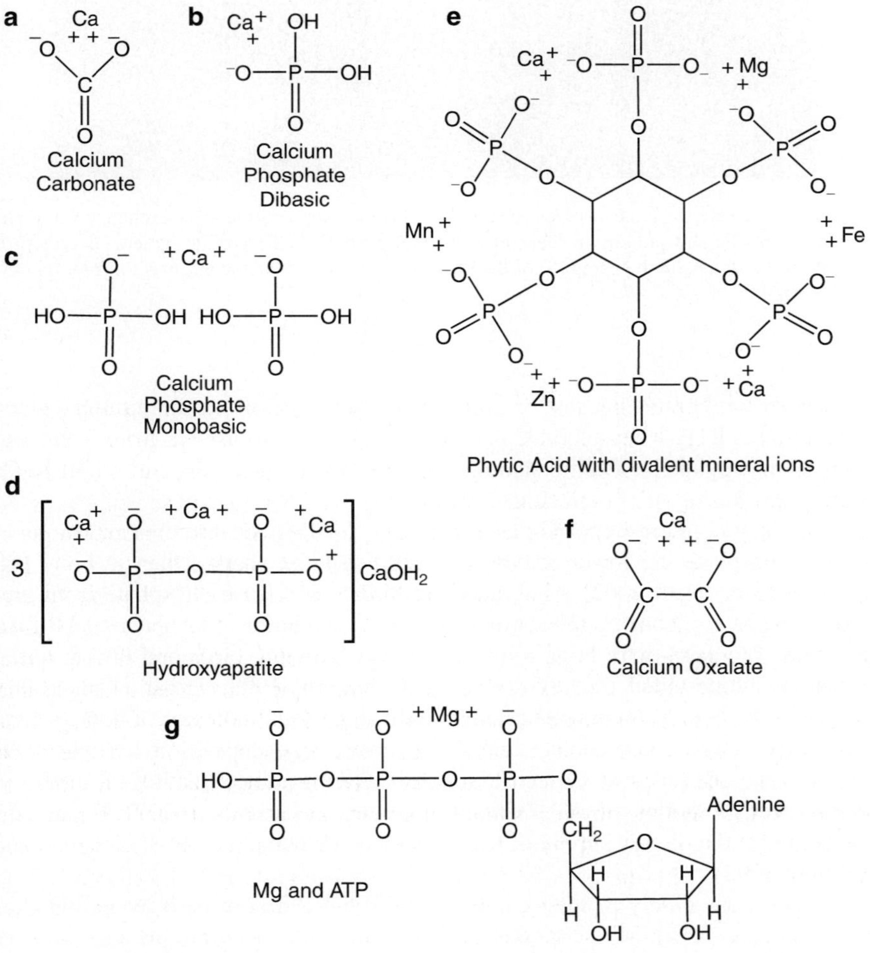

Fig. 9.8 Chemical forms of Ca and P and their interactions with other divalent mineral ions. **a** Calcium carbonate from shells of birds and mollluscs. **b** and **c** Calcium phosphates found in foods. **d** Hydroxyapatite from bone. **e** Phytic acid from seeds with associated mineral ions. **f** Calcium bound to oxalic acid from browse. **g** Phosphate linkages of ATP interact with magnesium

Fig. 9.9 Bones are dynamic systems of cells and channels that continuously exchange Ca, P and protein. Minerals and protein are deposited in rings as bones enlarge; the pattern of growth is evident in the scutes on the carapace of the Galapagos tortoise, which continues to grow for several decades

from other vertebrates because vitamin D and hormones from the pituitary gland substitute for PTH when blood Ca declines. Fish can also use Ca from water and from their integument; Pacific salmon can absorb Ca at the gills and mobilize Ca from scales during migration (Lall 2002).

Bone may be more important as a store of P for fish because concentrations of plasma phosphate are lower in cartilaginous fish (e.g., sharks) than in bony fish (e.g., flounder) (Lall 2002). Mammals and birds can release phosphate from glucose-phosphate in bone by the action of the enzyme alkaline phosphatase (AP); AP in plasma increases with bone mineralization in growing birds and during antler growth in white-tailed deer (Grasman and Hellgren 1993; Tilgar et al. 2008). Similarly, the bone protein osteocalcin is used as a serum indicator of bone growth and antlerogenesis. Osteocalcin and AP indicate bone deposition whereas PTH indicates the adequacy of Ca and P supplies. During mineralization of antlers in fallow deer, activities of AP increase and the concentration of PTH remains unchanged if the dietary supply of Ca is adequate (Eiben et al. 1984; Grasman and Hellgren 1993).

Tissues vary widely in their contents of Ca and P and in their availability for digestion (Fig. 9.10). Milk contains Ca and P in high concentrations and in forms that are readily absorbed by infant mammals; Ca availability is enhanced by interactions between lactose and Ca^{2+}, while milk phospholipids provide a ready source of phosphate (Linder 1991). Phosphate is also available in high concentrations from the membranes of microbes, muscles and egg yolks (Fig. 9.10). Acid digestion releases soluble Ca and phosphate from within cells and protein complexes. Insoluble Ca in bones and shell of whole prey and eggs, however, may require prolonged retention in the stomach, which would limit the digestion of other components of the diet and the supply of energy and protein. Consequently, a large proportion of ingested Ca and P is lost in the feces or regurgitated as pellets by carnivores such as owls that ingest whole prey. Grinding bone into small particles increases the surface area for digestion of both the protein and its minerals. Hyenas, Tasmanian devils and timber wolves chew bones and leave only small fragments and the hardest parts of the jaws and claws. Cape Griffon vultures scavenge small fragments of bone left by large predators to feed their chicks; the eradication of large predators in some areas of Africa lead to bone anomalies in young vultures (Richardson et al. 1986). Herbivores such as voles and porcupines also obtain Ca and P from bones by gnawing on antlers dropped from moose and elk.

Concentrations of Ca and P differ markedly in plants because the two minerals are not always linked by function or structure as they are in the bone matrix of animals. For example, P varies seasonally and is highest in growing plants. High concentrations of Ca^{2+}, however, can bind phosphate and reduce the availability of dietary P. The optimum ratio of dietary Ca: P is between 1:1 and 2:1 for growing bones in most animals. Adults may tolerate much higher ratios (up to 7:1) of Ca: P as long as the diet provides enough P because Ca absorption is regulated in relation to blood Ca^{2+}. Birds may require high ratios of Ca:P (up to 4:1) during egg laying because Ca is needed for calcium carbonate in egg shell, although both Ca and P are required for the yolk and albumen (Fig. 9.10). Although not common, low Ca:P ratios result in an excess of P in the body pool because there are few places to shunt excess P, in contrast to Ca that can be moved to bone or eggs. The resulting condition is known as nutritional secondary hyperparathyroidism (NSH); levels of PTH increase in response to low plasma Ca^{2+} and cause mobilization of Ca from bones, which can deform growing animals (National Research Council 2006, 2007a).

The availability of Ca and P is reduced by interactions with charged elements of the plant fiber matrix and by PSMs such as oxalic acid in leaves and phytic acid in seeds (Fig. 9.8). Microbial phytases in the digestive tract can release P and its associated mineral from phytates in herbivores such as ruminants. Fecal contamination of water near dense aggregations of animals such as cattle feedlots, piggeries and poultry farms provides phosphates and N for rapid growth of algae that may reduce stream flow and water quality for fish and other wildlife.

Inadequate dietary supplies of Ca and P usually result in the net loss of mineral from bone, and in elevated concentrations of PTH or vitamin D metabolites in plasma. Bone demineralization reduces the ability to resist the physical stresses of gravity and activity, which may result in enlarged joints and bent limbs or rib cages in growing animals (rickets) and a high incidence of fractures during capture and handling. Malformation of bones may be more easily discerned in growing animals with long limbs such as cranes (Olsen and Langenberg 1996). Skeletal bones are also a temporary store of Ca and P for production of milk, antlers and eggs. Red deer and mule deer accumulate bone Ca and P from the diet through spring and

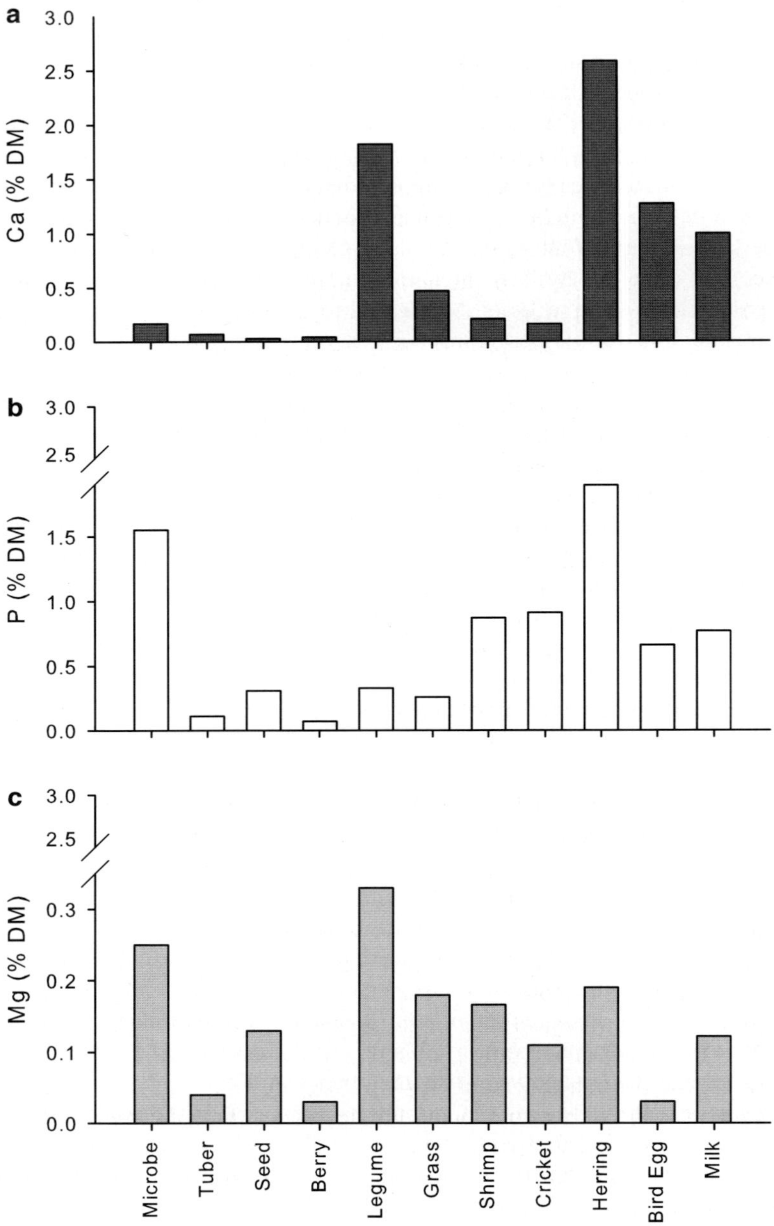

Fig. 9.10 Concentrations of three macrominerals in the dry tissues of selected microbes, plants and animals. **a** Calcium (Ca). **b** Phosphorus (P). **c** Magnesium (Mg). See Fig. 9.6 for details of the selected organisms

summer and translocate those stores to the hardening antler in the last phase of antlerogenesis (Hillman et al. 1973; Muir et al. 1987).

The development of the avian egg shell requires a steady delivery of Ca to the oviduct, which is provided by medullary bone at the core of leg and pelvic bones (Klasing 1998). Vitamin D stimulates absorption of Ca and P from the diet, both of which are deposited in medullary bone. The protein calbindin carries Ca from bone to the oviduct where bicarbonate is produced to combine with Ca as $CaCO_3$ in the egg shell (Fig. 9.8). Net loss of bone is minimal during egg formation in chickens, even though Ca deposited in the egg is equivalent to more than five times the amount of exchangeable Ca^{2+} in the body (Clunies et al. 1993). Granivorous songbirds rely on dietary Ca in insects and snails to support egg formation (Graveland 1996). Pollutants that alter the abundance of invertebrate prey therefore may alter Ca availability and egg production in songbirds such as pied flycatchers (Eeva and Lehikoinen 2004). In tree swallows, Ca availability affects the onset of laying, the size of eggs and the growth of nestlings (Bidwell and Dawson 2005; Dawson and Bidwell 2005). Severe depletion of Ca, P, Na and other minerals may cause pica (the consumption of items that are not usually considered food, such as wood, bone and hair) or soil (geophagia). These unusual materials may increase tooth wear, physical damage to the soft tissues of the mouth and digestive tract and increase the risk of infection (Underwood and Suttle 2001).

9.2.3 Magnesium and Sulfur

Magnesium and sulfur are widely distributed in animals because they form linkages with minerals and amino acids in bone, connective tissue and the enzyme complexes within cells. The largest amount of Mg in animals is associated with bone, where Mg^{2+} binds to hydroxyapatite. Mg^{2+} plays a wide variety of catalytic roles as a cation in the cells of both plants and animals. For example, Mg^{2+} ions stabilize the high-energy phosphate bonds in ATP and interact with enzymes to control the use of this energy store (Fig. 9.8) (Linder 1991). Plants use Mg as a central component of chlorophyll in their photosynthetic systems. Consequently, Mg concentrations are high in leaves and typically lower in plant storage organs such as tubers (Fig. 9.10). The absorption of Mg from plants may be reduced by binding to phosphates in phytic acid or by high concentrations of K (Figs 9.6, 9.8 and 9.10). Grass tetany is a condition found in high-producing livestock when high demands for lactation coincide with a switch to fertilized pastures where plants are high in K and N (Underwood and Suttle 2001). Low Mg availability apparently limits the supply of ATP for muscular contraction and animals quickly become immobilized, with rigid limbs, staggered gait, uncoordinated movements, convulsions and often death. This condition is uncommon among wild ruminants, even though lactating cervids often shift to highly digestible forbs or browse soon after birth. Mg tetany may be rare because those foods are high in Mg or because susceptible females would be quickly predated.

Animal tissue provides Mg in a readily available form for carnivores (Fig. 9.8). Excess Mg is excreted in urine with phosphate and other solutes. Urinary minerals can crystallize into 'stones' or uroliths that are large enough to obstruct the renal system, especially the narrow urethral duct of male mammals. Struvite is a urolith composed of Mg, ammonium and phosphate that is often found in domestic cats. The formation of struvite and other uroliths depends on the concentrations of the component ions, urinary pH and urinary flow rate. Urinary pH is influenced by the dietary load of ions and by the endogenous production of acid for digestion. After large meals, bicarbonate and ammonium ions titrate the acid in an alkaline tide in the urine. Water restriction or dehydration increases the concentration of solutes and reduces urinary flow, which favors the formation of uroliths. Captive cats and dogs may be predisposed to formation of uroliths because their low energy expenditures result in low water turnovers and urinary volumes, and because their food may be consumed in large, regular meals that result in concomitantly large fluctuations in urinary pH. The incidence of struvite uroliths in domestic cats has been reduced by formulating diets that are low in Mg and result in an acid urine (National Research Council 2006). Uroliths also may result from chronic metabolic conditions in wild animals. Uroliths of the sulfur amino acid cysteine (Fig. 8.1) are prevalent in both captive and wild maned wolves because cysteine is insufficiently resorbed at the kidney (Boveé et al. 1981). The genetic defect persists because uroliths mainly shorten the lives of adult males, but do not prevent the animals from reproducing. Captive breeding programs maintain maned wolves with diets that minimize excessive intakes of sulfur amino acids (Barboza et al. 1994; Childs-Sanford and Angel 2006).

Sulfur is found in all proteins, especially the keratins in skin, horns, claws, nails, hair and feathers. Animals therefore ingest most S as the amino acids cysteine and methionine in proteins from both plants and animals (Figs. 8.1 and 8.2). Microbes in the foregut of ruminants and kangaroos may use sulfate and non-protein N (e.g., urea) to synthesize S amino acids for microbial proteins that are subsequently absorbed by the animal (Chapter 8). Cysteine forms the sulfhydryl cross links (—S—S—) between proteins in hair, feathers and connective tissues whereas taurine is used as a charged sulphite group ($—SO_3^-$) in bile salts (Fig. 8.1). S compounds are used for numerous reactions of the vitamins biotin, thiamine and pantothenic acid, and in the antioxidant glutathione. Oxidation of S amino acids produces sulfate ions that contribute to the cation balance within cells. Sulfate excretion at the kidney contributes to decreases in urinary pH that counterbalance the alkaline tide following a large meal of protein.

Aberrations in S metabolism are primarily due to inadequate intakes of methionine and cysteine that can result in malformation of hair, feathers and connective tissue, especially in growing animals. Increases in requirements for S are usually accompanied by much greater increases in the requirements for N and energy to support tissue synthesis. For example, only 3% of the energy expended by molting birds is deposited in feathers (Lindström et al. 1993). Signs of limited S availability such as poor fur and feathers in wildlife therefore may arise from primary limitations on energy and protein, which are discussed further in Chapter 10.

9.2.4 *Trace Metals*

A prominent group of trace nutrients consists of trace metals such as Mn, Zn, Cu and Fe. Imbalances of Mn and Zn are not likely in free-ranging populations of wildlife, but cases of both Cu deficiency and Cu toxicity sometimes are reported. A deficiency of Cu results most commonly in wildlife that consumes low-Cu foods or when other trace minerals in excess inhibit Cu uptake from the gut. Cu toxicity may occur when animals have access to polluted water bodies and settling ponds. In the case of Fe, a deficiency usually occurs when animal health is already compromised, such as by blood-sucking ticks on moose, but Fe toxicity in wildlife is rare.

Ionic forms of metals such as Mn^{2+}, Fe^{2+}, Cu^{2+} and Zn^{2+} are highly reactive and are present in low concentrations both in extracellular fluids and within cells. These trace metals are usually bound to proteins that serve as transport molecules or enzymes. Trace metals are used to catalyze an enormous number of reactions that involve the loss (oxidation) or gain (reduction) of electrons (Table 9.3). Consequently, imbalances in trace metals result in numerous symptoms that range from impaired O_2 uptake for physical activity (Fe) to infertility (Zn) and increased susceptibility to infection (Cu). Reactions that transfer small functional groups include Mn-containing enzymes involved in the production of urea from arginine (arginase), and the coordinated oxidation of C from carbohydrate (pyruvate carboxylase) and lipid in the TCA cycle (Fig. 7.9) (Brody 1999). Interactions between trace metals and macromolecules include those between Zn and nucleotides that stabilize the DNA helix during transcription of RNA. Reversible electron-transfer reactions are used for gas exchange and respiration; the red heme complex at the center of the hemoglobin and myoglobin proteins contains Fe, which forms a readily reversible bond with O_2 and allows the gas to be transported in blood and stored in muscle (Fig. 9.11). A similar complex with Cu in hemocyanin carries O_2 in the blood of mollusks, albeit with a change in color to blue (Mathews and Van Holde 1996).

Protein complexes involving both Fe and S form parts of the electron transport chain for respiration in mitochondria and for photosynthesis in plant chloroplasts (Mathews and Van Holde 1996). Enzymes with trace metals catalyze reactions that defend against toxic products of respiration. The respiratory chain uses O_2 and H^+ to produce water (H_2O), but these reactions also can release toxic superoxide ions (O_2^-) that randomly destroy macromolecules such as lipids. Superoxide dismutase is a family of enzymes that contain one of the trace metals (Mn, Cu, Zn or Fe; Table 9.3) and convert superoxide to the less toxic hydrogen peroxide by the reaction $2O_2^- + 2H^+ \rightarrow H_2O_2 + O_2$. Peroxides are rapidly degraded to harmless water and oxygen by the enzyme catalase, which involves Fe in heme for the reaction $2H_2O_2 \rightarrow 2H_2O + O_2$. Trace metals such as Cu are also involved in production of peroxides

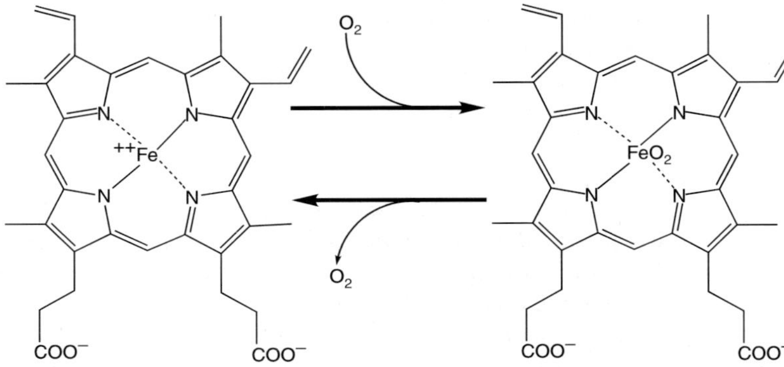

Fig. 9.11 Iron (Fe) readily interacts with oxygen (O_2) to form oxides that are too robust to release O_2 under cellular conditions. However, when combined with heme, an organic complex, Fe binds with O_2 in a controlled and reversible fashion that makes possible the transport of O_2 between lungs and tissues by hemoglobin, and storage of O_2 in muscle by myoglobin (Mathews and Van Holde 1996)

and superoxides for immune defense (Weiss and Spears 2006). White blood cells such as neutrophils, macrophages and lymphocytes use a rapid burst of O_2 consumption to produce toxic superoxides and hydrogen peroxide that destroy invading bacteria (Bonham et al. 2002).

Mucosal cells of the intestine absorb trace metals in the ionic state; Fe is also absorbed as heme but degraded within the mucosal cells or enterocytes (Brody 1999). Metal ions are transferred to intracellular proteins such as ferritin for Fe storage within the mucosal cells. Carrier proteins such as plasma albumen, ceruloplasmin (Cu) and transferrin (Fe) mediate the transfer of nutrients from mucosa to blood and target tissues such as liver, muscle or bone marrow. Blood samples are used to assess Fe status by measuring the total iron-binding capacity of plasma and serum ferritin. These parameters of Fe status have been used to assess the development of O_2 storage and diving ability in harbor seals, which fast on shore after they are weaned (Burns et al. 2004). Ceruloplasmin activities in blood samples are used to assess Cu status; activities of ceruloplasmin reflect concentrations of Cu in the liver of moose, muskoxen and caribou (Barboza and Blake 2001). Enzymes and proteins that indicate Cu, Zn and Fe status should be used to complement other measures because many of the enzymes are affected by development and stressors. For example, ceruloplasmin activities do not reflect liver Cu in muskox calves less than 90 days of age or liver Cu in male reindeer during rut (Barboza and Blake 2001; Rombach et al. 2002b). The liver mediates the uptake and storage of metals in proteins such as hemosiderin (Fe) and metalothionein (Cu and Zn). Concentrations of metals in the liver are therefore the most common method for assessing the status of Cu, Zn and other metals in fish and wildlife. Apparent digestibilities of trace metals are usually low, and some trace metals such as Mn can be used as a digestibility marker (Chapter 4). Low digestibilities of metals reflect both unavailable forms and fecal losses of absorbed metals in sloughed intestinal mucosal cells, and unresorbed secretions from the liver and pancreas.

Animals contain Fe, Cu and Zn in readily available forms that reach their highest concentrations in internal organs such as the liver, kidney and gonads (Fig. 9.12). Milk contains highly available forms of trace metals that are highest in the colostrum, which is produced soon after birth, but concentrations of these minerals are low for the rest of lactation (Underwood and Suttle 2001). Mothers transfer Fe, Cu and Zn to their young during fetal development in utero; suckling muskoxen and caribou rely on stores of Fe, Cu and Zn in the liver until they begin consuming forages (Rombach et al. 2002a; Knott et al. 2004). In marsupials, for which milk is used to support most of the development of the offspring in the pouch, milks are more concentrated in Fe, Cu and Zn than those of placental mammals (Green and Merchant 1988; Krockenberger 2006). Plants usually contain lower concentrations of trace metals than animals. High metal concentrations in plants are often associated with less available forms that are complexed with plant fiber or with PSMs such as phytate (Fig. 9.12). Steady consumption of soil may increase the uptake of trace metals. Drought is associated with increases in soil ingestion by hairy-nosed wombats, which leads to increases in liver Fe, Cu and Zn (Gaughwin et al. 1984). Lack of access to soils, as with Fe-deficient diets, by captive animals may result in a variety of deficiencies, including the Fe-deficient depigmentation of underfur known as cotton-fur that decreases the value of mink from fur farms (Robbins 1993).

Chronic exposure to trace metals can slowly increase their concentrations in liver, kidney and even hair as the metal ions bind to weakly negative groups such as the sulfhydryl group (—SH) on cysteine. Metalothionein is a S amino acid-rich storage protein which binds Cu, Zn and toxic heavy metals such as Cd (Bremner and Beattie 1990). High concentrations of Cd from invertebrate prey accumulate on metalothionein in the kidney of beluga whales (Dehn et al. 2006). Similarly, moose accumulate Cu in the liver from plants growing in acidic soils in Finland (Hyvarinen and Nygren 1993). Metals become toxic when large amounts of the ion are released; Cu is released to the blood when the storage capacity of the liver is exceeded, resulting in the oxidation and lysis of blood cells in sheep and hairy-nosed wombats on high-Cu diets (Barboza and Vanselow 1990; Underwood and Suttle 2001). Tolerances to metals vary with the ability to control absorption, safely store the metal in protein and excrete excesses. Trout can tolerate a diet of 665 ppm Cu whereas domestic sheep may exhibit signs of Cu toxicity on diets as low as 10 ppm Cu (Lall 2002; National Research Council 2005). Metal tolerances and requirements sometimes differ among closely related species and among subpopulations of the same species of domestic animals (Oldham 1999). Browsing rhinos such as the Sumatran rhino and black rhino are susceptible to Fe overloads and hemolytic crises in captivity, whereas captive African white rhinos and Indian rhinos are grazers that tolerate diets based on agricultural forages and products (Paglia et al. 2001).

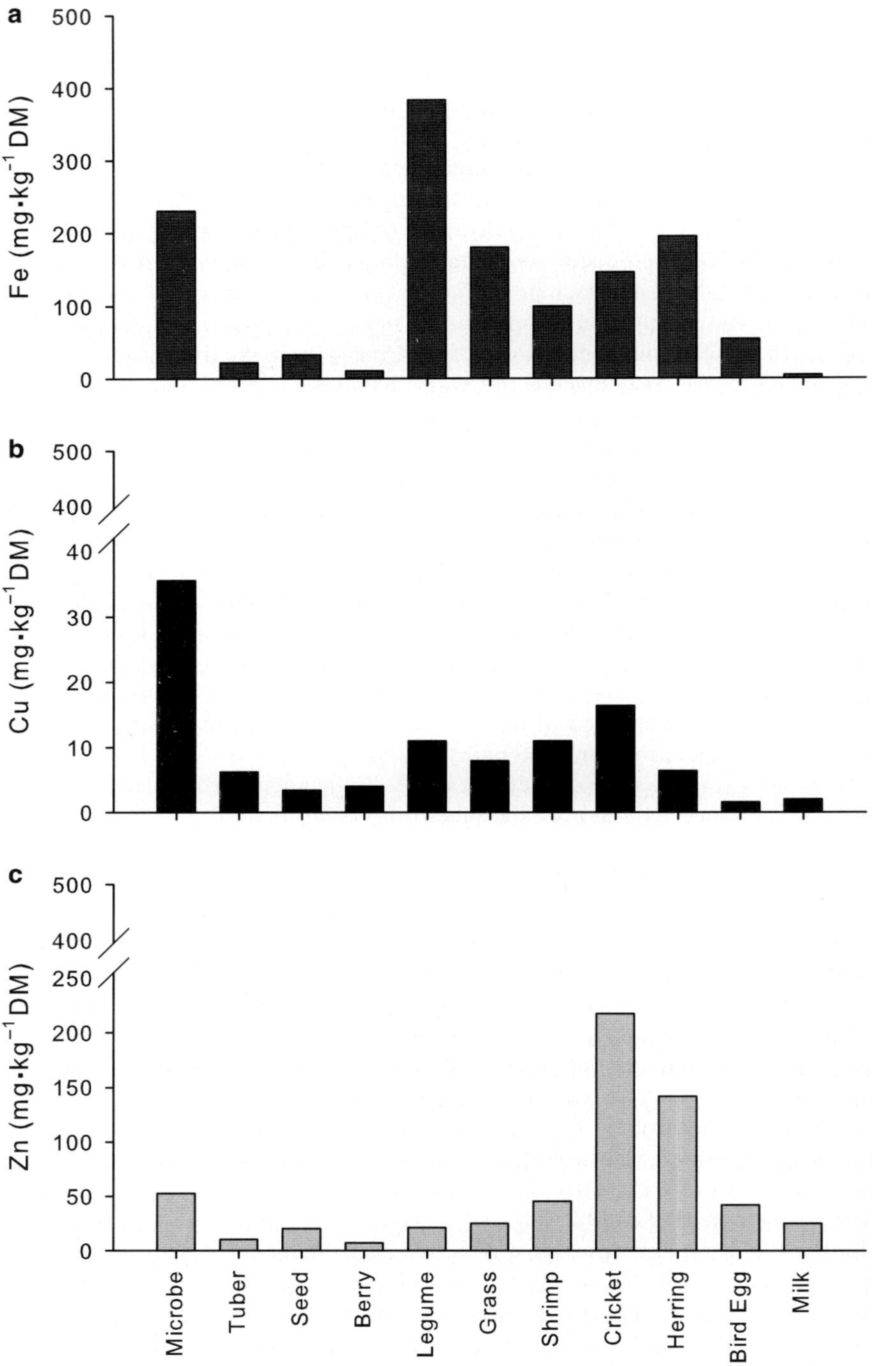

Fig. 9.12 Concentrations of three trace minerals in the dry tissues of selected microbes, plants and animals. **a** Iron (Fe). **b** Copper (Cu). **c** Zinc (Zn). See Fig. 9.6 for details of the selected microbes, plants and animals

Regional deficiencies of trace metals are probably more common than toxicities in populations of wild animals. Soils with low Cu availability for plants result in low Cu concentrations in the livers of Californian elk and Canadian muskoxen, and in the hair and hooves of Alaskan moose (Flynn et al. 1977; Gogan et al. 1989; Barboza et al. 2003). Alkaline soils or mine spoil areas with high concentrations of molybdenum (Mo) and sulfate reduce the availability of Cu by forming an insoluble copper thiomolybdate ($CuMoS_4$) in the forestomach of ruminants and kangaroos. Deficiencies are exemplified by defective keratinization of hooves in ungulates, which decreases mobility and increases the risk of predation. In ruminants such as moose, Cu deficiency can be induced on diets containing normal levels of Cu but relatively high levels of Mo and inorganic sulfate (Suttle 1991; Frank et al. 1994, 2000). Thiomolybdates are also associated with apparent Cu deficiency in the quokka, a small wallaby living on the limestone soils of Rottnest Island off the coast of Western Australia. Rottnest Island quokkas exhibit seasonal anemias whereas the same species living on better soils on the adjacent mainland show no signs of Cu deficiency (Barker 1961a,b; Hume 1999). Regional differences in Cu availability may become more apparent as populations grow. High densities of muskoxen increase the exposure of calves to intestinal infection, which increases their use of liver Cu and impairs their growth (Swor 2002). Intestinal infections debilitate and eventually kill calves by reducing food intake and increasing water loss through diarrhea. Trace metal imbalances therefore may intensify the adverse effects of disease and weather on young animals and thus reduce the rate of population growth (Barboza and Reynolds 2004) (Chapter 2).

9.2.5 Iodine and Selenium

The trace elements iodine and selenium serve critical roles in biochemical pathways and development. They also interact to affect the animal's metabolism and thermoregulation. In free-ranging wildlife, iodine toxicities are rare; deficiencies occur in iodine-deficient areas and in response to some plants and pesticides. Se toxicities result from the consumption of seleniferous plants and Se deficiencies are most common in regions leached by precipitation.

I is readily available as the iodide ion (I^-) in salts of Na or K, especially in the tissues of marine organisms such as seaweeds and sponges. The primary role of I in the body is in the function of the thyroid gland. Iodide is absorbed by the thyroid and attached to a derivative of the amino acid tyrosine to form the hormone thyroxine (T_4; Fig. 9.13). The thyroid gland stores T_4 as the protein thyroglobulin, which is hydrolyzed to form the hormone secreted into the blood. Removal of iodide from T_4 produces a more active form of the hormone, tri-iodo-thyronine (T_3). Both T_4 and T_3 are absorbed into target cells and bind to receptors in the nucleus and mitochondria.

Thyroid hormones stimulate transcription of nuclear genes, as well as protein synthesis and respiration in mitochondria (Norris 1997). Increasing concentrations

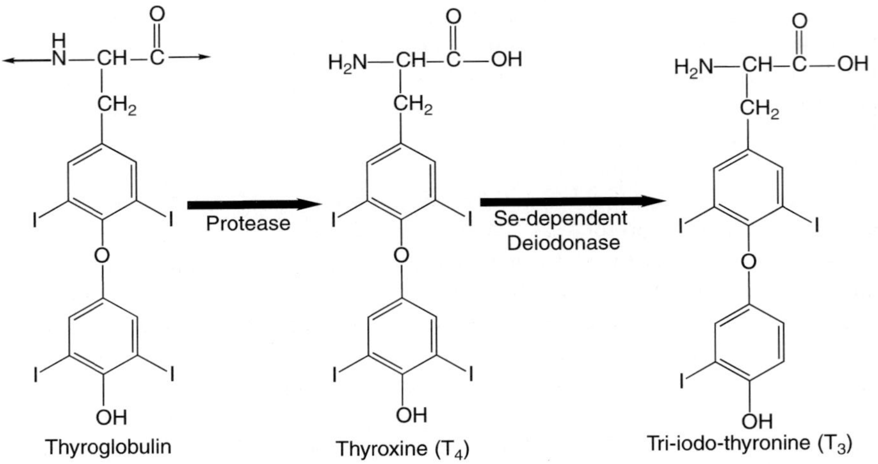

Fig. 9.13 Iodine (I) modifies the amino acid tyrosine to form a hormone (thyroxine; T_4) that regulates energy metabolism. The hormone is stored in the thyroid gland as a polymer (thyroglobulin) and released as T_4 when the peptide bonds are broken (Groff et al. 1995). Hormonal activity is further increased by removing one I to convert T_4 to tri-iodo-thyronine (T_3) with an enzyme that uses Se. Inadequate Se intakes therefore cause secondary symptoms of I deficiency

of plasma T_4 and T_3 therefore are associated with increases in resting energy expenditure and the production of metabolic heat for thermoregulation (Chapter 10). High concentrations of thyroid hormones in newborn muskoxen and caribou reflect high rates of heat production used to combat low ambient temperatures of late winter and early spring in the Arctic (Knott et al. 2005). Plasma T_4 increases with seasonal changes in daily food intake and mass gain of red deer and reindeer during summer (Ryg and Langvatn 1982; Ryg and Jacobsen 1983; Barboza et al. 2004) and with the increased activity of migration in Pacific salmon (Norris 1997). Thyroid hormones activate genes that cause the asymmetry of the eyes in flounder and the resorption of fins and tails in frogs during metamorphosis (Norris 1997). Similarly, T4 is also associated with muscular development in birds such as barnacle geese (Deaton et al. 1997). The suppression of thyroid hormones in I deficiency produces many symptoms that include impaired development of the brain (e.g., cretinism in humans), skin anomalies and suppressed rates of metabolism that vary with the role of the T_4 and T_3 in each species and life stage.

Thyroid secretions are stimulated by hormones released from the central nervous system at the hypothalamus and the pituitary gland; these hormones integrate thyroid function with seasonal changes in growth and reproduction. Plasma concentrations of T_3 and T_4 are influenced by the sensitivity of receptors and the clearance rate of the hormones from the blood, both of which may change with season (Tomasi and Mitchell 1994). Increasing concentrations of thyroid hormones in blood suppress the secretion of the thyroid gland by negative feedback. Iodine deficiency does not allow enough thyroid hormone to be released for negative feedback. Consequently, the thyroid continues to produce protein and enlarges into a tumor or goiter. Goiters also are produced by PSMs called goitrogens. Plants in the Family Brassicacea, which include turnips and cabbages, contain a PSM called progoitrin that is converted to goitrin when the plant is damaged; goitrin suppresses iodination and thus production of T_4 by the animal (Norris 1997). Goiters can be caused by cyanogenic glycosides from a variety of plants, including clover and cassava; the release of thiocyanate (CNS^-) ions from the hydrolysis of the PSM blocks iodide absorption by the thyroid (Chapter 3) (Harborne 1993). The pesticide DDT causes enlargement of thyroid glands in ring doves, which is associated with thinning of egg shells and low rates of survival in chicks. The widespread effects of DDT on egg shell thickness in peregrine falcons and other birds may be due to both a direct effect on Ca deposition in the oviduct and a secondary effect of thyroid hormones on this energetically demanding process (Moriarty 1999).

Deficiencies of Se impair thyroid function because the conversion of T_4 to T_3 requires a Se-dependent iodinase (Fig. 9.13). Se deficiency therefore is accompanied by some symptoms of I deficiency such as low T_3 and impaired thermogenesis in newborn ruminants exposed to the cold (Underwood and Suttle 2001). Se is incorporated into enzymes and other proteins as the modified S amino acids Se-methionine and Se-cysteine (Table 9.3). These amino acids are produced by substitution of S with Se immediately before translation of mRNA to protein (Brody 1999). Animals absorb the Se-amino acids with the same transporters for methionine and cysteine in the small intestine (Chapter 8). Inorganic Se is passively absorbed as selenite (SeO_3^{2-}) and actively absorbed as selenate (SeO_4^{2-}) in a carrier shared with sulfate. Plants that accumulate Se can discriminate between sulfate and selenate and can store Se-methionine and Se-cysteine in vacuoles, where the modified amino acids cannot interfere with protein synthesis (Harborne 1993). In animals, Se toxicity is associated with sloughing of hooves and nails and with rough hair coats because sulfhydryl cross links (—S—S—) between keratin proteins are lost when cysteine is replaced with Se-cysteine. In ungulates, the chronic form of Se toxicity is known as alkali disease; acute Se toxicity leading to respiratory failure is called blind staggers. Seabirds may be exposed to high levels of Se from sediments and invertebrate prey. Emperor geese, common eiders and spectacled eiders that winter in the seas off Alaska have high levels of blood Se when they arrive at the spring breeding grounds but those levels decline through the summer. Eiders and emperor geese may be more tolerant of Se than other waterfowl because blood Se concentrations of these waterfowl were 10 times greater than those of freshwater birds such as mallards at the start of spring (Franson et al. 2002; Grand et al. 2002).

The cells of the liver, heart and blood are rich in Se and its associated proteins. Liver Se and the activity of the Se-dependent enzyme glutathione peroxidase are used to assess Se status (Underwood and Suttle 2001). Glutathione is a derivative of cysteine that serves as a substrate for detoxification of peroxides of hydrogen (H_2O_2) and lipids (—ROOH) in the cytosol and mitochondria of cells throughout the body (Fig. 9.14). Peroxides are highly reactive byproducts of O_2 utilization that can damage proteins and membranes; oxidative defenses such as glutathione peroxidase are therefore crucial to maintaining tissues with intense O_2 consumptions such as exercising muscles, liver and kidney (Hulbert et al. 2007). Glutathione and glutathione peroxidase may serve as stores of S and Se respectively during feather synthesis in birds such as white-crowned sparrows when demands for S-amino acids and energy for protein synthesis are high (Murphy and King 1990; Underwood and Suttle 2001). Se deficiency is associated with fragile red cells and muscle damage or myopathy in ruminants and birds, especially after the intense exercise of capture and handling (Spraker 1993). White-muscle disease, characterized by dystrophy and white streaks in muscle tissue, results from low levels of Se and vitamin E, which are augmented by capture stress, in species such as mountain goats.

Demands for Se are affected by complementary systems of oxidative defense such as superoxide dismutase and vitamin E; inadequate supplies of trace metals and vitamin E may intensify the role of Se in oxidative defense and constrain the supply of Se for thyroid function. Signs of Se deficiency therefore may vary with environmental conditions and with the supply of trace metals and vitamin E (Underwood and Suttle 2001). Populations of wild animals often can tolerate chronically low supplies of Se because other nutrients ameliorate the need for Se and because stores of Se in the liver can satisfy the low demands for Se in adults. The constraints of low intakes of Se and other trace minerals become apparent when high demands for growth and reproduction coincide with high environmental demands. These problems are more readily discerned in livestock that are confined to small managed areas than in wild animals with large home ranges. However, the subtle effects of trace nutrients can be revealed by experimental approaches such as supplementation. For example, Se supplementation of wild mule deer increased the survival of fawns even though mothers did not exhibit any symptoms of Se deficiency (Flueck 1994). Trace mineral status can contribute to high mortality and low recruitment in a population when home range and age structure are altered by changes in the environment (Chapter 2).

9.3 Vitamins

Vitamins are comprised of structures that cannot be synthesized by animals. Many vitamins contain N, which led to the original name of 'vital amines' for this group of essential nutrients (Linder 1991). Some vitamins also contain essential elements such as cobalt (Co) in cyanocobalamin (vitamin B_{12}) and S in biotin (vitamin B_8) and thiamine (vitamin B_1; Table 9.4). The structures of vitamins include C rings and C chains. Rings are used to carry electrons on niacin and riboflavin (vitamin B_2) for respiration, and to absorb light on retinal for vision (vitamin A; Table 9.5). Chains are used to incorporate tocopherol (vitamin E) into membranes and to transfer functional groups between larger molecules in the TCA cycle with thiamine, transaminate amino acids with pyridoxine (vitamin B_6), and synthesize amino acids and bases with folic acid (vitamin B_9).

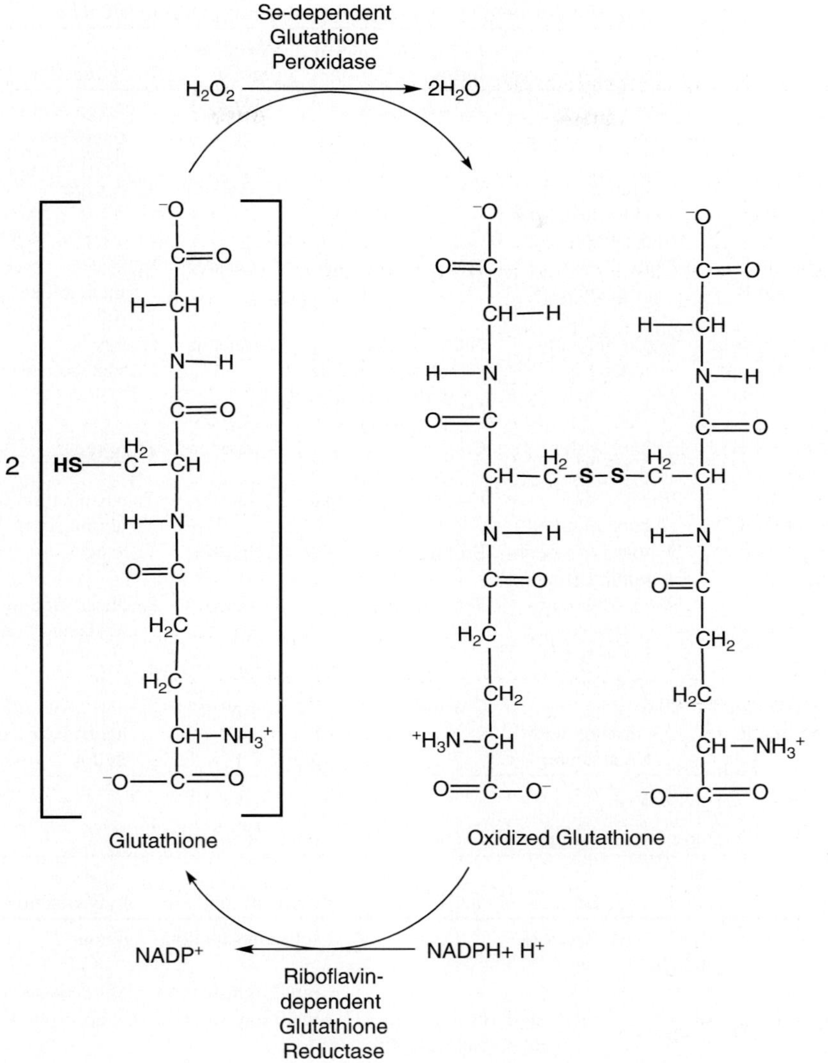

Fig. 9.14 The amino acids cysteine, glycine and glutamate are combined in glutathione, which serves as a substrate for reactions with peroxides of hydrogen (H_2O_2) and lipids (—ROOH) in the cytsol and mitochondrial matrix of cells (Underwood and Suttle 2001). Glutathione peroxidase is a Se-dependent enzyme that destroys peroxides of hydrogen and free fatty acids. Glutathione reductase is an enzyme that depends on riboflavin to recycle glutathione by using electrons from NADPH (Brody 1999)

Vitamin deficiencies are difficult to identify in populations of wild animals because symptoms may be subtle and require repeated testing that is often only feasible in captivity. Deficiencies are also slow to develop because these metabolites are conserved and recycled within the body. Seasonal or short-term deficits

Table 9.4 Common forms and functions of selected water-soluble vitamins in animals

Name	Chemical structures	Biologically active forms	Chemical function	Body function
Vitamin C	Sugar	Ascorbic acid	e^- transfer, reduction	Collagen synthesis, oxidative defense
Thiamine, Vitamin B_1	Pyrimidine and side chain with S and N	Thiamine-pyrophosphate	C oxidation	TCA cycle
Riboflavin, vitamin B_2	Triple C ring containing N	Flavin nucleotides (e.g., FMN, FAD)	e^- transfer	Energy metabolism
Niacin, nicotinic acid	Single ring containing N	Nicotinamide adenine dinucleotide (e.g. NADH)	e^- transfer	Energy metabolism
Pantothenic acid, vitamin B_5	C chain with N	Coenzyme A (e.g., acetyl-CoA)	C transfer	TCA cycle
Pyridoxine, vitamin B_6	Single C ring containing N	Pyridoxal-phosphate	C transfer	Transamination of amino acids
Biotin, vitamin B_8	Double C ring containing S and N	Biotin-AMP	C transfer	TCA cycle
Folic acid, vitamin B_9	Two C ring complexes containing N with C side-chain	Tetrahydrofolates	C transfer	Synthesis of amino acids and bases
Cyanocobalamin, vitamin B_{12}	Multi-ring complex with Co at center	Cobalamines	C transfer	Synthesis of amino acids, bases and SCFA

Table 9.5 Common forms and functions of fat-soluble vitamins in animals

Name	Chemical structures	Biologically active forms	Chemical function	Body function
Vitamin A	Single C ring and chain	Retinoids	Light transduction	Vision
			Signal transduction	Cell division
Vitamin D	Sterols	Ergocalciferol (D_2 from plants), cholecalciferol (D_3 from animals)	Signal transduction	Ca absorption
Vitamin E	Double C ring and chain	Tocopherols	e^- transfer	Membrane integrity
Vitamin K	Double C ring and chain	Phylloquinone (K_1 from plants), menaquinone (K_2 from bacteria)	C transfer	Activation of blood clotting enzymes

in vitamin intake are also offset by stores in tissues such as the liver. Consequently, symptoms of vitamin deficiency may only appear briefly in a portion of the population. For example, folic acid deficiencies produce specific anemias that are associated with impaired development, which is most likely to appear among neonatal animals when the demands for cell synthesis are highest. Anomalies of vitamin nutrition are most frequently reported in animals that are restricted to a small area with a relatively small number of dietary choices for long periods of time, such as populations that are isolated on islands by water, or in fenced reserves. Animals are less vulnerable to these problems if they are able to consume a diversity of foods that compensate for the low concentration or availability of vitamins among food items.

Vitamins are categorized on the basis of their extraction from tissues in animals and plants. Water-soluble vitamins include the B vitamins and ascorbic acid (vitamin C) that are mainly associated with the aqueous contents of cells (Table 9.4). The fat-soluble vitamins A, D, E and K are found in lipids and membranes (Table 9.5). Fat-soluble vitamins can be absorbed and transported with dietary lipids, whereas water-soluble vitamins are usually absorbed by specific carriers at the intestinal mucosa. Water-soluble vitamins are deposited in cells throughout the body, whereas fat-soluble vitamins can be accumulated in body lipids and may lead to toxic effects. Fat-soluble vitamins often are excreted in bile from the liver, whereas water-soluble vitamins are usually excreted in urine from the kidney. High doses of B vitamins in captive animals are readily discerned by urinary excretion of fluorescent cyclic compounds. The specific analysis of vitamins and their associated metabolites is often complex. Samples are usually frozen immediately in liquid N_2 or dry ice (solid CO_2), stored in ultra-low freezers (−80°C), and shielded from light in opaque containers to minimize degradation. Extractions of samples may involve several steps to remove extraneous compounds (e.g., cell proteins) before the specific metabolites can be isolated and quantified by one of the many chromatographic methods available. The activity of vitamins in foods is measured with bioassays on cell cultures or laboratory animal models (e.g., chickens, mice or rats). As with mineral analyses, the quality of vitamin analyses should be monitored closely by including certified standards and controls in each batch of samples and by comparing results between laboratories.

9.3.1 Water-Soluble Vitamins

9.3.1.1 B Vitamins

Vitamins in the B group are involved in several pathways of intermediary metabolism (Table 9.4) that interact with the functions of other trace nutrients. For example, riboflavin is required for maintaining glutathione reductase, the enzyme that recycles glutathione for oxidative defense with Se-dependent peroxidase (Fig. 9.14) (Brody 1999). Inadequate supplies of B vitamins can therefore impair the production and maintenance of many tissues and are most evident as anomalies of the skin, muscle and blood cells. Symptoms vary between species and life stages with the relative demands for each vitamin.

Plant storage organs such as tubers and the endosperm of seeds are poor sources of B vitamins and trace minerals, but the synthetic portions of the plant such as the photosynthetic leaves and the germ of seeds are rich sources of these trace nutrients. Many B vitamins are synthesized by microbes. Microbial fermentation provides most of the requirements of foregut fermenters (e.g., ruminants and kangaroos) for these vitamins. The cecotrophs ingested by rabbits, rats and ringtail possums are important sources of these microbial vitamins for these small hindgut fermenters (Stevens and Hume 1995). The presence of mucosal transporters in the colon indicates that some microbial vitamins also may be absorbed directly from the hindgut (Said 2004). Prolonged antibiotic treatments of captive animals can reduce the supply of these endogenous sources of vitamins by altering the microbial fermentation. Muscle and organs of animal prey are rich sources of vitamins for carnivores. However, plant and animal tissues begin to oxidize B vitamins soon after injury. These vitamins continue to be oxidized with exposure to air, light and increasing temperature. The loss of vitamins in stored food is therefore a major concern for the nutrition of captive wildlife in zoos and aquaria.

Some foods contain compounds that destroy vitamins or impair their absorption. Thiamin is degraded by enzymes called thiaminases in fish and some plants (e.g., horsetail, nardoo and bracken fern). Similarly, much of the biotin in eggs is bound to the protein avidin (Brody 1999). Conversely, vitamins such as niacin are partly produced from other dietary components. Niacin can be synthesized from tryptophan in a series of reactions that require mineral ions (Cu^{2+}, Fe^{2+}, Mg^{2+}) and another B vitamin (pyridoxine); this source of niacin may be only important when intakes of protein are high, because only 3% of the tryptophan from dietary protein is converted to niacin by this pathway (van Eys 1991). Obligate carnivores such as cats receive both tryptophan as well as niacin from NAD and NADP in the muscles of their prey (Morris 2002). Dietary changes also may affect requirements for B vitamins; requirements for pyridoxine may increase with protein intake and carnivory because the demands for transamination increase as amino-acid C is used for energy (Leklem 1991). Similarly, changes in the fermentation of dietary carbohydrate may increase the demand for vitamin B_{12} when propionate production is increased; high starch diets and low B_{12} have been associated with chronic wasting and inflammatory bowel disease in captive moose (Shochat et al. 1997). Supplements of cobalt, which is essential to the formation of B_{12}, and increased offerings of dietary browse are used to partially remedy these problems.

9.3.1.2 Vitamin C

Unlike the other water-soluble vitamins, vitamin C (ascorbic acid) is a carbohydrate and does not contain N (Table 9.4; Fig. 9.15). Ascorbic acid can be synthesized from glucose in a pathway that includes the enzyme glucono-lactone oxidase (GLO). Sturgeons express the gene for GLO, but most other fish do not

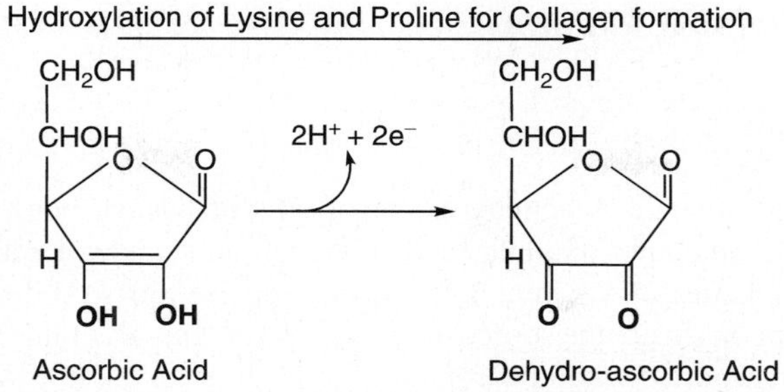

Fig. 9.15 Ascorbic acid (vitamin C) donates electrons for modifying amino-acid side chains that are involved in the cross-linking of collagen in connective tissues (Linder 1991)

express the enzyme and therefore require dietary sources of vitamin C (Halver 2002). The expression of GLO follows a phylogenetic pattern that progresses from kidney to liver. In amphibians, reptiles and monotremes (echidna, platypus), the enzyme is only expressed in the kidney. GLO is expressed in both the kidney and liver in marsupial bandicoots, but only in the liver of other marsupials and placental mammals (Hume 1999). A similar progression of GLO expression from the kidney to the liver is also reported in passerine birds. Ascorbic acid is required by some songbirds, some rodents (e.g., guinea pigs) and by most primates and bats. Fruits and flowers provide the highest concentrations of ascorbic acid for herbivores, whereas organs such as the liver may provide a rich source of the vitamin for carnivores.

Vitamin C deficiency causes weak connective tissues because ascorbic acid is used to cross-link collagen. Vitamin C deficiency causes skeletal abnormalities in growing fish (Halver 2002) and weak membranes in the mouth and blood vessels of other vertebrates. Symptoms of vitamin C deficiency include poor wound healing, increased rates of oral infection and tooth loss, hemorrhages and bruising. Vitamin C deficiency in humans is called 'scurvy', a condition frequently incurred by explorers when supplies of citrus fruits or preserved vegetables were unavailable. Requirements for ascorbic acid may increase with infection, energy use and exposure to toxins, as the vitamin is part of the oxidative defenses of the body (Moser and Bendich 1991; Halver 2002). Ascorbic acid may be used to protect the nervous system of hibernators from oxidative damage during the intense metabolic activity of arousal from torpor. In Arctic ground squirrels, the concentration of ascorbic acid in the blood plasma increases during torpor when body temperatures are low, but declines quickly as the animal increases O_2 consumption during rewarming (Drew et al. 1999). Glutathione and its Se-dependent peroxidase also are used to protect the fragile mucosa of the intestine from oxidative damage during rewarming of thirteen-lined ground squirrels (Carey 2005).

9.3.2 *Fat-Soluble Vitamins*

9.3.2.1 Vitamin A

Compounds with vitamin A activity are involved with vision, bone remodeling of young, reproductive output of adults, and epithelial tissue and coloration. Both deficiencies and toxicities of vitamin A have been reported in wildlife. Lipid-soluble retinoids of plant origin are the source of vitamin A for animals (Table 9.4; Fig. 9.16). Retinoids are transported with lipoproteins to lipid stores in the liver and adipose tissue. The liver exports retinol on retinol-binding protein, which is recognized by the light-sensitive cells on the retina at the back of the eye. The retinal cells convert retinol (an alcohol) to retinal (an aldehyde), which is bound to the protein opsin to form rhodopsin. Rhodopsin is degraded by light in a reaction that releases retinol, but also transfers the energy to a Na pump in the cell membrane, allowing the transmission of a nerve impulse that is perceived as vision (Linder 1991). Small amounts of retinol are lost from this system, but most of the retinol is recycled to retinal and returned to

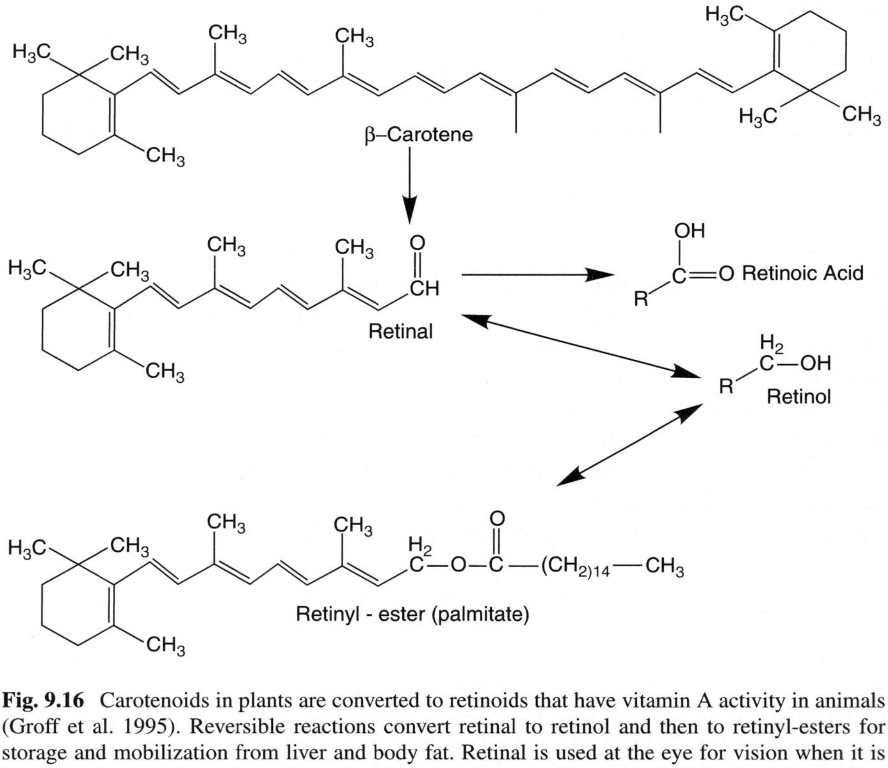

Fig. 9.16 Carotenoids in plants are converted to retinoids that have vitamin A activity in animals (Groff et al. 1995). Reversible reactions convert retinal to retinol and then to retinyl-esters for storage and mobilization from liver and body fat. Retinal is used at the eye for vision when it is bound to the protein opsin to form the light transducer rhodopsin. Retinal is also converted to retinoic acid, which serves as a hormone for cell differentiation

rhodopsin. Although the supply of retinol is maintained by small stores of retinyl-ester in the retina, depletion of that store in vitamin A deficiency impairs vision, especially in low light. Night-blindness is the earliest sign of vitamin A deficiency in humans, but this sign may only be apparent as subtle changes in behavior of animals such as domestic pigs and horses (Olson 1991).

Retinol is also absorbed at other tissues and converted to retinoic acid, which is used to control differentiation and division of cells (Fig. 9.16). Vitamin A deficiency affects cells in the outer layer (cornea) of the eye and the associated mucus glands; dry eyes accumulate keratin which eventually causes inflammation of the eye and blindness. Insufficient retinoic acid impairs growth and reduces reproductive output of adults by impairing development of both sperm and the fetus (Olson 1991). Cell division is also impaired by vitamin A toxicity; cartilage and bone overgrow and fuse in fish and mammals (Halver 2002). Carnivorous mammals tolerate higher concentrations of retinyl-esters in their blood than omnivorous and herbivorous mammals (Schweigert et al. 1990) and safely accumulate these retinyl-esters in specialized cells within the liver (Leighton et al. 1988). Toxicities of vitamin A may occur in people consuming liver from marine carnivores such as cod, whales, seals and polar bears. High concentrations of vitamin A in the liver of whales and polar bears are an example of biomagnification, that is, the accumulation of a compound in each successive level of the food chain. The concentration of vitamin A or any other potential toxin (e.g., organochlorines, heavy metals) in an organ depends on the rates of uptake and elimination by each prey and predator in the trophic chain, and by the population dynamics of each species (Moriarty 1999). Long-lived apex predators such as whales that consume large quantities of prey with short lives are more likely to accumulate lipid-soluble compounds such as vitamin A because they undergo several years of lipid storage and depletion during their lifetime.

Herbivores derive most vitamin A from carotenoids in the diet (Fig. 9.16). Carotenoids are converted to retinal by carotenoid-dehydrogenases in the intestinal mucosa and to a lesser extent in the liver. This conversion varies widely with the structure of the carotenoid, the lipid content of the diet and the digestive function of the species. Carotenoids are converted to retinal in domestic dogs but not in domestic cats, which require retinoids in the diet (Morris 2002). Obligate carnivores therefore rely on prey for vitamin A as well as amino acids (Chapter 8). Vitamin A activity of carotenoids is expressed as the equivalent activity of the same mass of retinol. In humans and domestic herbivores, β carotene has a retinol equivalent of 1/6 (167 μg retinol = 1,000 μg β carotene), whereas mixed carotenoids are equal to only 1/12 the same mass of retinol (Olson 1991; National Research Council 2007a). These conversions vary widely across the 50 carotenoids known to provide vitamin A activity; consequently the carotenoid content of the diet may be poorly related to the vitamin A activity derived by the animal.

Carotenoids also provide the functions of light absorption and oxidative defense in organisms. The large planar structure of carotenoids (Fig. 9.16) reinforces membranes in bacteria and fungi. In plants the linear structure and cyclic ends of carotenoids trap light on membranes during photosynthesis. These same attributes produce the colors of fruits and flowers that attract animals, in which the same compounds produce colors in skins and feathers. Vertebrate animals cannot synthesize carotenoids de novo, but they can modify carotenoids in the liver, kidney and at the skin to change the colors they produce. Enzymes for modification may be expressed differently as plumage coloration changes from chicks to adults and between adult males and females. Female scarlet tanagers use dietary carotenoids with little modification to produce yellow plumage, whereas males modify the same carotenoids extensively to produce red plumage (McGraw 2006a).

Animals require a dietary source of carotenoids to produce colors that range from red to yellow, depending on the mixture of carotenoids. Common dietary carotenoids are α and β carotene, lutein, zeaxanthin and canthaxanthin (McGraw 2006a) (Fig. 9.17a). The orange feet of mallards and greylag geese are a combination of β carotene, lutein, zeaxanthin, astaxanythin and canthaxanthin. Flamingos, canaries, goldfish and rainbow trout may lose their colors in captivity unless the diet includes appropriate carotenoids (Choubert 2001; Hill 2006) because carotenoid coloration depends on a continuous supply of carotenoids in the blood serum to the skin during seasonal production of new skin, scales or feathers. Dietary carotenoids therefore may influence the intensity of the red–yellow colorations of plumage on some birds such as American goldfinches (McGraw 2006a). Consequently, some of the variation in carotenoid coloration of wild songbirds and fish is associated with the foods selected in their foraging area or territory (Hill 2006). Carotenoids in tissues and serum reflect diet, but they also reflect selective uptake and retention of compounds as well as their conversion (McGraw 2006a). Carotenoids could serve as dietary markers for some animals, as indigestible markers similar to the alkanes

Fig. 9.17 Carotenoids and melanins are the two most common color compounds in birds. Carotenoid coloration is directly influenced by diet; dietary carotenoids are incorporated into the lipid of fat depots, skin and feathers. Carotenoids produce red, orange and yellow hues in feathers. **a** The pink pigmentation of flamingos is derived from the carotenoids in invertebrates that obtained carotenoids from algae. Feather coloration signals mate quality in greater flamingos and many other birds. **b** Eumelanins produce black and blue colors in the feathers of common ravens. Phaeomelanins produce brown and rust colors in the feathers of ptarmigan and mallards. Melanins are produced endogenously from the amino acids tyrosine and cysteine

in plant cuticle, or as assimilable markers such as the essential fatty acids in prey items (Chapter 4).

Carotenoids may indicate the condition of an animal when these compounds are used for oxidation reactions in immune responses. Infections affect the concentrations of carotenoids in liver and bursal (immunogenic) tissue of growing chickens, and supplementation with carotenoids enhances the immune response of zebra finches (McGraw 2006a). Carotenoid coloration of feathers declines with the intensity of bacterial, protozoal and ectoparasitic infections in several songbirds (Hill 2006). These colors therefore may indicate the quality of mates for some birds because they integrate characteristics of foraging ability as well as disease resistance.

Not all colors are based on carotenoids; melanins are also an important group of pigments in animals. Eumelanins produce black and blue colors, whereas phaeomelanins produce brown to rust colors. Melanins combine with structural proteins such as keratin and collagen in feathers and skin. Ultrastructural properties of pigmented feathers produce the iridescent feathers on the heads of peacocks and male mallards (McGraw 2006b). Melanins are synthesized de novo from the amino acids tyrosine and cysteine that are derived from the essential amino acids phenylalanine and methionine (Fig. 9.17b, Chapter 8). Melanophores synthesize melanin in the skin and control its deposition in feathers; melanophores can produce bands of pigment in mottled feathers that provide camouflage for ground birds and waterfowl. Although the amino acids in melanin are conditionally essential, diet does not affect melanization of feathers until dietary protein is so severely restricted that it limits feather synthesis (Hill 2006). In cats, the enzymes that synthesize eumelanin have a greater affinity for tyrosine than those that route tyrosine to other lean tissue; imbalanced diets that restrict the supply of tyrosine and phenylalanine therefore may affect coat color in cats (Morris 2002). The structural colors of melanized feathers are decreased by intestinal infections with coccidia in turkeys during molt (Hill et al. 2005). Coat color of mammals and plumage quality of birds may be adversely affected by diet and disease that limit supplies of energy, amino acids and carotenoids to the skin during molt.

9.3.2.2 Vitamin D

The principal function of vitamin D is the homeostatic control of Ca and P (Table 9.5); imbalances in vitamin D result in deficiencies or toxicities in wildlife. Light also plays a role in the metabolism of vitamin D; ultraviolet (UV) wavelengths (290—315 nm) provide the energy to break specific bonds in sterols produced by both plants and animals (Fig. 9.18). Vitamin D from animals is called cholecalciferol (D_3) whereas ergocalciferol (D_2) is produced in terrestrial plants and phytoplankton (Table 9.5). Light exposure has the greatest effect on concentrations of D_2 when plants senesce because less energy is captured by other light-sensitive compounds such as carotenoids (Collins and Norman 1991). Skin pigments (e.g., melanin), feathers and fur all reduce the UV light that is available for synthesis of D_3.

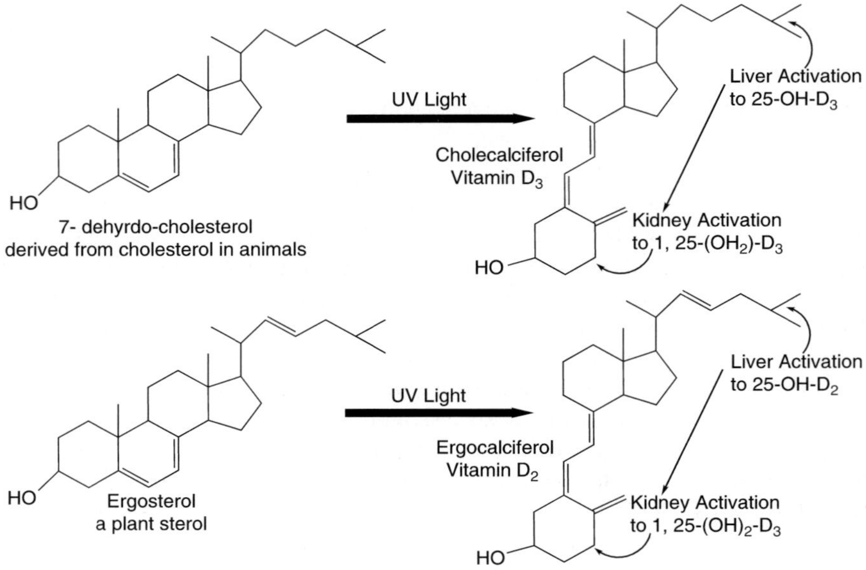

Fig. 9.18 The sterols of plants and animals can be converted to compounds with vitamin D activity in reactions that use ultraviolet (UV) radiation from sunlight at the surface of the leaf or the skin (Linder 1991; Groff et al. 1995). Hydroxyls are added at the *curved arrows* to activate vitamin D to 25-OH-D, which circulates in the blood, and to 1,25-$(OH)_2$-D, which stimulates Ca uptake from the intestine and release of Ca from bone

Full sunlight for only a few minutes each day, however, is often sufficient to produce the daily requirement of D_3 from a small portion of the skin surface at the face, legs and forelimbs of many animals, including humans. Consequently, dietary sources of vitamin D are not essential for ruminants and horses that are active in full sunlight during the day (National Research Council 2007a,b).

The light reaction for synthesis of D_3 can vary in its sensitivity to UV among animals with different diurnal behaviors; a comparison of sympatric species of lizards from the genus *Anolis* indicates that a shade-loving species produces more D_3 than a basking species when both animals are given the same UV exposure (Ferguson et al. 2005). Burrowing animals with crepuscular or nocturnal patterns of activity may still be able to use brief periods of UV exposure to produce D_3 if they bask at the entrances of their burrows to thermoregulate. Production of D_3 at the skin is also dependent upon the supply of the steroid precursor. Basking cats do not produce D_3 because the precursor 7-dehydro-cholesterol is catabolized before it can be used for D_3 synthesis (Morris 1999). Similarly, UV exposure is not a significant source of vitamin D_3 synthesis in fish because water blocks the radiation and because the concentration of the precursor is low (Holick 1989). Transfer of D_3 from skin to blood is regulated by vitamin D binding protein (DBP) from the liver; D_3 is catabolized by further UV exposure at the skin if the vitamin is not absorbed into the blood. Consequently, UV exposure cannot cause toxicity of vitamin D because the accumulated vitamin is degraded; UV protection may have been the principal role of these steroids in the evolution of plants and animals (Holick 1989).

Dietary vitamin D is absorbed at the small intestine and transported with either lipoproteins or DBP. The lipoproteins deliver vitamin D to stores in fat depots and DBP delivers the vitamin to the liver where it is converted to 25-OH-D_3 and 25-OH-D_2 and returned to the blood with DBP (Fig. 9.18). Serum 25-OH-D therefore is used as an indicator of vitamin D stores (Collins and Norman 1991; Klasing 1998). Vitamin D is activated further to 1,25-$(OH)_2$-D_3 and 1,25-$(OH)_2$-D_2 by the cells of the proximal tubules in the kidney (Collins and Norman 1991). More 1,25-$(OH)_2$-D_3 enters the circulation than 1,25-$(OH)_2$-D_2 in fish, cats, dogs, primates and birds because D_2 may be less efficiently absorbed and activated than D_3, and excreted more rapidly than D_3 when activated to 1,25-$(OH)_2$-D (Klasing 1998; Halver 2002; National Research Council 2003, 2006). Consequently, vitamin D_2 from plants may provide less than 10% of the activity of the same mass of vitamin D_3 (Klasing 1998).

Production of 1,25-$(OH)_2$-D is controlled by PTH when serum Ca is low. Absorption of Ca from the diet is increased by the action of 1,25-$(OH)_2$-D on mucosal cells in the small intestine, which increase production of Ca transporters. Blood Ca is elevated by the complementary effects of 1,25-$(OH)_2$-D and PTH on bone and kidney; bone releases both Ca and phosphate while the kidney retains Ca and excretes phosphate to increase blood Ca (Groff et al. 1995). In birds, growth hormone also increases production of 1,25-$(OH)_2$-D when blood P is low. Blood phosphate is elevated by the complementary effects of 1,25-$(OH)_2$-D and growth hormone on bone and kidney; bone releases both Ca and P, while the kidney excretes Ca and retains phosphate to elevate blood phosphate (Klasing 1998). Serum concentrations of 1,25-$(OH)_2$-D increase during mineralization of antlers, eggs and the growing skeleton (Eiben et al. 1984; Klasing 1998). Serum concentrations of 1,25-$(OH)_2$-D probably depend on the kinetics of clearing the hormone as well as the sensitivity of target tissues to the hormone. New World primates such as common marmosets from the Family Callitrichidae have serum concentrations of 1,25-$(OH)_2$-D that are five times higher than other primates such as rhesus monkeys (National Research Council 2003). The apparently low sensitivity of New World moneys to 1,25-$(OH)_2$-D may reduce their susceptibility to a plant analog of the hormone that is produced as a glycoside or PSM by some plants in the Family Solanaceae. These calcigenic plants cause vitamin D toxicity in ruminants, which includes malformations of bone and mineralization and malfunction of vital organs such as the heart, lungs and kidney (Mello 2003).

Deficiency of vitamin D results in the same symptoms as inadequate Ca and P, that is, weak bones and egg shells. The seeds, fruits and leaves of most plants are poor sources of vitamin D, as are milk and the lean tissues of animals (Collins and Norman 1991). Animals therefore must derive vitamin D from UV exposure or from stores of vitamin D in fat. Maternal vitamin D is transferred to the fetus of mammals and to the yolk of birds for development. Fat depots also provide most of the vitamin D required by terrestrial animals in winter because fat gained in the bright light of summer is mobilized in the dark of winter when D_3 production is low or absent. Dietary vitamin D for terrestrial predators varies with the fat content, light exposure and diet of their prey. For example, the vitamin D content of wild rodents available to cats varies from 0.06–0.33 IU•kg^{-1} DM (Morris 1999). The livers of terrestrial prey are low in vitamin D (< 4 IU•kg^{-1}), but the livers of marine fish such as cod, tuna and shark are rich in the vitamin (> 600 IU•kg^{-1}) because the liver is also a principal fat depot in these fish (Halver 2002). Vitamin D is usually not limiting for marine predators, but may be limiting for terrestrial predators with a low diversity or availability of prey. Captive carnivores such as lions that are fed muscle (meat) which is low in Ca and vitamin D are vulnerable to bone fractures, especially during growth and lactation (Allen et al. 1996).

The need for vitamin D in Ca absorption may be obviated in some animals. Fish can absorb Ca across the gills; signs of vitamin D deficiency and inadequate absorption of dietary Ca are only evident in fish kept in water with low concentrations of Ca (Halver 2002). Some animals can absorb Ca by transport systems that are not dependent on vitamin D, especially during growth and lactation, but also during normal maintenance of the body. Mole rats (Family Bathyergidae) are completely subterranean animals that feed on tubers; these animals do not produce D_3 from UV exposure or receive D_2 from their diet of tubers (Skinner et al. 1991) even though they are able to activate D_3 in a similar fashion to other mammals (Buffenstein et al. 1993). Fortunately, their favored diet of gemsbok cucumbers is high in Ca. Damara mole rats passively absorb Ca independently of vitamin D (Pitcher et al. 1992).

9.3.2.3 Vitamin E

Vitamin E is critical to the maintenance of all cell membranes. Although it is found in plant and animal fats, deficiencies are most likely in carnivores with access to only young prey without large fat reserves and in herbivores consuming forages with low vitamin E activity. Toxicities of vitamin E are unlikely in free-ranging animals, but could have severe consequences associated with bleeding because they antagonize the absorption of vitamin K.

Vitamin E is part of the oxidative defence of cells that is provided by other trace nutrients including glutathione (S) and its Se-dependent peroxidase, ascorbic acid (vitamin C), superoxide dismutases (Mn, Cu, Zn) and catalase (Fe; Tables 9.2–9.5). Superoxides (O_2^-), peroxides (H_2O_2) and free radicals ($OH^\bullet$) are reactive species of oxygen that randomly damage membranes. Antioxidant systems prevent damage to fatty acids and can even reverse some of the changes to lipids. Antioxidant systems that resist and repair oxidative damage to cell membranes are associated with the prolonged life spans of some mammals including naked mole rats and humans (Hulbert et al. 2007; Buffenstein 2008). Unlike the other antioxidant systems, vitamin E provides an oxidative defense within membranes (Table 9.5). The role of vitamin E therefore increases in importance with the risk of producing strong oxidants and with the vulnerability of membrane lipids to oxidative damage. The most vulnerable membranes are those that contain high concentrations of polyunsaturated fatty acids (PUFA), including the essential fatty acids (Chapter 7). All reactions that involve O_2 present some risk of producing highly reactive free radicals ($OH^{\bullet}$), which lead to spontaneous degradation of double bonds (Fig. 9.19). Membranes with high risk of oxidative damage during respiration include the mitochondrial membranes of muscles and the plasma membranes of red blood cells and lungs. Immune cells such as neutrophils and lymphocytes also incur oxidative risks when they produce peroxides to destroy invading pathogens.

The tocopherols are a group of eight compounds in plants that have vitamin E activity (Machlin 1991). The most active is α-tcopherol, which is the reference for all other natural and synthetic forms of the vitamin (Fig. 9.19). All forms of the vitamin have a hydrocarbon tail that dissolves in the lipid bilayer of membranes. The reactive part of the molecules is the chromanol ring. Vitamin E is a sacrificial substrate because it reacts with a free radical to form a stable radical on the chromanol ring that does not damage the surrounding lipids (Fig. 9.19). Subsequent reactions with glutathione and ascorbate, however, can regenerate the vitamin. The concentration of vitamin E in membranes is similar to that of the essential fatty acids (Chapter 7); the minimum concentration is defined by the function of the

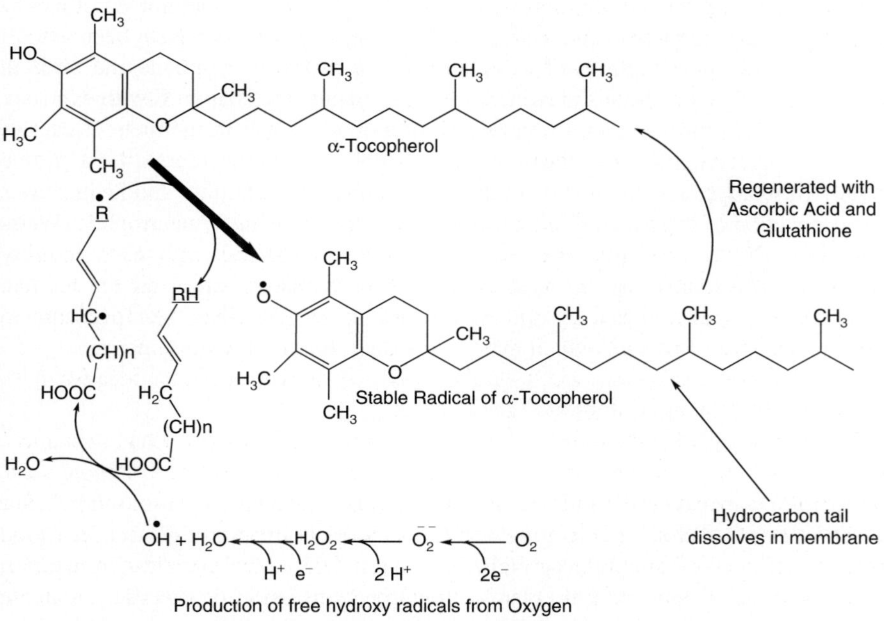

Fig. 9.19 The use of O_2 as an electron acceptor for respiration spontaneously forms superoxides (O_2^-), peroxides (H_2O_2) and free radicals ($OH^{\bullet}$). These compounds can start chain reactions in unsaturated fatty acids by forming fatty acid radicals ($R^{\bullet}$) that randomly exchange H and lose double bonds. Tocopherols (vitamin E) discharge free radicals and protect membrane lipids from disruption by returning the fatty acid to the stable form (RH) (Linder 1991). The oxidized tocopherol can be recycled with ascorbic acid (vitamin C) and glutathione

membrane, but additional vitamin E and PUFA may be incorporated when available from the diet. Tocopherol concentration of lungs and muscle increases with vitamin E consumption in rats, chickens and sheep (Machlin 1991; National Research Council 2007b). Similarly, tocopherol concentrations in plasma and red cells increase with vitamin E intake as the vitamin accumulates in adipose tissue, liver and muscle (Machlin 1991).

In plants, concentrations of tocopherols also follow those of PUFA, both being highest in the germ of seeds and lowest in tubers. Animal lipids with high PUFA content such as fish are also good sources of tocopherol for carnivores. Tocopherol contents of foods may vary by a factor of 10 as plants and prey change seasonally in lipid content and membrane composition. Vitamin E content of foods can also decline quickly, especially as PUFA are oxidized. Stored diets for captive wildlife are often fortified with vitamin E to offset the loss of tocopherol during storage.

Deficiencies of vitamin E are associated with muscle degradation or myopathy in fish, birds and mammals. These lesions are most prevalent when wild animals struggle intensely during capture and restraint (Spraker 1993). Myopathies associated with vitamin E have been reported in wallabies, white-tailed deer and domestic ruminants (Tramontin et al. 1983; Hume 2006; National Research Council 2007b). Low vitamin E concentrations also are associated with vascular problems such as leakage of fluids (exudative diathesis) and hemolysis (Machlin 1991). Signs of vitamin E deficiency include more general problems with growth and reproduction that have been associated with low concentrations of tocopherol in the blood of mammals and birds in captivity. Captive conditions may increase requirements for vitamin E with exposure to disease and other stressors; supplementation of vitamin E in the diets of captive wildlife can alleviate some of these problems (Robbins 1993; Dierenfeld 1994; Hume 1999). Similarly, supplementation of domestic cattle with vitamin E and Se increases the production of superoxides and the ability to kill pathogens by neutrophils (Weiss and Spears 2006). The requirement for vitamin E therefore varies with the availability of other antioxidants and the type of disease or oxidative stress on the animal. Changes in population density and movements of animals alter their exposure to disease and other stressors, which may in turn alter vitamin E requirements.

9.3.2.4 Vitamin K

Vitamin K was named by its Danish discoverer for 'koagulation' of blood because it serves as a cofactor for activating 4 of the 13 steps in the clotting cascade. Blood clots are formed from soluble proteins and small cells called platelets. Activation of the soluble proteins and the platelets requires conversion of specific glutamate residues to γ-glutamate that can bind Ca^{2+} (Fig. 9.20). The clotting cascade is similar to the serial activation of zymogen enzymes in protein digestion (Chapter 8). Each step of the clotting cascade activates the next level, which results in rapid precipitation of the soluble proteins in a matrix with platelets; the resultant clot blocks further blood loss from a damaged vessel. Vitamin K also is involved in similar reactions with glutamate residues in osteocalcin, which facilitates the release of Ca^{2+} from bone (Linder 1991).

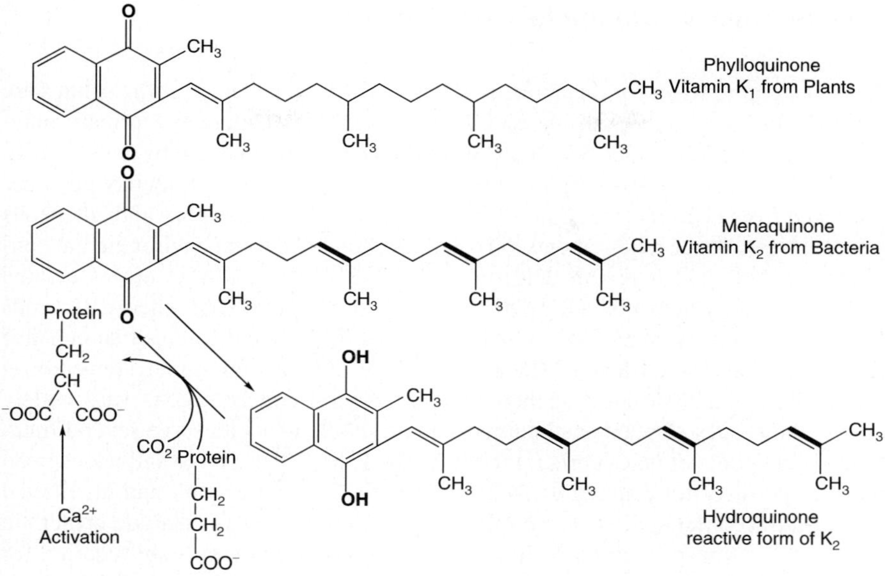

Fig. 9.20 Quinones with vitamin K activity are produced by plants and bacteria. Vitamin K is used to add a carboxyl group (—COO⁻) to the end of glutamate residues in proteins (Groff et al. 1995). The divalent negative charge on γ-glutamate allows the protein to interact with Ca^{2+}, which activates a cascade of proteins in the formation of a blood clot. Dicoumarol and Warfarin impair blood clotting by blocking the reactions that convert vitamin K to active hydroquinone and back to quinone

Vitamin K activity is derived from quinones that are involved in electron transfer reactions in photosynthesis and oxidation in plants and bacteria (Fig. 9.20). Plant leaves are a rich source of phylloquinone (vitamin K_1), and intestinal bacteria provide menaquinone (vitamin K_2). Phylloquinone is absorbed actively from the small intestine and transported on lipoproteins, whereas bacterial menaquinone is absorbed passively from the colon (Groff et al. 1995). Both forms of vitamin K accumulate in the liver. Excessive bleeding and slow clotting times are the principal signs of vitamin K deficiency. Neonatal mammals may be vulnerable to vitamin K deficiency because they have meager stores of the vitamin and because they are born without the intestinal bacteria that produce menaquinone (Linder 1991). Poor clotting reactions are also the result of impaired cycling of vitamin K between the reactive form of hydroquinone and the original quinone. Fungal damage and spoilage of sweet clover releases the anticoagulant dicoumarol, which blocks the vitamin K cycle in grazing ruminants (Harborne 1993; Van Soest 1994). Animals that are exposed to rodenticides such as Warfarin may show signs of poor clotting because the poison also blocks the vitamin K cycle (Groff et al. 1995). Supplemental intakes of vitamin K can offset these toxins. The effects of stressors such as toxins on nutrient requirements and body condition are discussed further in Chapter 11.

9.4 Summary: Metabolic Constituents

Animals require water for metabolic reactions. Water is channeled while solutes are filtered and pumped across gills, salt glands, rectal glands, kidneys, urinary bladders and the digestive tract. The body water pool can be maintained by a wide range of complementary influxes (free water, preformed water in food, metabolic water produced from chemical reactions) and effluxes (excreta, respiratory and cutaneous water losses). Water requirements vary with age, body size and environmental conditions including temperature and drought. High water turnovers are associated with high energy demands during growth and reproduction. High energy demands increase respiratory water loss as well as intakes of food and solutes that enhance fecal and urinary water losses. The availability of minerals for animals depends on the geology and hydrology of their habitat. Mineral concentrations vary widely among the tissues of plants and animals, and each element can serve several functions. Body pools of macrominerals (Na, K, Ca, P, Mg, S) are large and widespread whereas pools of trace minerals (Mn, Cu, Fe, Zn, I, Se) are small and distributed among intracellular spaces. Trace minerals are mainly used as catalytic centers in enzymes and transport proteins. Both trace minerals and vitamins are required for metabolism of energy and for protecting cells against oxidative damage. Water-soluble vitamins include the B vitamins and ascorbic acid (vitamin C), which mainly are associated with the aqueous contents of cells distributed throughout the body. The fat-soluble vitamins (A, D, E and K) are found in lipids and membranes; accumulations in body lipids may be at high enough concentrations to have toxic effects. Requirements for trace minerals and vitamins are affected by interactions among nutrients and by stressors (such as disease), which may change with population density and animal movements.

Part III
Energy and Integration

Chapter 10
Energy: Carbon as a Fuel and a Tissue Constituent

The measurement of energy use is of fundamental importance to wildlife biologists because energy flux through the animal dictates the use of food resources from the habitat. Energy use is related to the use of nutritional constituents such as protein because animals use both energy and protein to maintain and grow tissues. The energy stored in carbohydrates, lipids and proteins is used to maintain the body and to fuel both exercise and reproduction when food is limited.

10.1 Energy Flow and Balance

Dietary carbon enters an exchangeable pool of metabolizable carbon in the body that can be oxidized as fuel for chemical reactions (Fig. 10.1). All chemical reactions produce heat as well as new chemical bonds. Bonds may be formed and broken repeatedly as muscles contract to do work. Longer-lasting bonds between amino acids, fatty acids and glucose produce body tissues from the metabolizable pool of carbon. Core tissues are maintained for the minimum or basal function of the animal; additional bonds and carbon are used to produce more tissue for growth and to deposit stores of glycogen, fat and protein. Stores are mobilized to return carbon to the metabolizable pool when dietary carbon is insufficient for work or synthesis of tissue during fasting or reproduction (Fig. 10.1).

10.1.1 Digestible Energy

Gross energy is the heat released by burning samples of food, tissue or excreta in a bomb calorimeter. The bomb calorimeter is a robust stainless steel chamber (bomb) that is filled with oxygen and immersed in water. A sample is ignited within the bomb, which releases heat that is measured as an increase in temperature of the water. Gross energy is therefore the maximum yield of energy from oxidizing all carbon compounds in a sample to CO_2. Only a small faction of the gross energy in a food may be available to the animal because all carbohydrates, lipids and proteins

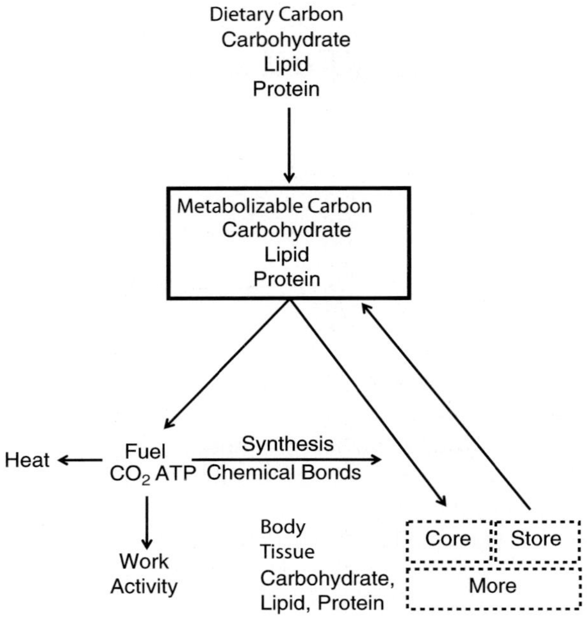

Fig. 10.1 Carbon from dietary carbohydrates, lipids and proteins enters a pool of metabolizable carbon that serves as fuel for work and synthesis as well as a constituent of body tissues. Carbon flows maintain core organ systems (e.g., liver glycogen, cell turnover) before depositing stores (e.g., fat) or growing more of the body (e.g., protein). Body stores return carbon to the metabolizable pool when food intakes are insufficient to maintain energy and constituents for core functions or for synthesis of sperm, eggs, fetuses and milk during reproduction

are not completely digested by the animal. Fibrous grasses and starchy tubers have the same concentrations of gross energy (kilojoules per gram of dry matter, $kJ \bullet g^{-1}$ DM) because both cellulose and starch are glucose polymers with the same gross energy content. However, grasses and tubers may not provide the same concentrations of digestible energy (DE) for geese and small mammals. The DE absorbed by the animal is the difference in gross energy between the food ingested and the feces produced (Fig. 10.2; Table 10.1). The proportion of energy lost in feces increases with the concentration of indigestible components in the food, such as the waxes and cell walls of plants or the skins and shells of prey. Digestibilities of gross energy and organic matter are closely related for large herbivores because plant cell walls contribute most of the organic matter in both food and feces (Table 10.1).

10.1.2 Metabolizable Energy

The metabolizable energy (ME) content of a food was introduced in Chapter 1 as fuel values, which were the average concentrations of ME for carbohydrates ($16.7\,kJ \bullet g^{-1}$ DM), lipids ($37.7\,kJ \bullet g^{-1}$ DM) and proteins ($16.7\,kJ \bullet g^{-1}$ DM) in a

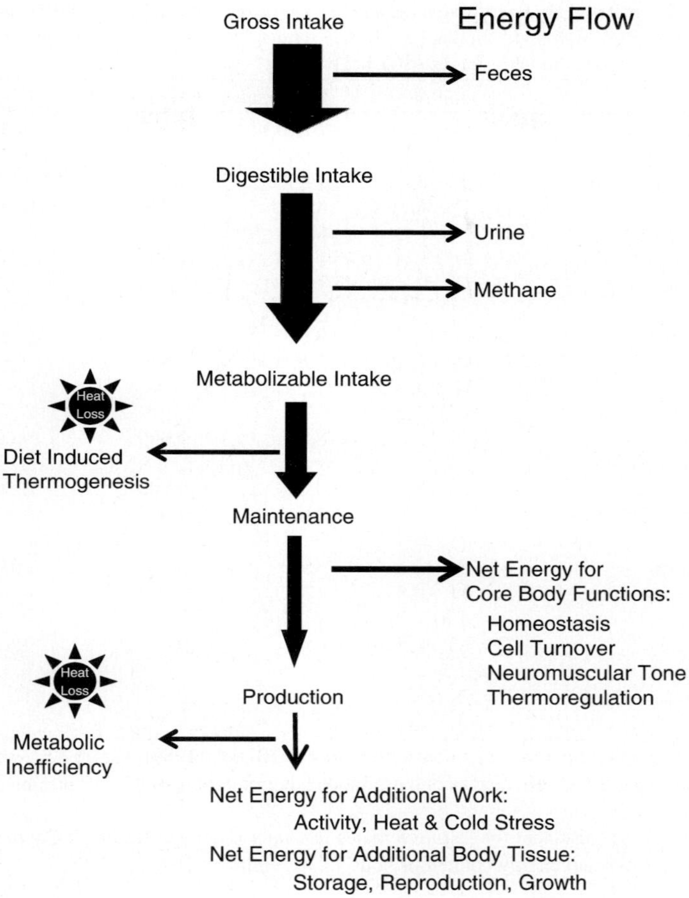

Fig. 10.2 Dietary energy flows to body tissues and supports the activity of the animal. The energy retained by the animal is estimated by the difference between the energy in food and energy in excreta. The energy available for maintenance or productive functions can be estimated as the difference between metabolizable intake and the loss of heat, or by accounting for the energy deposited in tissues or expended in work

range of foods (Blaxter 1989) (Fig. 1.5). The actual concentration of ME in a food is affected by both its digestibility and the losses of energy in urine and methane (Fig. 10.2; Table 10.1). Methane losses primarily affect the ME content of carbohydrates for herbivores because methane is a loss of carbon from fermentation. Methane production depends on the substrates fermented and the community of microbes (Chapter 6). Fermentation of fiber tends to produce less methane than that of starch. Consequently, methane production is lower in hindgut fermenters (e.g., horses) than in foregut fermenters (e.g., ruminants) because very little starch is exposed to fermentation in the colon.

Table 10.1 Estimated daily energy flow ($kJ \cdot d^{-1}$) in a female white-tailed deer (40 kg body mass) at low, medium and high food intakes (1.11, 1.19 and 1.78 kg $DM \cdot d^{-1}$). Energy balances are equated to daily changes in body fat ($g \cdot d^{-1}$) at $39.3\,kJ \cdot g^{-1}$

Parameter	Calculation	Low intake	Medium intake	High intake
Gross energy intake	A	20,000	21,429	32,000
Fecal energy loss	B	6,000	6,429	9,600
Digestible energy intake	C = A − B	14,000	15,000	22,400
Digestibility	D = C ÷ A	0.700	0.700	0.700
Urinary energy loss	E	600	643	960
Methane energy loss[1]	F = 0.07 × A	1,400	1,500	2,240
Metabolizable energy intake	G = C − E − F	12,000	12,857	19,200
Metabolizability	H = G ÷ A	0.600	0.600	0.600
Energy lost to DIT[2]	I = 0.3 × G if M < 0; 0.5 × G if M > 0	3,600	3,857	9,600
Net energy for maintenance	J = G − I	8,400	9,000	9,600
Net efficiency	K = J ÷ G	0.700	0.700	0.500
Maintenance requirement for NE[3]	L	9,000	9,000	9,000
Energy balance	M = J − L	−600	0	600
Net energy from stores[4]	N = M ÷ 0.82 (if M < 0)	−732	0	0
Net energy for production[4]	O = M × 0.7 (if M > 0)	0	0	420
Body fat change[4] ($g \cdot d^{-1}$)	P = (N or O) ÷ $39.3\,kJ \cdot g^{-1}$	−19	0	11

[1]Typical methane loss for ruminants consuming grasses (Blaxter 1989).
[2]M > 0 when G > L ÷ 0.7; efficiencies of diet-induced thermogenesis (DIT) for ruminants (Blaxter 1989).
[3]Net energy (NE) requirement for mature females in winter (National Research Council 2007b).
[4]Energy conversions and energy content of body fat in ruminants (Blaxter 1989).

Urinary energy losses are related to digestible nitrogen (N) intakes because most of the carbon in urine is associated with the principal products of waste N such as urea or uric acid. The ME content of protein is affected by the route of N excretion because urea-N is less costly ($22.6\,kJ \cdot g^{-1}N$) than uric acid-N ($34.3\,kJ \cdot g^{-1}N$); the gross energy from digested muscle protein (myosin $24.89\,kJ \cdot g^{-1}$) is 86% metabolized after urea excretion and 78% metabolized after uric acid excretion. Amino acid composition and the routes of metabolizing both C and N therefore affect the actual ME content of a protein used by mammals and birds (Chapter 8, Fig. 8.6). Urinary N also includes endogenous excretions of purine metabolites and creatinine (Chapter 8) that further alter the apparent ME content of the diet. Truly metabolizable intakes of energy are calculated by adding endogenous excretions of energy in urine to ME intake (see Table 8.2 for calculations of truly metabolizable N). Concentrations of both apparent and true ME in foods are used to estimate the quality of habitats for animals such as waterfowl in winter (Jorde and Owen 1988; Ballard et al. 2004).

True ME estimates are not always available and thus apparent metabolizabilities are more commonly used.

10.1.3 Net Energy

Net energy (NE) is the ME that is available from the diet for activities and synthesis of tissues (Fig. 10.2). Animals must first use NE for maintenance of the body core before they meet the demands for additional activities and growth or reproduction (Fig. 10.2). Maintenance functions include homeostatic control of fluid balance and body temperature, cell replacement, physical posture and processing food. Additional work includes meeting environmental challenges such as cold or heat, changing location and building new tissues. The process of metabolizing carbon produces heat, which is a loss of ME that can detract from the energy available for maintenance and production (Figs 10.1 and 10.2).

All metabolic pathways produce heat because all reactions are inefficient. Heat is therefore lost continuously from the animal and increases with all metabolic activities including feeding. Diet-induced thermogenesis (DIT) is the increase in heat production of an animal from the start of a meal. This form of heat production has also been referred to as the 'heat increment of feeding', 'specific dynamic action' and 'specific dynamic effect' (Blaxter 1989; McCue et al. 2005). The additional heat is released from metabolic reactions associated with secretion, gut motility, absorption and excretion as the food is digested and metabolized (Bureau et al. 2002). DIT varies with the composition of the food and with the digestive physiology of the animal. In herbivores such as muskoxen, heat is produced after a meal by fermentation of carbohydrate (Lawler and White 2003). In carnivores such as Burmese pythons, protein consumption increases heat production through acid secretion, mucosal transport of amino acids and protein synthesis (McCue et al. 2005; Wang et al. 2005). DIT increases with the size of the meal and with the time required for its digestion and metabolism; in pythons and fish, large meals increase heat production and energy expenditure by 100% above resting metabolism for several days (Secor and Diamond 1997). Similarly, in foraging Weddell seals, digestion increases energy expenditure by 45% (Williams et al. 2004b).

The heat lost from the ME in food varies from less than 10% to more than 50% because DIT increases with the number of metabolic pathways for the absorbed carbon (Blaxter 1989). Low DIT is associated with the use of seeds for immediate energy demands to maintain omnivores such as songbirds or grouse because most of the starch is oxidized soon after it is digested and absorbed in the small intestine (Figs 6.8 and 10.1). Fermentation of the same seeds in a deer increases the heat loss because the carbon from starch is first metabolized by microbes to SCFA before it is absorbed and metabolized by the animal (Fig. 6.8). More heat is lost when NE intake exceeds the requirement for maintenance because absorbed carbon flows further through intermediary pathways that allow interconversion of the metabolizable carbon among carbohydrates, lipids and proteins (Fig. 10.1). Animals must therefore increase their ME intake to cover the additional heat lost to DIT during mass gain.

In the example in Table 10.1, a deer must increase ME intake by 49% (from 12,857 to 19,200 kJ • d^{-1}, i.e. by 6,343 kJ • d^{-1}) to achieve a positive NE balance of just 3% ME (600 kJ • d^{-1}). Lower DIT at low food intakes reduces the difference between the changes in ME and NE; a reduction in ME intake by 7% (from 12,857 to 12,000 kJ • d^{-1}, i.e. by 857 kJ • d^{-1}) results in a negative NE balance (–600 kJ • d^{-1}), which is 5% of ME intake (Table 10.1). Heat loss or metabolic inefficiency also reduces the energy deposited in tissues, and also the energy returned from body stores (Fig. 10.2) (Blaxter 1989). The efficiency of using stored energy is typically higher than the efficiency of depositing the same amount of energy because the pathways are not the same for mobilization and synthesis of carbohydrates, lipids and proteins (Figs 8.6 and 10.1). A negative energy balance of 600 kJ • d^{-1} NE for a deer in winter is equivalent to a loss of 19 g fat, but a positive balance of 600 kJ • d^{-1} NE only gains 11 g fat (Table 10.1). Similarly, the efficiency of routing carbon during mass loss and mass gain affects the rate at which birds lose mass during migration as well as the amount of food and the time required to recover that body mass at a staging ground or nesting area.

Mass gains that include protein require both energy and amino acids. Protein synthesis may be limited by the availability of amino acids even though metabolizable carbon is available as dietary carbohydrate and lipid. Under these conditions intermediary metabolism may spill even more energy as heat. Bacteria and animals can generate heat through pathways that discharge concentration gradients; for example, a controlled proton (H^+) leak in mitochondria generates heat without ATP in brown fat in some mammals (Hochachka and Somero 2002; Russell 2002). This has particular adaptive value for young neonates such as caribou born during very cold ambient conditions. Although these mechanisms may serve to regulate ion balance or body temperature, they can divert NE from growth. Energy spilling in ruminal microbes may increase the heat of fermentation when N supplies are low but diets are rich in digestible carbohydrate; spilled energy ultimately reduces the bacterial protein and SCFA available for absorption by the ruminant (Chapter 6). Energy spilling is also implicated in bears because mass gains are highest when supplies of both metabolizable energy and protein are adequate (Felicetti et al. 2003a). Foraging behavior and food selection of bears and other animals are therefore influenced by the metabolizable contents of both energy and protein in food (Raubenheimer and Simpson 1997; Robbins et al. 2007). Associative effects among foods allow animals to balance intakes of energy and nutrients and improve the metabolizability of energy. Ruminants improve fiber digestion and increase DE intake when small amounts of dietary protein meet the requirements of microbes for fermentation. Bears grow fastest when they combine the carbohydrates from berries with the protein from salmon.

10.2 Measuring Energy Expenditure

The energy expended by an animal is the sum of heat released and NE invested in chemical bonds (Fig. 10.2). The net gain or loss in chemical energy of the body is measured by changes in mass and body composition. Heat production includes

DIT, metabolic inefficiency and the heat released by breaking bonds during the contraction of muscles and the catabolism of body stores. Calorimetry is the measurement of heat; direct calorimetry measures the heat released by an animal whereas indirect calorimetry measures the consequences of metabolism. Direct calorimetry is limited to measures of animals in controlled-temperature chambers or tanks with systems that can monitor minute changes in temperature of the air or water. Consequently, indirect approaches to measuring energy use are much more common and appropriate for animals under natural conditions. Indirect calorimetry is usually associated with gas exchange measures such as oxygen (O_2) consumption and carbon dioxide (CO_2) production. Gas exchanges also are used to validate other correlates of energy use such as changes in body temperature or heart rate (Butler et al. 2004). Gas exchanges can be measured with a wide variety of approaches depending on size of the animal, its behavioral patterns and the habitat. Studies of captive animals measure the energetic costs of behaviors and environmental conditions that are observed in wild animals. These data from captive animals are combined with measures of wild animals collected by direct observation, markers and telemetry devices (Chapter 4). The combined approach relates the behavior and environmental conditions of wild animals to the energy they expend on survival, growth or reproduction.

Chambers and tanks are most often used to measure gas exchanges of small animals such as ducks, songbirds, rodents and fish (Kaseloo and Lovvorn 2003; Millidine et al. 2006). The flow (liters per minute, $L \bullet min^{-1}$) and composition of air into and out of the chamber is monitored for O_2 and CO_2 (both in milliliters per liter of air, $mL \bullet L^{-1}$ air) to measure the rate at which the animal consumes O_2 and produces CO_2 (both in milliliters per minute, $mL \bullet min^{-1}$). Chambers are well suited to measuring energy expended during periods of inactivity in burrowing moles, hibernating bears and developing eggs in both captive and field situations (MacCluskie et al. 1997; Withers et al. 2006). The energy needed to maintain body temperatures is measured with captive animals accustomed to chambers in which the temperature, humidity and flow of the air can be altered with fans and heat exchangers (Parker 1988). Energy required for swimming, walking and running, and flying has been measured with small animals that are accustomed to circulating water flows, treadmills or wind tunnels in their chambers. Larger animals that are trained to breathe through a mask have been used to measure gas exchanges while walking and running on a treadmill or outdoors (Parker and Robbins 1984; Fancy and White 1985a). The cost of diving has been measured in chambers above dive tanks for captive ducks and in chambers on the breathing hole for Weddell seals diving under the Antarctic ice (Castellini et al. 1992). Chambers or respiratory masks also are used to validate markers such as hydrogen and oxygen isotopes in water to measure gas exchanges. Labeled water dilution is used to measure energy expended in the field for a wide variety of animals and habitats, ranging from desert reptiles and arctic reindeer to marine mammals and seabirds (Lifson and McClintock 1966; Nagy 2005).

All methods of gas exchange are based on the role of O_2 and CO_2 in the pathways for metabolizing carbon (Figs. 6.1 and 10.1). Oxidation of carbon produces CO_2 and electron carriers (NADH, $FADH_2$) from the TCA cycle. Oxidative phosphorylation uses O_2 to discharge those carriers and to produce ATP. Carbohydrates, lipids and proteins have different fuel values because they produce different amounts of CO_2 and ATP when they are oxidized. Oxidation of lipid produces more CO_2 with a higher energy equivalent (27.8 kJ • L^{-1}) than the CO_2 from carbohydrate (21.2 kJ • L^{-1}; Table 10.2). Energy equivalents for O_2 consumption are less variable than those for CO_2 production because the yield of ATP varies with the point at which carbon enters the oxidative pathways (Fig. 6.1; Table 10.2). The ratio of CO_2 to O_2 is the respiratory quotient (RQ); this ratio indicates the substrate oxidized (Table 10.2). The RQ also indicates the energy equivalent for each gas when calculating the energy expenditure of an animal (Table 10.3). For example, a fasted duck that is resting in a metabolism chamber has an RQ of 0.71, which indicates that the animal is oxidizing lipid from the body (Table 10.3). The RQ increases to 0.81 as carbohydrate and protein are oxidized from the diet when birds are allowed to feed (Table 10.3).

The RQ estimate is an important component of evaluating energy expenditures from CO_2 production. CO_2 production can be measured in both captive and wild animals using doubly labeled water. This marker technique uses water (H_2O) that is labeled on both the hydrogen and the oxygen; hydrogen is labeled with either the radioactive isotope tritium (^3H) or the stable isotope deuterium (^2H) and oxygen is labeled with the stable isotope oxygen-18 (^{18}O). Following administration of a dose of doubly labeled water into the animal, the hydrogen label is lost from the body as

Table 10.2 Indirect measures of energy produced by oxidation of carbohydrate, protein and lipid based on the ratio of carbon dioxide (CO_2) produced to oxygen (O_2) consumed, which is the respiratory quotient (RQ) for an animal (Blaxter 1989)

Substrate	RQ (mol CO_2 • mol O_2^{-1})	O_2 energy equivalent (kJ • L^{-1})	CO_2 energy equivalent (kJ • L^{-1})
Carbohydrate	1.00	21.2	21.2
Protein	0.81	19.2	23.8
Lipid	0.71	19.7	27.8

Table 10.3 The calculation of energy expended by a duck in a metabolism chamber. The composition and flow of air into and out of the chamber is used to measure oxygen (O_2) consumption and carbon dioxide (CO_2) production. The basal metabolic rate (BMR) is the energy expended by fasted birds at rest in their thermoneutral zone (Kaseloo and Lovvorn 2003)

			Experimental conditions		
Parameter	Units	Calculation	Fasted at rest (BMR)	Fed at rest	Fed and swimming
CO_2	mL • min^{-1}	A	8.68	13.54	22.22
O_2	mL • min^{-1}	B	12.23	16.72	27.43
RQ	mL • mL^{-1}	C = A ÷ B	0.71	0.81	0.81
O_2 equivalent[1]	kJ • L^{-1}	D	19.70	19.20	19.20
Energy expended	kJ • min^{-1}	E = (D ÷ 1000) × B	0.24	0.33	0.54
BMR multiple			1.00	1.37	2.24

[1]Energy equivalents from Table 10.2.

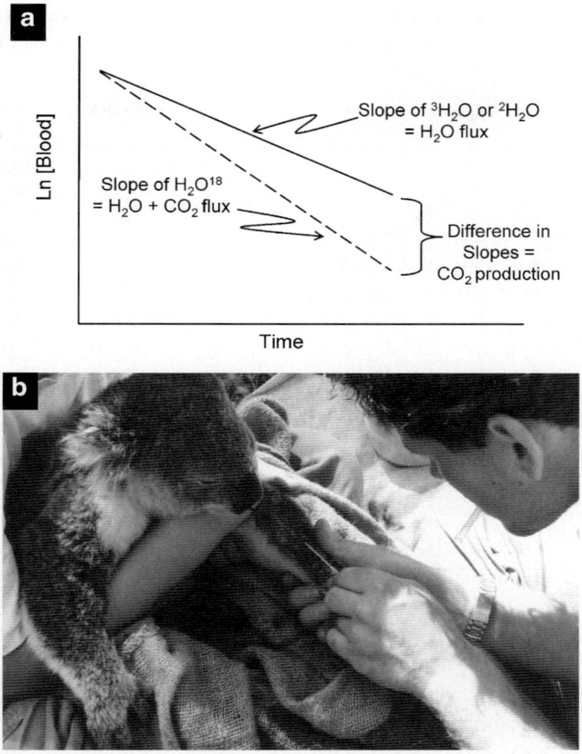

Fig. 10.3 **a** Water dilution with labels of both hydrogen (^{2}H or ^{3}H) and oxygen (^{18}O) can be used to estimate carbon dioxide (CO_2) production, which is subsequently converted to energy expended by the animal. **b** Water dilution techniques are based on samples of blood or other body fluids. These approaches have been used to measure body water space and energy expenditures of several species including koalas (Krockenberger and Hume 1998)

water whereas the oxygen label is lost as both water and CO_2 (Fig. 10.3). The difference in the rates of disappearance of the two labels is used to calculate CO_2 production (Fig. 10.3). The method is most sensitive to an accurate delivery of a single dose and to the recovery of the animal for a second sample before the isotopes are eliminated from the body (Speakman 1993). Doses can be administered into the mouth or stomach with a tube, which is most appropriate for large volumes, or injected into the blood or muscle. The concentration of the labels in the body can be monitored in blood, urine and even saliva as long as extraneous water is excluded. Recaptures of animals are facilitated by using telemetry devices such as radio-collars or homing behaviors that return the animal to a roost, nest or burrow.

In the example in Table 10.4, CO_2 production is used to estimate energy expenditure and the corresponding consumption of food by a fox. The calculation uses the RQ estimated from the composition of the diet and the ME content of each food.

Table 10.4 Calculation of energy expended by a fox (5 kg body mass) on a diet of berries (high carbohydrate) or voles (high protein and lipid) by using the doubly labeled water method to measure CO_2 production. RQ and energy equivalents for CO_2 are calculated from the composition of the diet and from the values in Table 10.2

Dietary substrates	Units	Calculation	Berries	Voles
Carbohydrate	$g \bullet g^{-1}$ DM	A	0.920	0.010
Protein	$g \bullet g^{-1}$ DM	B	0.044	0.710
Lipid	$g \bullet g^{-1}$ DM	C	0.022	0.170
Ash	$g \bullet g^{-1}$ DM	D	0.014	0.110
Carbohydrate	$kJ \bullet g^{-1}$ DM	$E = A \times (4\,kcal \bullet g^{-1} \times 4.184 kJ \bullet kcal^{-1})$	15.40	0.17
Protein	$kJ \bullet g^{-1}$ DM	$F = B \times (4\,kcal \bullet g^{-1} \times 4.184 kJ \bullet kcal^{-1})$	0.736	11.88
Lipid	$kJ \bullet g^{-1}$ DM	$G = C \times (9\,kcal \bullet g^{-1} \times 4.184 kJ \bullet kcal^{-1})$	0.828	6.40
Whole diet	$kJ \bullet g^{-1}$ DM	$H = E + F + G$	16.96	18.45
Estimated RQ		$I = ((E \times 1.00) + (F \times 0.81) + (G \times 0.71)) \div H$	0.98	0.78
CO_2 equivalent	$kJ \bullet L^{-1}$	$J = ((E \times 21.2) + (F \times 23.8) + (G \times 27.8)) \div H$	21.64	25.16
CO_2 produced	$L \bullet d^{-1}$	K	100	100
Energy expended	$kJ \bullet d^{-1}$	$L = J \times K$	2,163	2,516
ME content	$kJ \bullet g^{-1}$	M	2.62	4.48
Food intake	$g\ DM \bullet d^{-1}$	$N = L \div M$	826	562

High concentrations of carbohydrate increase the RQ for berries (0.98) whereas protein and lipid reduce the RQ for voles (0.78). The lower RQ of the voles increases the energy equivalent for CO_2 and the estimate of energy expended for the same volume of CO_2 ($100\,L \bullet d^{-1}$). The ME content of voles is much greater than that of berries; the fox therefore requires a smaller mass of voles (562 g) than berries (826 g) to balance the energy expended.

10.3 Basal Metabolism and Maintenance of the Body

Basal metabolic rate (BMR) is used as a reference for energy expenditures of animals because it is the minimum energy required to maintain the function of core tissues, including electrochemical gradients, respiration, circulation and muscle tone. Energy expenditures are often expressed as multiples of BMR for calculating the added costs of activity, thermoregulation, reproduction or growth. The BMR of an endotherm is measured as the O_2 consumption of a fasted non-reproducing adult animal at rest in a chamber within its thermoneutral zone (see Section 10.4.2). If it is not feasible to fast animals, the measure is called the resting metabolic rate (RMR), which is usually higher than BMR (Hudson and Christopherson 1985). A similar measurement of the metabolic rate of fasted ectotherms is called the standard metabolic rate (SMR) when resting animals are held in a chamber at a specified temperature within their preferred thermal range (Bennett 1982; McNab 2002). The O_2 consumption of a fasted duck resting in a chamber is used to calculate BMR in Table 10.3; O_2 consumption and the corresponding energy expenditure increase when the birds are fed because DIT adds to energy use at rest (1.37 × BMR), which is further increased by muscle activity during swimming (2.24 × BMR; Table 10.3).

The cost of maintaining the body depends on the energy required to replace membranes and proteins of cells (cell turnover) and to regulate the internal environment with transporters (homeostasis). The rates of cell turnover and transport vary widely between tissues but are typically highest for neural tissues (brain and nerves) and the digestive organs (liver, small intestine) (Summers et al. 1988). The proportionately large size of the brain in developing fetuses, hatchlings and neonates contributes to their high BMR. Similarly, the relatively large amount of small intestinal tissue in seals and dolphins (Stevens and Hume 1995) is associated with high BMR and thus a high demand for food (Williams et al. 2001). Conversely, tissues with lower rates of cell turnover such as bone, skin and adipose have relatively low energy requirements. BMR is low in animals with a proportionately large mass of skin and bone in their bodies such as sloths and tortoises (Barboza 1996; Grand and Barboza 2001). Metabolic rates vary with both the mass of tissues and the biochemical activity of their cells (Vézina et al. 2006). Some birds reduce the mass of liver and intestine during migration, which may serve to reduce the energy required to carry and maintain those tissues in flight (Piersma et al. 1996; Battley et al. 2001). Conversely, increases in reproductive tissue as well as liver and gut contribute to increases in the metabolic rate of lactating rodents and laying birds (Vézina and Williams 2003; Ksiazek et al. 2004). Seasonal changes in RMR in northern animals such as moose, muskoxen, sheep and foxes allow animals to reduce their requirements for food in winter (Schwartz et al. 1988; Lawler and White 1997; Argo et al. 1999; Fuglesteg et al. 2006). Winter reductions in metabolism may be associated with a variety of tissues including a reduction in ruminal secretions and absorption in muskoxen, and reduced rates of skin and hair replacement in bears (Barboza et al. 1997; Crater et al. 2007). Summer increases in metabolic rates are associated with greater activity and prolonged effects of DIT (Mautz et al. 1992).

Basal metabolic rate varies widely among species because the number and activity of cells change with behaviors and body form. Predators such as mink that actively pursue prey have higher BMR than insectivorous anteaters; bats that forage widely for nectar from flowers have higher BMR than those that feed on insects (McNab 2002). The surface area exposed to the environment affects the exchange of heat, water and solutes and thus the cost of homeostasis. Animals with proportionately large surface areas in relation to body volume use more energy for homeostasis; the ratio of surface area to volume is smaller for spheres than for cylinders and decreases with body size. Ground squirrels and bears both reduce their surface area by rolling into a ball during winter torpor, but the larger volume of the body allows bears to maintain a higher body temperature than the small squirrels (Guppy 1986).

Body size also affects the diffusion and distribution of nutrients and gases through the animal, and thus BMR generally scales to kilograms to the power of 0.75 ($kg^{0.75}$) from bacteria to birds and mammals (Savage et al. 2004). The specific scalar and the constant in the allometric equation for BMR changes among taxonomic groups (Hayssen and Lacy 1985). The average for all mammals (eutherians, marsupials and monotremes) is $246\,kJ \cdot kg^{-0.73} \cdot d^{-1}$ ($3.45\,mL\ O_2 \cdot h^{-1}$ at RQ 0.81) (McNab 2002).

However, a somewhat higher value of $293\,kJ \cdot kg^{-0.75} \cdot d^{-1}$ is traditionally used for eutherians (placental mammals) and a lower value of $204\,kJ \cdot kg^{-0.75} \cdot d^{-1}$ is used for marsupials even though different groups of marsupials vary from 130–$310\,kJ \cdot kg^{-0.75} \cdot d^{-1}$(Kleiber 1947; Dawson and Hulbert 1970; Hume 1999). Similar taxonomic differences prevail among birds, but these measures are also affected by the daily cycle of activity, even though the birds are inactive during BMR measurements. The BMR of songbirds (passerines) is typically higher than those of non-passerines in both the active ($590\,kJ \cdot kg^{-0.70} \cdot d^{-1}$ vs. $381\,kJ \cdot kg^{-0.73} \cdot d^{-1}$) and roosting (480 vs. $308\,kJ \cdot kg^{-0.73} \cdot d^{-1}$) periods of the day (Aschoff and Pohl 1970a). Ambient temperatures directly affect rates of metabolism in ectotherms such as fish; SMR of rainbow trout increases from 14 to $37\,kJ \cdot kg^{-0.82} \cdot d^{-1}$ between 5 and 15°C (Bureau et al. 2002). These rates are only a small fraction of those for mammals and birds because ectotherms expend less energy on thermoregulation. Buoyancy in water minimizes the costs of maintaining posture in aquatic ectotherms and further reduces their RMR.

10.4 Temperature

Temperature affects all biochemical reactions and therefore the total energy expended by an animal. Progressive cooling reduces the rate of reactions by reducing the activity of all molecules, and the flexibility of enzymes. Conversely, heating increases molecular activity and reaction rate until the structure of enzymes becomes unstable and the proteins denature. The operational range of an animal is therefore defined by the thermal limits of its enzymes; animals can tolerate a zone of high and low temperatures on either side of a preferred range that includes the optimal temperatures for its metabolic pathways (Fig. 10.4). The effect of temperature on reaction rates is measured as Q_{10}, which is the proportional change in the rate of a reaction (R) over 10°C:

$$Q_{10} = \left(\frac{R_{High}}{R_{Low}}\right)^{10/(T_{High} - T_{Low})} \qquad \text{(Eq. 10.1)}$$

where R_{High} and R_{Low} are the rates of reaction at high and low temperatures (T) respectively. At a Q_{10} of 2, reaction rates double over a temperature range of 10°C. Endotherms attenuate the effects of ambient temperature on internal reactions at their preferred temperatures; this is also called the thermoneutral zone because energy expenditures are not changed by ambient temperature when Q_{10} is 1 (Fig. 10.4). Reactions and energy expenditures increase with temperature in ectotherms, that is, Q_{10} is always greater than 1.0. Bacteria, fish, amphibians and reptiles usually double energy expenditures when temperature increases by 10°C ($Q_{10} = 2$) in their zone of preferred temperatures (Fig. 10.4).

Thermal responses to ambient temperatures should be quantified relative to an effective environmental temperature experienced by the animal. Air temperature alone often does not adequately reflect the combinations of thermal variables to

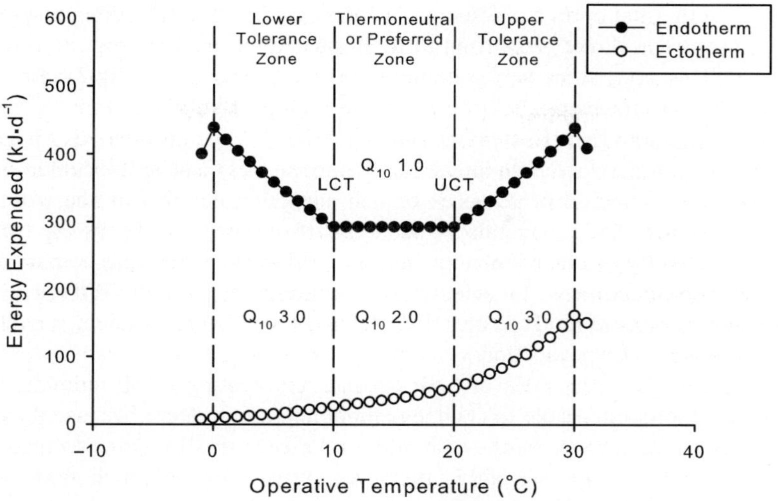

Fig. 10.4 Energy expenditure of ectothermic (*open circles*) and endothermic (*closed symbols*) animals in relation to operative temperature. Operative temperature includes the combined effects of air temperature, convection and radiation experienced by the animals (Bakken 1980). Energy use at 10°C for an animal of 1 kg body mass was 293 kJ • d^{-1} for a mammal (endotherm) and 27 kJ • d^{-1} for a fish (ectotherm). The rate of energy use in each zone of temperatures (0–10; 10–20; 20–30°C) is calculated on the basis of Q_{10}, that is, $R_{High} = e^{\{Ln R_{Low} + [Ln Q_{10} \cdot (T_{High} - T_{Low}) \div 10]\}}$, where R_{High} and R_{Low} are the respective rates of energy expended at high and low temperatures (*T*). The Q_{10} is 2.0 for ectotherms in the preferred zone and 1.0 for endotherms in the thermoneutral zone. Metabolic rates increase outside the thermoneutral zone for endotherms. At temperatures below the lower critical temperature (*LCT*) and above the upper critical temperature (*UCT*), the mammal increases energy expenditures to maintain body temperatures within acceptable limits. Declining metabolic rates below 0°C and above 30°C in this model reflect tissue damage and loss of function

which free-living animals respond. A standard operative temperature for endotherms more appropriately integrates the influences of air temperature, wind speed and incident radiation into one thermal index describing the animal's thermal environment (Bakken 1980), and has been used to model the thermal environments of numerous species including mule deer (Parker and Gillingham 1990). Without the effects of wind and radiation, operative temperature is equivalent to air temperature.

10.4.1 Ectothermy

Ectotherms such as fish increase food intake as ambient temperatures increase in their preferred zone (Bureau et al. 2002). Carnivorous fish can maintain digestibility of proteins over a wide range of temperatures by using more than one form of the same enzyme. Atlantic salmon are able to maintain protein digestion at low temperatures by producing proteases such as trypsin in four forms with overlapping thermal optima (Clements and Raubenheimer 2006). Net energy from a diet increases in direct proportion to food intake because the proportion of heat lost as DIT is consistent across temperatures (Bureau et al. 2002). However, maximum food intakes are achieved at higher temperatures than those for maximum growth rate in fish (Kestemont and Baras 2001). The greatest NE for growth for fish is achieved at moderate intakes and temperatures close to the center of their preferred zone. Thermal preferences of fish and other ectotherms therefore may vary with the metabolic demands associated with season and life history (Madrid et al. 2001). Lizards such as green iguanas use basking sites because both digestion and growth are maximized by selecting temperatures in the preferred zone; some species aggressively defend basking sites because the behavior is selectively beneficial for growth and reproduction.

Body size affects the rates of both cooling and heating of all animals. Small animals are more vulnerable to changes in ambient temperature because they have a large surface area to exchange heat and a relatively small volume to retain that heat. Juvenile desert tortoises of 25 g body mass warm and cool much more quickly than adults at 2,000—4,500 g body mass; small tortoises forage earlier in the day than adults because the juveniles can warm to active body temperatures by basking in the morning, but they also reach their upper tolerance zone more quickly in the desert sun.

Food intakes of fish decline outside of the preferred temperature range when temperatures enter the upper or lower tolerance zones (Fig. 10.4), even though reaction rates increase more rapidly with temperature (i.e., Q_{10} increases above 2) (Fig. 10.4). This is because net energy is diverted from growth to metabolic inefficiency (heat loss) as well as additional maintenance in these thermal tolerance zones. The costs of maintenance are increased by the synthesis of small proteins and solutes such as trehalose and glycerol that maintain the intracellular environment and the structures of nucleic acids and proteins in the lower and upper tolerance zones (Hochachka and Somero 2002). These proteins and metabolites are therefore used as indicators of ‘cold shock’ or ‘heat shock’ in many microbes and aquatic ectotherms. Antarctic and Arctic ice fish survive in sea water below 0°C because long-chain glycoproteins serve as anti-freeze agents in their cells. Reaction rates and Q_{10} decrease precipitously when animals fail to regulate their internal chemistry and begin to die at the outer limits of their tolerance zones (Fig. 10.4). The combined effects of high temperature and low O_2 concentration can depress feeding in fish and cause sudden deaths when active animals are isolated in stagnant water or even slow-flowing segments of rivers (Kestemont and Baras 2001).

10.4.2 Endothermy

Endothermic animals maintain body temperatures in the optimal range for their metabolic pathways. Endotherms can therefore tolerate a wider range of ambient temperatures than ectotherms; polar bears can live at 20°C on the shores of Hudson

Bay in summer and below –70°C on the open sea ice in the high Arctic (Blix 2005). The apparent costs of thermoregulation are lowest in the thermoneutral zone because animals can alter heat flux internally without increasing energy expenditure. In addition to behavioral changes, in most animals blood flow can be redirected to dump excess heat at the surface or conserve heat at the core without substantially increasing metabolic rate. Layers of fat, feathers and fur limit the exchange of heat with the environment; these layers minimize the gain of heat by desert quail and the loss of heat from Arctic muskoxen (Fig. 10.5). The manes of African lions vary with the annual range of temperatures; male lions grow longer, denser manes in cool climates than in hot regions (Patterson et al. 2006). Seasonal changes in feathers and fur alter the thermoneutral range; deposition of underfur or

Fig. 10.5 The exchange of heat between an animal and its environment depends on the difference in temperature between the body and the environment (radiant exchange), contact with the environment (conduction to air, ground or water), the movement of the animal or the medium (convection of air and water), and evaporative water loss. **a** Gambel's quail experience high radiant and conductive loads of heat in the desert; shade and cover in shrubs provide thermal refuges that lower the energy and water needed to maintain body temperature within acceptable limits. **b** Muskoxen experience high losses of heat through radiation, conduction and convection on the Arctic tundra; a deep layer of insulation by subcutaneous fat and pelage minimizes the surface temperature of the animal and reduces the rate at which body heat is lost to the environment

down and new seasonal pelage shifts the thermoneutral range to lower temperatures in autumn whereas molting the insulation returns the range to higher temperatures that are experienced in summer. The loss of these thermal insulative layers dramatically increases heat loss, especially when the skin is in contact with water; seabirds and otters lose the insulative layer of air next to the skin when they are contaminated with oil spills (McEwan and Koelink 1973). Similarly, terrestrial animals such as black-tailed deer lose insulation value when the pelage becomes soaked from rain at cold temperatures (Parker 1988).

Energy expenditures rise when endotherms work to keep body temperature relatively constant below a lower critical temperature and above an upper critical temperature (10 and 20°C respectively in the model in Fig. 10.4). Evaporative cooling uses energy for secretion of sweat or saliva and for muscle activity during panting. High operative temperatures increase the expenditures of both energy and water; thermal refuges in shrub cover are therefore crucial to small animals such as quail in arid and semi-arid habitats (Guthery et al. 2005). Ungulates that dissipate excess heat by sweating can forage simultaneously, in contrast to those that rely primarily on panting for body cooling (Parker and Robbins 1984). Endotherms and ectotherms both avoid high radiant heat loads in deserts by using shade and burrows. The low energy expenditures of ectotherms are an advantage in these dry habitats, where plant production is low. Water conservation and its relationship with sodium and other solutes are discussed in Chapter 9.

Thermogenesis or heat production is a large component of the metabolic rate of endotherms. It involves controlled leaks of ions across membranes; this form of metabolic inefficiency is related to the lipid composition (Chapter 7) and the activity of uncoupling proteins in the membranes of mitochondria (Hulbert and Else 2000). Animals generate more heat as operative temperatures decline below their lower critical temperature by increasing metabolic inefficiency in brown fat (non-shivering thermogenesis) and by contracting muscles without moving (shivering). Non-shivering thermogenesis is most common in mammals from cold climates, such as neonatal reindeer and plateau pika (Soppela et al. 1991; Wang et al. 2006). Heat released by other muscle activities and by feeding (DIT) may also contribute to maintaining body temperature. Feeding on fir needles reduces the lower critical temperature of blue grouse during winter (Pekins et al. 1994). In white-tailed deer and pronghorns, a 40–50% increase in heat production in fed versus fasted animals can lower the lower critical temperature by up to 16°C (Welsley et al. 1973; Jensen et al. 1999). In muskoxen, the heat of fermentation also can offset the cost of warming water when animals ingest snow (Crater and Barboza 2007).

The food required to maintain animals in cold temperatures can be estimated from measures of energy expended by captive animals. Table 10.5 estimates the energy expenditure and corresponding ME intake of a duck within the thermoneutral zone at 20°C and in the lower tolerance zone at −1°C. The BMR is used to estimate maintenance costs that include radiant and conductive losses at thermoneutrality. The minimal cost of thermoregulation is estimated for conductance to air without any additional effects of air currents (convection). Thermoregulation at −1°C is estimated to increase daily energy expenditure by 74% above BMR, which would require a similar increase in food intake for the bird to maintain body mass.

Table 10.5 The effect of air temperature on energy expended by a black duck at rest within (warm; 20°C) and below (cold, −1°C) the thermoneutral zone in still air

Parameter	Units	Calculation	Warm	Cold
Ambient T	°C	A	20	−1
Body mass	kg	B	1.00	1.00
BMR	$kJ \bullet d^{-1}$	C	339	339
T difference[1]	C	D = 20 − A	0	21
Heat loss active phase[2]	$kJ \bullet C^{-1} \bullet 12h^{-1}$	$E = ((33.4 \bullet B^{-0.484}) \bullet 12) \bullet 0.01968$	7.89	7.89
Heat loss passive phase	$kJ \bullet C^{-1} \bullet 12h^{-1}$	$F = ((16.9 \bullet B^{-0.583}) \bullet 12) \bullet 0.01968$	3.99	3.99
Added cost of heat loss	$kJ \bullet d^{-1}$	G = D × (E + F)	0	249
Total energy expended	$kJ \bullet d^{-1}$	H = C + G	339	588
Food intake[3]	g whole	I = H ÷ 2.94	115	200

[1]Thermoneutral reference temperature.

[2]Predicted increases in oxygen (O_2) consumption for each degree of temperature (Aschoff and Pohl 1970a; Mason et al. 2006). O_2 consumption is converted to energy equivalent ($0.01968\,kJ \bullet mL^{-1}$) based on RQ of 0.83 for carbohydrate, protein and lipid in a diet of amphipods. Duration of each phase is $12\,h \bullet d^{-1}$. That is, half the day is spent active.

[3]Based on the ME content of amphipods of $2.94\,kJ \bullet g^{-1}$.

Thermal costs directly affect foraging patterns of mallards and the ability of brant geese and black ducks to over-winter at the northern extremes of their migratory range (Jorde et al. 1984; Mason et al. 2006).

Some endotherms can reduce their energy expenditures by using torpor, which maintains a lower body temperature and minimizes radiant heat loss. A wide variety of mammals and birds use torpor to conserve energy stores when food availability is low and when thermal demands are high. Short bouts of torpor are used to reduce the costs of cold nights in birds, bats and primates, whereas longer bouts are used by seasonal hibernators among the rodents and bears. Energy expenditures of torpid Arctic ground squirrels at a body temperature of 5°C are reduced to 2.4% of the rate of the same animals at a body temperature of 37°C (Buck and Barnes 2000). Thermogenesis to warm the body during arousal from torpor uses more energy than the active state. Nonetheless, the average energy expended during hibernation may be as little as 20% of the energy required to maintain the body at the active temperature through the same period. The costs of rewarming may be reduced in animals at temperate latitudes by using basking, which substitutes radiant heat from the sun for some of the thermogenic effort (Geiser and Pavey 2007). Bears only reduce body temperature to 35°C during hibernation, which reduces their average energy expenditure to 73% of BMR. Similarly, small reductions in body temperature allow red deer to conserve energy during the night while remaining active throughout winter (Arnold et al. 2004). Warm body temperatures may also allow wintering animals to maintain metabolic pathways; unlike smaller hibernators, bears are able to recycle urea

N via microbial urease in the digestive tract and also give birth and begin milk production during winter (Farley and Robbins 1995; Barboza et al. 1997).

10.5 Activity

Energy expenditure for activity can be estimated by behavioral observations of wild animals. The cost (kilojoules per minute, $kJ \cdot min^{-1}$) of each behavior is determined in captive animals held in chambers for measuring gas exchange (Table 10.3). A broader range of activities can be measured for animals marked with radio-telemetry devices. Heart-rate monitors have been used to estimate the energetic costs of swimming, running and flying following calibration against O_2 consumption of the same individuals in captivity (Butler et al. 2004). Doubly labeled water has also been used to measure the energy used for complex foraging behaviors such as plunge diving in seabirds (Jodice et al. 2003). The energy expended (kilojoules, kJ) by an animal is the product of the cost (kilojoules per minute, $kJ \cdot min^{-1}$) and the duration (minutes, min) of each activity. The example in Table 10.6 uses an activity budget to calculate the energy expended by a duck and the corresponding food intake to maintain body mass. In mallards, resting at 8°C increases their metabolic rate to 1.5 × BMR measured at 23°C, but the cost of feeding at the two temperatures is similar (3.0 × BMR) (Kaseloo and Lovvorn 2003) (Table 10.6). The daily energy expenditure estimated for a black duck (1,090 $kJ \cdot d^{-1}$) is 2.4 × BMR, because most time is spent in the low-energy activities of resting or feeding rather than swimming or flying (Table 10.6).

Intense activities such as flying are usually limited in duration because of their high costs, which can exceed seven times the resting rate of energy expenditure

Table 10.6 Activity budget and estimated energy expenditure of a female black duck (1 kg body mass) at 8°C (lower tolerance zone)[1]

Activity	Cost ($kJ \cdot min^{-1}$)[1]	Duration ($min \cdot d^{-1}$)[2]	Energy expended (kJ)[3]	Equivalent food intake (g)[4]
Resting	0.49	750	368	125
Preening	0.98	60	59	20
Feeding	0.99	540	535	182
Swimming	1.04	78	81	28
Flying	3.98	12	48	16
Total (24 h)	—	1,440	1,090	371
BMR[1]	0.32	1,440	458	156

[1]Estimated from expenditures of mallards resting, preening, feeding and swimming in a metabolism chamber at 8°C and resting at 23°C (BMR) (Kaseloo and Lovvorn 2003). The cost of flying was estimated from heart rate at 12.5 × BMR (Wooley and Owen 1978).

[2]Times are based on observed activities of female black duck on a pond (Wooley and Owen 1978).

[3]Energy expended = cost ($kJ \cdot min^{-1}$) × duration ($min \cdot d^{-1}$).

[4]Based on the ME content of amphipods of 2.94 $kJ \cdot g^{-1}$.

(Table 10.6). Similarly, pursuit predators such as cheetahs and lions spend only a small part of their day in energy-costly high-speed chases; more time is spent resting or searching for prey. Ambient temperature may affect the costs of swimming and flying through conductive and convective exchanges of heat. Low temperatures may remove the excess metabolic heat generated by intense activity that would otherwise limit activity at higher operative temperatures. Captures of large mammals such as kangaroos and pronghorns are best performed under cool conditions because they can more easily dissipate excess heat generated during pursuits and struggling during handling. Exertion in wildlife under high thermal loads increases the risk of muscle damage if their oxidative defenses are reduced by low concentrations of some trace nutrients (Chapter 9).

Activity costs vary widely among animals because body form affects the costs of locomotion (kilojoules per meter, $kJ \bullet m^{-1}$) in a particular medium for a limited range of actions (Fig. 10.6). Bounding kangaroos, striding ostriches and running gazelles all achieve low costs for moving large distances over level land. Shorter legs allow marmots to climb and to burrow more effectively than to cover long distances. The webbed feet and cylindrical bodies that allow muskrats and ducks to swim and dive are poorly suited to walking. Streamlining of the body increases the ability of penguins to swim underwater and to approach some of the locomotion efficiencies of fish, but this adaptation precludes them from flight. Forest birds use short wings for agile flapping flight whereas albatross use long wings to glide over open oceans (Fig. 10.6).

Maintaining a resting posture by lying or roosting is an activity that is included in BMR. Standing requires more energy than lying because muscles are used to oppose gravity and hold the position of the animal; in ruminants, the energy cost of standing is 1.1 to 1.3 × BMR (Fancy and White 1985b). The cost of moving the body over a level surface (kilojoules per minute, $kJ \bullet min^{-1}$) increases with speed (meters per second, $m \bullet s^{-1}$; Fig. 10.7). Ungulates increase their speed by changing from walking to trotting to galloping. Each gait optimizes stride length to provide the maximum propulsive force from each muscle contraction; stride frequency enhances the elastic return of energy from ligaments. Kangaroos are very inefficient when moving on all four legs during feeding but become highly efficient when they can bound bipedally between foraging areas (Dawson 1983).

The net cost of moving a distance does not depend on speed; the same energetic increment above standing is required whether the animal walks or runs because the same mass is moved over the same distance. The cost of moving each kilometer declines with increasing body size because the longer strides of larger animals cover more distance at $10.75\,kJ \bullet kg^{-0.32} \bullet km^{-1}$ (Taylor et al. 1982; Robbins 1993). A reindeer calf (10 kg body mass) that follows its mother (100 kg body mass) for 5 km each day at a speed of $3\,km \bullet h^{-1}$ will expend $139\,kJ \bullet d^{-1}$ on locomotion while the mother expends $384\,kJ \bullet d^{-1}$ if the net cost of movement is added to the cost of standing (1.2 × BMR) (Fancy and White 1985b; Robbins 1993). Traveling the same distance is proportionately more expensive, however, for the small calf (8% of BMR) than the mother (4% of BMR). The cost of horizontal travel increases when traction or stride lengths are reduced by sand or snow (Fig. 10.7). Increased foot

Fig. 10.6 Costs of locomotion vary with the body form of the animal and with the medium. **a** Walking and running in ostriches; **b** hopping in kangaroos; **c** climbing in hoary marmots; **d** flying and soaring in mew gulls; **e** swimming and diving in harlequin ducks; **f** cratering in snow by reindeer. Energy expenditures are reduced by adaptive changes in the body form such as limb length, which increases the distance covered in each stride or stroke. Costs increase for opposing gravity during hopping or flying and for opposing the medium (water, snow, soil) during walking, swimming and burrowing

area enables better traction and may reduce sinking in marshes and snow, where energetic costs increase exponentially with increasing sinking depths of the animal (Parker et al. 1984). Deep snows that exceed the breast height of animals can prevent caribou, bison and elk herds from traveling and restrict their winter range (National Research Council 2002). Moose often walk along roads and trails because it is less costly than traveling through snow, although there may be increased risk of mortality from vehicle collisions. Wolves likewise use trails made by humans, which may increase their success in killing caribou (McLoughlin et al. 2003).

The costs of locomotion increase when animals oppose gravity or their medium. The average cost of vertical movement is $25.1\,\mathrm{kJ \cdot kg^{-1} \cdot km^{-1}}$ for several mammals

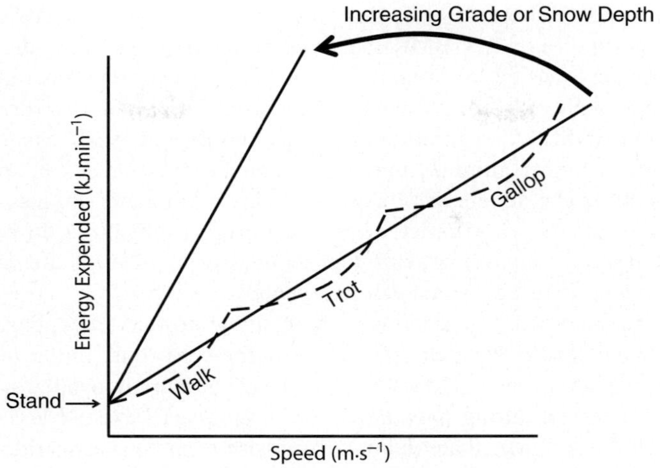

Fig. 10.7 The energy expended by elk increases with speed. Changes in gait reduce the time spent in contact with the ground as the animal progresses from walking to trotting to galloping (Parker and Robbins 1984)

that walk on all four legs including red squirrels, elk and mule deer (Robbins 1993). Walking uphill is the combined cost of horizontal and vertical movement. If a female reindeer (100 kg body mass) expends 384 kJ to walk 5 km over level ground (see above), a vertical gain of 0.25 km over this distance will increase energy expenditure by 628 kJ (25.1 kJ • kg^{-1} • km^{-1} × 0.25 km × 100 kg) to 1,012 kJ. The same journey for her calf (10 kg body mass) would increase energy expenditure by 63 kJ from 139 to 202 kJ. The cost of climbing is only 32% (126 ÷ 264) of the total energy expended by the calf but 62% (1,255 ÷ 1,640) of the total expenditure of its mother. The added cost of climbing is low (kilojoules per kilojoule, kJ • kJ^{-1}) for small animals because they already expend proportionately more energy (kilojoules per kilogram, kJ • kg^{-1}) on horizontal travel and basal metabolism than large animals. Herbivores balance the costs of vertical and horizontal movement as they move between foraging areas on a hillside (Robbins 1993). Downhill movement is assisted by gravity, that is, animals may recover up to 9.79 kJ • kg^{-1} for each kilometer of vertical descent. The amount of potential energy recovered in descending depends on the slope and the body form of the animal; mountain goats and sheep moving down steep slopes can recover most of this potential energy from gravity (Dailey and Hobbs 1989). Reindeer recover 59% of the potential energy when descending a 5% slope (e.g., 250 m vertical ÷ 5,000 m horizontal = 5 m/100 m), which reduces the total cost of travel (White and Yousef 1977). Therefore, a female reindeer (100 kg body mass) would recover 144 kJ (9.79 kJ • kg^{-1} • km^{-1} × 0.25 km × 100 kg × 0.59 efficiency) by descending 0.25 km over 5 km and reduce the cost of travel from 384 to 240 kJ.

Flight is much more expensive (kilojoules per second, kJ • s^{-1}) than vertical locomotion on the ground because vertical position must be gained quickly and maintained in air (Robbins 1993). Flapping flight (e.g., at 23 × BMR) is more costly than gliding (e.g., 3 × BMR) because the latter mainly uses the energy of wind currents and vertical position to stay aloft. Albatross may therefore forage over vast areas with long flight times (Weimerskirch et al. 2005) whereas smaller birds use short flights at lower altitudes (Tatner and Bryant 1986). Small birds use wind currents at higher altitudes for long migratory flights that can span oceans (Gill et al. 2005). The net energetic costs of flying a kilometer are less than those for walking or running a kilometer because during flight birds do not have to support their weight against gravity. Consequently, small birds are more likely to migrate longer distances than small mammals.

Aquatic animals spend less energy on maintaining posture and opposing gravity if they use buoyancy. In Weddell seals, RMR at the surface is similar to the metabolic rate during short dives (Castellini et al. 1992) because buoyancy supports the body both at rest and during movement. Swimming seals expend less energy on each stroke of their flippers than running mammals expend on each stride (Williams et al. 2004b), and net energetic costs per kilometer are less for species adapted to swimming than for species that fly or run the same distance. The specific gravity of the body is reduced by body lipids and by gas in the lungs or swim bladder. Marine iguanas therefore expend less energy swimming than walking at the same speeds (Bennett 1982).

The energy used for swimming increases exponentially with speed, and also with ambient temperature for both reptiles and fish; in reptiles, maximum energy use increases from 5 × BMR at 20°C to 10 × BMR at 40°C (Bennett 1994; Jobling 1994). Cold water may limit activity. Low temperatures limit dive times and dive depths of marine iguanas foraging on algae (Wikelski and Carbone 2004).

Fish expend energy at 3 to 7 × SMR when swimming at their optimal stroke rates. In contrast, swimming endotherms expend energy at only 2 to 3 × BMR, but this is mainly because of their relatively high rates of basal metabolism (Prange and Schmidt-Nielsen 1970; Robbins 1993; Jobling 1994; Williams et al. 2004b). Because water provides greater resistance (drag) and heat conductance than air, it selects for streamlined body forms as well as for insulation in aquatic endotherms. Leaping out of water by dolphins at high speeds actually costs less energy than swimming because of reduced drag. Surface turbulence that increases drag favors diving, especially as the body length of the animal increases. Maximum swimming speeds also increase with body length, which probably selects for the largest body sizes in wide-ranging pelagic predators such as white sharks.

Burrowing is equivalent to swimming through an extremely dense medium. Although the energy used is equivalent to only 4 to 5 × RMR, speeds are extremely slow and very little distance is traveled underground (Lovegrove 1989). Consequently, animals burrow for only short periods of time and make extensive use of established burrows (Reichman and Smith 1987). Red foxes rarely dig their own burrows, but utilize the excavations of other animals. Desert wombats that burrow under layers of limestone may continue to use and expand the same burrow complexes for many years. Burrowing is similar to diving because exercise is combined with limited gas exchange. Burrowing animals therefore often have low metabolic rates, which increase their tolerance of low O_2 concentrations and high concentrations of CO_2 (Frappell et al. 2002; McNab 2002).

10.6 Energy Budgets and Field Metabolic Rate

The added costs of activity and thermoregulation may increase the total daily energy expended by wild animals to over 7 × BMR (Bryant and Tatner 1991). In summer, activity costs are highest for birds that fly to catch prey and commute back to the nest to feed their chicks several times each day (Drent and Daan 1980). In winter, thermoregulatory costs dominate the energy use of small birds at high latitudes and altitudes and limit their distribution (Root 1988; Doherty et al. 2001; Sharbaugh 2001). Conversely, in fish and reptiles, low rates of activity and SMR at low temperatures dramatically reduce rates of daily energy use in winter.

Rates of daily energy use by most endotherms and ectotherms are sustained between 1 and 5 × BMR (Peterson et al. 1990). The maximum rate of energy expenditure probably reflects a variety of controls on energy input (e.g., maximum food intake, digestion and intermediary metabolism) and energy output (e.g., maximum endurance for exercise or for production of eggs or milk) (Hammond 1998) that vary with life-history pattern. Total energy use increases with body mass in all animals, albeit with different allometric scalars to those for BMR or SMR (Nagy 2005); field metabolic rates are higher for birds ($10.5\,kJ \bullet g^{-0.68} \bullet d^{-1}$) than for mammals ($4.82\,kJ \bullet g^{-0.73} \bullet d^{-1}$) and are lowest for reptiles ($0.196\,kJ \bullet g^{-0.89} \bullet d^{-1}$) (Nagy et al. 1999).

Food intakes reflect total energy expended when animals maintain body mass, that is, when the cumulative costs of BMR, activity and thermoregulation are equivalent to ME intake (Table 10.7). Energy from body stores can substitute for dietary energy when energy expenditures are increased above ME intake (Table 10.7). Fat stores are used to meet energy demands when food availability is limited in winter or drought and when gut capacity and retention time limit food intake during reproduction (Chapter 5). The amount of ME derived from body stores depends on the metabolic efficiency of converting carbohydrate, lipid or protein to metabolizable carbon (Figs 10.1 and 10.2). In wintering waterfowl and deer, low temperatures or increased activity can increase mass loss when foraging is limited by snow and ice, and when food quality and abundance are low (Whyte and Bolen 1984; Lovvorn 1994; Garroway and Broders 2005).

10.7 Body Condition

Body condition of individuals is often used to assess the potential for growth in a population during the year (Fig. 2.3) because the amount of stored energy and nutrients in the body is often related to the ability to survive inadequate food intakes

Table 10.7 Energy budget (kJ • d^{-1}) and corresponding food intake (g) for a black duck consuming amphipods. Costs of activity, cold and basal metabolism (BMR) are derived from Tables 10.3, 10.5 and 10.6

Parameter	Calculation	Body mass held (low activity)	Body mass lost (high activity)
NE from body (kJ • d^{-1})	A	0	122
ME from body[1] (kJ • d^{-1})	B = A × 0.82	0	100
Activity cost (kJ • d^{-1})	C	87	187
Cold cost (kJ • d^{-1})	D	249	249
BMR cost (kJ • d^{-1})	E	339	339
Total energy cost (kJ • d^{-1})	F = C + D + E	675	775
ME from diet (kJ • d^{-1})	G = F − B	675	675
Dietary ME (kJ • g^{-1})	H	2.94	2.94
Food intake (g)	I = G ÷ H	230	230

[1]Efficiency of converting NE to ME at 0.82 kJ NE • kJ^{-1} ME.

Table 10.8 Advantages and disadvantages of some common methods of measuring the body condition of wildlife

Method	Measure	Advantages	Disadvantages
Morphometric	Photogrammetry	Repeatable, non-invasive	Non-linear, site dependent, needs validation
	Body mass indices	Repeatable, quick, easily standardized for field	Non-linear, non-specific mass gains (fat vs. eggs or water), needs validation
	Subcutaneous fat and muscle	Repeatable by ultrasound	Limited range in relation to whole body composition
	Internal fat depots (kidney, omentum)	Quick	Single lethal measure, needs validation
Chemical	Whole body lipid, protein, ash	Validation standard, comprehensive	Single lethal measure, time consuming
	Internal imaging (X-ray tomography)	Repeatable	Site-based machine, expensive
	Electrochemical (BIA[1], TOBEC[2])	Repeatable, quick	Muscle tone and posture effects
	Water dilution	Repeatable, portable	Time for equilibration and sampling

[1]BIA = Bioelectrical impedance analysis.
[2]TOBEC = Total body electrical conductivity.

or to reproduce. Chemical measures of lipid, protein and mineral content of the body are directly related to energy or nutrient fluxes, but those measures typically require more effort than morphometric approaches to body condition that provide ranks or relative measures of body stores (Table 10.8).

10.7.1 Morphometry

Morphometry is the measurement of body form, usually as size and mass of the body or its components. The simplest approach to monitoring condition is the use of visual indicators, such as plumage color of breeding waterfowl or antler size of male deer. These indicators are often used to identify animals for harvest. Photogrammetry (i.e., analysis of a recorded image) can be used to quantify these visual cues when handling the animal is either undesirable or impractical. Photogrammetry has been used to estimate body size of killer whales from the size of tail flukes, body mass of bison from the length of the jaw, and fat stores of snow geese from the abdominal profile (Berger 1992; Feret et al. 2005).

Direct measures of body size and mass are simple to obtain during capture and release of small mammals and songbirds, or harvested waterfowl and ungulates. However, measures of body mass are not always feasible under field conditions; large animals such as moose, elephants and rhinos cannot be weighed easily either alive or dead, whereas body length, body circumference, jaw length and limb length can be measured quickly with relatively little training. The dimensions of jaws, femurs and antlers have been used to monitor changes in the body size of red deer, moose and reindeer (Meldgaard 1986; Solberg and Saether 1994; Loison and Langvatn 1998; Weladji et al. 2005). Structural size is routinely monitored in waterfowl as the length of the body, legs, wings and bill; these indicators have been used to evaluate changes in the winter distribution of snow geese and the value of agricultural crops for those populations (Alisauskas 1998).

A large part of the seasonal change in body mass of adult animals is associated with changes in their lipid stores (Allaye Chan-McLeod et al. 1999; Mason et al. 2007). Body mass is usually corrected for structural size to remove the variation that is associated with size alone and to focus comparisons on the changes in mass that are associated with body stores. However, changes in body mass are not always related to the mass of stored lipid or protein; changes in the contents of the digestive tract, urinary bladder and the reproductive tract also affect body mass. The body mass of desert tortoises may increase as digesta fill is gained on coarse grasses, as water is stored in the bladder and as eggs are formed in the oviduct (Barboza 1995). Conversely, the loss in body fat and protein of female muskoxen during pregnancy is masked by the growth of uterine and fetal tissue (Rombach et al. 2002a).

Specific tissues such as muscles, bones and fat depots provide a more direct measure of body condition than total body mass (Ringberg et al. 1981). Body fat is scored by observing the subcutaneous fat around the base of the throat (furculum) of songbirds, by palpating the flanks and rump of ungulates, and by gauging the thickness of the tail in lemurs and platypus (Gerhart et al. 1996; Schmid and Kappeler 2005). Scoring provides a quick but subjective measure that can be improved by using devices such as calipers to measure the thickness of skin folds or ultrasound devices that can visualize and measure the depth of fat and muscle. Ultrasonic measures have been used to evaluate fat stores of moose, deer, elk, caribou and seals (Stephenson et al. 1998, 2002; Cook et al. 2004, 2007; Mellish et al. 2004; Gustine et al. 2007). Dissection and scoring of internal fat depots have also been used to evaluate body condition. Kidney fat indices have been used in several ruminants (Dauphine 1975; Torbit et al. 1988; Forsyth et al. 2005) and marrow fat has been used for a large range of species including deer, coyotes and birds (Ringelman et al. 1992; Thouzeau et al. 1997; Sacks 2005; Mech 2007).

Fat depots vary widely among species (Pond 1978). Depots within an animal are also variable in their rate of loss and gain (Pond et al. 1995b); consequently, the relationship between whole body lipid and the size of fat depots may be linear for only part of the total range of body fat. Fat in bone marrow best indicates very low levels of body fat, whereas kidney fat and subcutaneous depots are better indicators of moderate and high stores of total body fat in deer and waterfowl (Torbit et al. 1988; Ringelman et al. 1992). Morphometric measures and indices therefore require validation against direct measures of the chemical composition of the body to define the range and accuracy of the estimates for the species (Huot et al. 1995; Cook et al. 2001a,b, 2007; Hayes and Shonkwiller 2001).

10.7.2 Chemical Composition

The chemical composition of the body is used to determine the amount of gross energy and nutrients stored in the body. Direct measures of whole body composition are lethal and are therefore used only for comparison between groups of animals (Table 10.8). This approach is thorough but labor intensive, especially with large animals (Reynolds and Kunz 2001). The carcass is usually dissected into components that are analyzed separately to partition the stores of lipid and protein in the body; the components are subsequently summed to calculate the composition of the whole animal (Figs. 1.2–1.5). Partitioning of body mass and composition is used to validate indirect measures or indices and to describe the response to energy deficits or gains during fasting and growth. For example, approximately 50% of the lipid in black duck and Pacific black brant is associated with subcutaneous fat depots whereas breast muscles account for 20% of body protein. Lipid is used from both the subcutaneous depots and the musculo-skeleton during winter and migration when protein and breast muscle are conserved (Barboza and Jorde 2002; Mason et al. 2006).

Chemical measures divide the body into fat mass and lean (fat-free) mass. Fat mass is the lipid stored within adipose cells whereas lean mass is all the cellular spaces and extra-cellular tissues that contain protein, nucleic acids, minerals and water. Fat contents vary widely among animals from 50% to less than 2% of body mass. Animals such as penguins and seals that rely on stored energy for fasting during winter or incubation typically have the highest fat stores (Cherel et al. 1994a; Noren and Mangel 2003). Fat stores are small in ectotherms such as turtles that can minimize energy demands, and in agile prey such as black-tailed jack rabbits and sage grouse that can ill-afford the burden of fat when evading predators (Derickson 1976; Remington and Braun 1988; Henke and Demarais 1990). Neonatal animals are usually low in fat content, that is most of their mass is lean. For this reason their body condition and development are best evaluated by water and protein in relation to structural size (Starck and Ricklefs 1998).

Non-lethal methods of body composition permit repeated measures that can be used to calculate net changes in body stores throughout the year, or as long as animals can be recaptured (Parker et al. 1993b). These methods can be divided into three general categories: internal imaging, electrochemical measures and water dilution (Table 10.8). Internal imaging methods calculate the volumes of fat depots and organs in animals such as seals to determine total composition (Nordoy and Blix 1985). A recent advance in this approach uses dual emission of X-rays (DEXA) to relate the absorption of radiation to tissue density and thus body composition (Korine et al. 2004; Stevenson and van Tets 2008). Imaging and DEXA require large, heavy machines that are not easily transported and are thus best used for small animals that can be brought out of the field to the machine rather than vice versa (Forbes 1999).

Electrochemical methods rely on the conductive properties of the lean components of the body that contain water and ions. For measurement of total body electrical conductivity (TOBEC) the animal is placed within a chamber with a standard magnetic field; lean mass increases the disturbance to the field and is therefore inversely related to fat mass. This method can be applied to animals up to 2 kg body mass, depending on the size of the measuring chamber (Asch and Roby 1995; Zuercher et al. 1997). The measurements are quick, but several determinations may be required if animals struggle and disrupt the field during each measurement. Bioelectrical impedance analysis (BIA) is also affected by muscle tone and activity because lean mass is calculated from the conduction of a small current between electrodes placed at the extremities of the body. Unlike TOBEC, BIA can be applied to larger animals such as bears and seals as well as fish (Farley and Robbins 1994; Cox and Hartman 2004). Both TOBEC and BIA require standardized procedures for handling and positioning the animals during measurements as well as careful calibration for each species (Unangst and Wunder 2001).

The lean mass of the body contains all the water and ions because lipids are hydrophobic. Consequently, measurements of body water have been used to determine the lean mass of animals across a wide range of body sizes and taxa, from small seabirds and rodents to large ungulates, bears and seals (Fancy et al. 1986; Bowen and Iverson 1998; Speakman et al. 2001). The wide use of body water measurements is due to its portability and ease of administration; in studies of infant mammals, the method only requires a syringe and a tube to administer an oral dose of water and a cup to sample urine after the dose (Dove 1988).

The amount of water in the body is estimated by dilution of a single dose of water labeled with either the radioactive isotope tritium (^3H) or the stable isotope deuterium (^2H; Fig. 10.3). The dose is allowed to equilibrate with body water for 60–180 minutes, depending on the size of the animal and the route of its administration; intravenous doses in small endotherms have the shortest times to equilibration. At the other extreme, 24 hours may be required to equilibrate the dose of labeled water in desert tortoises, because of their slow rates of metabolism and large pools of water in the digestive tract and the urinary bladder. Animals are usually held quietly without access to drinking water during the equilibration period. The method is usually not applied to animals during egg laying or late pregnancy because equilibration is prolonged by diffusion across multiple membranes and tissue pools in eggs or fetuses.

A representative sample of body water from blood or urine is collected to measure the concentration of label (E) (milligrams per milliliter, $mg \bullet mL^{-1}$, or counts per milliliter, counts • mL^{-1}) after equilibration of the dose (D) (milligrams, mg, or counts) to estimate water space (W) (milliliters, mL) as: W = D ÷ E. A dose of 3,000 mg ^2H_2O administered to a black duck will equilibrate to 5.454 mg • mL^{-1} in a water space of 550 mL (Table 10.9). The lean and lipid mass of the animal is associated with the ingesta-free or empty body. Body mass is therefore corrected for the digestive tract contents as well as non-exchangeable tissues such as feathers that can be estimated as a proportion of the total mass or structural size of the animal (Table 10.9). The contribution of water from the digestive tract and the urinary bladder may be measured with the extracellular marker sodium bromide (NaBr) when those components are likely to vary widely. The net water space is therefore the water associated only with lean body tissue. The average water content of lean tissue is 73%, but this value varies among species with the concentration of lipid, protein and minerals in the dry fraction of the tissue. Water contents are highest in the lean tissues of young birds and mammals (66–90% water) (Starck and Ricklefs 1998). Protein and other components of the lean mass are calculated from relationships determined by direct measurement of tissues, whereas fat is estimated as the difference between empty body mass and lean mass (Table 10.9). In the example in Table 10.9, repeated measures of body composition indicate that lean mass and body protein are lost at low rates in black duck over winter. Most of the energy

Table 10.9 Measuring body composition of black duck by dilution with isotopically labeled water.

Parameter	Units	Calculation	Day 0 Early winter	Day 60 Late winter	Average change[1] (g or kJ.d^{-1})
Body mass	g	A	1,000	800	–3.33
Water space	g	B	550	500	
Estimated digesta mass	g	C = A × 0.086	86	69	
Estimated feather mass	g	D	72	72	
Empty body mass	g	E = A – C – D	842	659	
Net water space[2]	g	F = B – (0.80 × C)	481	445	
Lean mass[2]	g	G = F ÷ 0.6994	687	636	–0.86
Fat mass	g	H = E – G	155	24	–2.18
Protein[2]	g	I = G × 0.260743	179	166	–0.22
Energy in fat	kJ	J = H × 39.3	6,075	925	–85.82
Energy in protein	kJ	K = I × 23.7	4,248	3,928	–5.33
Total body energy	kJ	L = J + K	10,323	4,853	–91.16

[1](Late winter – early winter) ÷ 60 days.

[2]Coefficients derived from direct measures of body composition of black duck (Barboza and Jorde 2001).

lost from the body is from fat, which provides 91 kJ • d^{-1} or 27% of the estimated BMR (339 kJ • d^{-1}; Table 10.7).

10.8 Survival

Body stores are mobilized when food intakes decline or when energy is expended more rapidly than ingested (Table 10.7). Energy deficits are encountered daily by songbirds when foraging bouts are limited by predation or inclement weather (Fig. 3.7). Migrating birds incur seasonal deficits of energy by fasting during long flights. The importance of each type of energy substrate depends on the duration of the energy deficit and the rate of energy use. Slow and sustained rates of energy use are fueled by aerobic metabolism mainly through the oxidation of lipid. Aerobic metabolism is used for most energy expended by animals because it is used at maintenance and for sustained activities such as steady swimming and feeding in ducks (Table 10.6) or long-distance migration (McWilliams et al. 2004).

Rapid rates of energy use are fueled by anaerobic metabolism through the oxidation of carbohydrates that are stored as glycogen in liver and muscle. Anaerobic metabolism is used for intense activities such as sprints or bursts of running and swimming in lizards and fish, or when tissue oxygen is exhausted during the long dives of seals (Hochachka and Somero 2002). Anaerobic pathways produce partially oxidized acids such as lactate that may limit the endurance of these activities, even though the carbon can still be metabolized once the animal slows down and oxygen becomes more available for tissue metabolism. The limits to aerobic metabolism determine the duration of chases by cheetahs and wolves, and of foraging dives by penguins and seals (Kooyman et al. 1992; Burns et al. 2006). Muscle glycogen is therefore conserved even during fasting in birds because it is an anaerobic fuel that allows the birds to respond to emergencies such as the pursuit by predators (Gaunt et al. 1990).

Body lipid is the principal fuel for maintaining animals during a prolonged energy deficit, such as a fast during the winter dormancy of bears or during the spring incubation of ducks (Fig. 10.8). Stores of glucose are conserved by deriving most of the energy from fat. However, conservation of glucose may result in the production of ketone bodies at high rates of lipid oxidation because of a shortage of glucogenic precursors (Fig. 7.9); concentrations of ketone bodies normally increase in the blood of fasting animals such as penguins during incubation, migrating shore birds and seal pups after being weaned by their mothers (Cherel et al. 1994b; Rea et al. 2000; Gugliemo et al. 2005). Glucose that is used during a fast may be derived from glycerol in fat and from amino acids; low glucose demands in dormant bears are mainly derived from glycerol, whereas the higher demands for glucose in seals that are fasting and producing milk are met from body protein (Barboza et al. 1997; Houser et al. 2007).

Depletion of energy stores may be accompanied by changes in activity and metabolic rate that slow the rate of body mass loss. Fasting animals also may enter

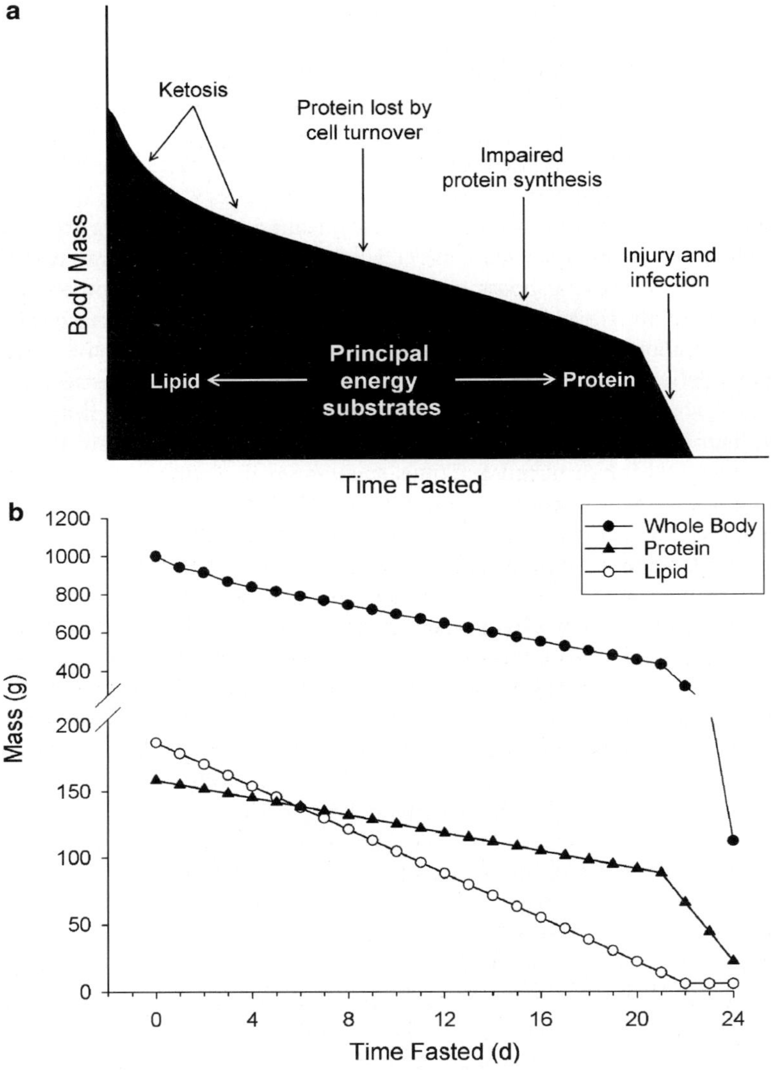

Fig. 10.8 **a** Fuel metabolism in a fasting animal. **b** Model of fasting in a bird with an initial body mass of 1,000 g comprised of 7% digesta and 93% empty body mass. Lipid is 20% of the empty body mass. Lean mass is 70% water and 21% protein. Energy is used at the basal metabolic rate of 308 kJ • d^{-1} (Aschoff and Pohl 1970b). Protein is lost at the maintenance N requirement of 3.49 g crude protein • d^{-1} (Klasing 1998). Conversions of net energy from body lipid and protein are the same as Table 10.9. Whole body mass declines as the animal loses digesta fill at the start of the fast. Mass is lost as both lipid and protein until body lipid declines to 5% of the initial body mass. Mass loss is fastest when body protein is the primary fuel at the end of the fast

torpor or reduce BMR (Ehrhardt et al. 2005; Moe et al. 2005). Protein is usually not used as a primary fuel until lipid stores are depleted (Fig. 10.8). However, body protein is lost at low rates during a fast through normal turnover of tissues and amino acid pools (Fig. 10.8). Labile stores of protein in muscle, liver and connective tissues allow animals to maintain protein synthesis during a fast without losing critical function; bears conserve muscle strength during dormancy even though they lose body protein during the fast (Harlow et al. 2001; Hershey et al. 2008). Animals lose body mass rapidly when protein is used as a fuel because they also lose body water when they break down lean mass (Fig. 10.8). These rapid depletions of body mass may compromise core functions such as homeostasis and resistance to infection (Oarada et al. 2002; Krasnov et al. 2005). Animals that are debilitated by fasting are more likely to succumb to disease and predation; wolves are more likely to kill young deer with low levels of marrow fat or moose with heavy infestations of tapeworms (Joly and Messier 2004a; Mech 2007). The effects of starvation and disease are often difficult to discern because animals may die in sheltered locations that are difficult to find, or their carcasses are quickly scavenged, which leaves little evidence of the cause of death (Cook et al. 2004; Woebeser 2006).

The ability to survive a fast is related to the amount of energy stored as lipid and protein. Small animals are more vulnerable to fasting because they have proportionately higher maintenance requirements for energy and N than large animals (Chapter 1). Those requirements scale to body mass at less than 1.0. For example, the model in Fig. 10.9 uses rates of $308\,kJ \bullet kg^{-0.734} \bullet d^{-1}$ and $3.8\,g$ crude protein $\bullet kg^{-0.58} \bullet d^{-1}$ for birds. Fasting duration may be enhanced in large animals if they can store proportionately more lipid; body fat scales to body mass $kg^{1.146}$ in the widest range of mammals from shrews to whales (Prothero 1995). Consequently, a bird with an

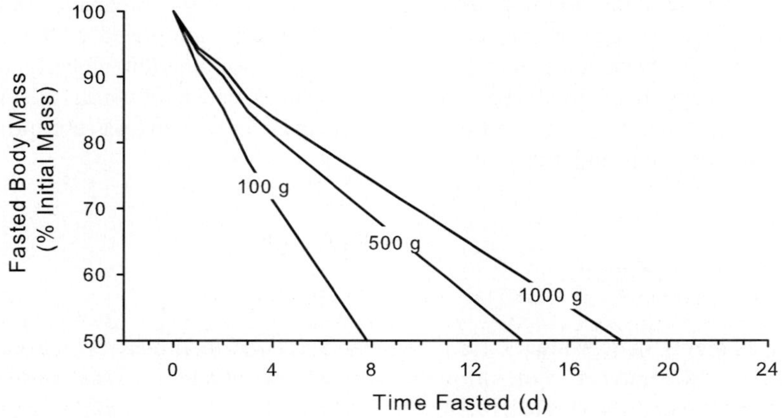

Fig. 10.9 Patterns of mass loss during a fast for birds of different initial body mass but the same lipid content (20% of initial empty body mass). Fasting duration decreases as energy is used more quickly with declining body mass. Calculations were the same as those for Fig. 10.8. Energy is expended at the basal metabolic rate of $308\,kJ \bullet kg^{-0.734} \bullet d^{-1}$ (Aschoff and Pohl 1970b). Protein is lost at the maintenance N requirement of $3.8\,g$ crude protein $\bullet kg^{-0.58} \bullet d^{-1}$ (Klasing 1998)

initial mass of 100 g would use its body stores of lipid and protein more quickly than a bird of 1,000 g even though both animals start fasting with a lipid content of 20% empty body mass (Fig. 10.9). Young animals with small lipid stores are particularly vulnerable to energy deficits. Small shorebirds and songbirds that winter in northern sites deposit more body fat in autumn to offset energy expenditures that are higher than those experienced by the same species at lower latitudes during winter (Rogers 1987; Castro et al. 1992). Waterfowl may also minimize their energy costs and mass loss during winter by moving among sites within the season or by using different sites within their wintering range each year (Lindberg et al. 2007). Deep snows increase the mortality rates of deer because thermoregulation and activity costs increase at the same time as food availability declines, which forces animals to use their body stores (Parker 2003; Garroway and Broders 2005).

10.9 Reproduction

Reproduction increases the daily use of energy and nutrients by animals because changes in activity, thermoregulation and synthesis of new tissues are added to the cost of maintaining the body (Fig. 10.2). The total cost varies widely with the pattern of investing in each stage of reproduction from the synthesis of gametes (sperm or ova), to mating and parental care. Males invest in many small gametes (sperm) to favor success of fertilization, whereas females invest in fewer ova that contain enough nutrients to begin development once fertilized (Table 10.10). Each reproductive stage may vary in duration. That is, the total investment is divided over long or short periods to provide small or large increments in daily demands. Female investment in fetal development may require several months of pregnancy in large ungulates, whereas egg production in waterfowl may be completed within a few days. Parental investments may be supported by a combination of body stores (nutritional capital) and food intake (nutritional income), which influences the timing of each stage of reproduction in relation to environmental demands and primary production of plants and thus prey.

10.9.1 Life History

All investments in reproduction are ultimately related to fitness, which can be measured as the number of offspring that survive to continue the transfer of genes to another generation (Table 10.10). The strong selection for lifetime output of mature offspring is reflected in the wide diversity of breeding and mating systems in vertebrates. Animals may breed once in a lifetime (semelparous) or repeatedly through their lives (iteroparous). They may mate with multiple partners (polygynandry) or with just one partner (monogamy). Parental care may be minimal, shared equally (monogamy) or biased to either females (polygyny) or males

Table 10.10 Hypothetical parameters for some life histories used by vertebrate animals with the corresponding commitments of energy and nutrients by males and females to each stage of reproduction

Representative animals		Salmonid fish		Ungulate mammal		Passerine bird		Ratite bird	
Breeding system[1]		Semelparity		Iteroparity		Iteroparity		Iteroparity	
Mating system[2]		Polygynandry		Polygyny		Polygyny		Monogamy	
Parameter	Calculation	Female	Male	Female	Male	Female	Male	Female	Male
Fecundity (gametes)	A	1,000	1×10^{10}	2	1×10^{8}	1	1×10^{8}	2	1×10^{8}
Proportion fertilized	B	0.5	5×10^{-8}	0.95	1×10^{-8}	0.95	1×10^{-8}	0.95	5×10^{-8}
Mating frequency (y^{-1})	C	1	1	1	8	1	1	4	1
Offspring (y^{-1})	$D = A \bullet B \bullet C$	500	500	1.9	8	0.95	1	7.6	5
Offspring survival ratio	E	0.01	0.01	0.20	0.20	0.40	0.40	0.20	0.20
Surviving offspring (y^{-1})	$F = D \bullet E$	5.00	5.00	0.38	1.60	0.38	0.40	1.52	1.00
Time to maturity (y)	G	5	5	4	8	1	1	3	2
Maximum life span (y)	H	5	5	15	10	10	10	6	6
Adult survival ratio	I	0.01	0.01	0.95	0.80	0.95	0.95	0.80	0.90
Breeding duration (y)	$J = 1 + I \bullet (H - G)$	1	1	11.45	2.6	9.55	9.55	3.4	4.6
Mature offspring produced over lifetime	$K = F \times J$	5	5	4	4	4	4	5	5
Commitments of energy and nutrients at each reproductive stage[3]	Gamete synthesis	Eggs, mass gain	Mass gain	Eggs, mass gain	Plumage, mass gain	Eggs, mass gain	Plumage, mass gain	Eggs, mass gain	Mass gain
	Mating	Activity	Activity	Low activity	Activity	Activity	Activity	Activity	Low activity
	Parental care	None	None	Pregnancy/ lactation, incubation, activity	None	Pregnancy/ lactation, incubation, activity	Incubation, activity	None	Incubation, activity

[1]Breeding systems: animals may breed once in their life (semelparity) or more than once (iteroparity).

[2]Mating systems: animals may exchange gametes with only one partner (monogamy) or multiple partners (promiscuity). Promiscuity may be practised by both genders (polygynandry) and either by males (polygyny) or females (polyandry).

[3]Activity is the combined costs of locomotion and thermoregulation for animals engaged in migration, displays and contests during mating, or in brood rearing and defense of offspring.

(polyandry). There are many ways to make an investment in reproduction that will result in the same reproductive success. The hypothetical parameters in Table 10.10 result in the lifetime production of 4–5 offspring that survive to maturity in four different patterns, ranging from polygynandrous Pacific salmon that breed once before dying (semelparous) to polyandrous emus that are iteroparous. More than one pattern of reproduction may be used within a species; males may be semelparous in fish such as capelin and in some carnivorous marsupials whereas females of the same species are iteroparous (Dickman and Braithwaite 1992; Moyle and Cech 2004). Among ungulates, polygynous mating results in fewer breeding years for males than females because males mature later and die younger than females (Table 10.10). Alternatively, monogamous males and females among songbirds have similar life spans and life-history parameters (Table 10.10).

Reproductive patterns may change with environmental conditions because the fitness of each generation is affected by food supplies and environmental demands that alter survivorship as well as the energy and nutrients available for production by each gender (Smith and Smith 2001). American shad produce fewer eggs but spawn more than once at northern latitudes whereas southern populations of the same species spawn only once in a lifetime (Moyle and Cech 2004).

The energy available for reproduction depends on the cost of maintaining the body. Endotherms are less able to meet their high demands for both maintenance and production from food supplies than ectotherms. Consequently, populations of ectotherms such as fish realize higher efficiencies of production than mammals and birds; 25% of the annual energy use is deposited as offspring in ectotherms compared with 2.5% in endotherms (Humphreys 1979; Reiss 1989). Hibernation reduces the annual cost of maintenance and therefore increases the production efficiency of populations of marmots in comparison with other small mammals (Kilgore and Armitage 1978).

Not all parents maintain their bodies throughout the reproductive period. Body lipid and protein are used to support activities such as migration for spawning in Pacific salmon and the defense of females in male moose and reindeer during the rut (Mommsen et al. 1980; Mysterud et al. 2004). Male reindeer may reduce their energy demands and extend their body stores during the rut by reducing the size of metabolically expensive tissues such as the digestive tract and by reducing homeostatic functions of the kidney and immune system (Barboza et al. 2004). Birds that are constrained from foraging during incubation of eggs or tending their chicks may likewise reduce their activity and resting metabolism to conserve their body stores (Buttemer and Dawson 1989; Ancel et al. 1998). Alternatively, high demands for egg production may be offset by lower activity in female zebra finches, which reduces energy expenditure and thus food intake (Vézina et al. 2006).

The new tissues produced for reproduction require both energy and nutrients such as protein. Body stores are deposited by both genders for use during mating and parental care (Table 10.10). These stores may be deposited over several weeks to minimize the increment in daily food intake. In the example in Table 10.11, daily gains of 0.3% body mass in black duck only require 18% greater daily food intake to deposit a store of 8% body mass over 28 days at the end of winter. Females are usually obliged to invest more energy and protein in gamete synthesis than males.

Females of the semelparous sockeye salmon deposit 20 times more energy in ovaries than males invest in testes (Diana 2004). Males may apportion more of their investments to secondary sexual characteristics such as plumage or antlers for mating, whereas females develop tissues for parental care such as the uterus for pregnancy, brood patches for incubation of eggs, and mammary glands for milk synthesis. Males may also develop brood patches and skin folds to share incubation in monogamous species (e.g., penguins) and polyandrous species (e.g., emus).

Males often invest proportionately more energy and nutrients than females in activity and secondary characteristics to attract and secure their investment in mates by displays and contests. The antlers of cervids are a large investment of protein that is deposited over several weeks as both body protein and fat are gained; growth of antlers and body mass in young white-tailed deer increases their N requirement by 74%, from 0.61 to 1.06 g $N \bullet kg^{-0.75} \bullet d^{-1}$ (Asleson et al. 1996). Antler size may reflect the overall condition of males and their likely fitness because antler growth may be constrained by poor food quality and availability and by high parasite loads at high population densities (Bowyer et al. 2001; Ditchkoff et al. 2001a; Mysterud et al. 2005a). Plumage colors produced by carotenoids and melanins may also indicate the quality of individuals as mates (Chapter 9). Feathers require large investments of amino acids (Chapters 8) and energy for protein synthesis (Lindström et al. 1993; Murphy and Taruscio 1995; Klasing 1998). In white-crowned sparrows, feather synthesis at peak molt requires constant supplies of essential amino acids such as threonine and methionine and increases the metabolic rate to 1.58 × BMR (Murphy and King 1990; Pearcy and Murphy 1997). The costs of molt depend on the number of active feather tracts and the duration of their activity (Heitmeyer 1987; Leafloor and Ankney 1991). Wing molts are most expensive because the costs of synthesis are compounded by impaired movement that may reduce the efficiency of foraging (Bridge 2004). Female waterfowl can minimize the functional constraints of feather replacement by synchronizing the molt of their wing feathers with brood rearing when they forage with their flightless hatchlings.

Females must commit more nutrients to gametes than males because the lipids and proteins are required for survival and at least part of the development of the fertilized egg. Mammals continue to support the egg by providing energy and nutrients across the placenta during pregnancy and in milk after birth. In placental mammals, pregnancy and lactation may require the same absolute amount of energy and protein for the offspring, but daily mass gains of the offspring are slower during pregnancy than during lactation. Consequently, in caribou, daily energy demands increase by 30% during pregnancy and by 60–100% during lactation, while demands for protein increase by 23–43% in pregnancy and by 110–130% in lactation (Allaye Chan-McLeod et al. 1994; Barboza and Parker 2008). High demands for milk production divert energy and protein from deposition of keratin in the horns of female mountain goats, Japanese serow and bighorn sheep (Festa-Bianchet and Côté 2008).

The costs of milk production vary with stage of lactation and milk composition (Oftedal 1984). Milks with high concentrations of lipid require concomitant investments of energy that are greatest for marine mammals with short periods of lactation and rapid growth of offspring; the hooded seal produces milk containing 87.5% lipid

Table 10.11 Calculation of the energy and protein budgets for a duck during winter (non-reproductive), spring laying and incubation. Food intakes are estimated for high-protein amphipods and for low-protein tubers.

			Late winter		Spring	
Parameter	Units	Calculation	Mass held (amphipod)	Mass regained (amphipod)	Mass held Laying (amphipod)	Mass lost Incubation (tuber)
Average body mass	kg	A	1	1	1	1
Body mass change	$g \cdot d^{-1}$	B	0	3	0	–3
Body fat change	$g \cdot d^{-1}$	C	0	2.1	0	–2.1
Lean body mass change	$g \cdot d^{-1}$	$D = B - C$	0	0.9	0	–0.9
Body protein change[1]	$g \cdot d^{-1}$	$E = 0.26 \times D$	0.0	0.2	0.0	–0.2
Body energy change[2]	$kJ \cdot d^{-1}$	$F = (39.3 \times C) + (23.7 \times E)$	0.0	88.1	0.0	–88.1
ME of change[3]	$kJ \cdot d^{-1}$	$G = F \div 0.7\ B > 0;\ F \div 0.82\ B < 0$	0.0	125.8	0.0	–107.4
BMR[4]	$kJ \cdot d^{-1}$	$H = 308 \times A^{0.75}$	308	308	308	308
Energy increment for activity, thermoregulation	$kJ \cdot d^{-1}$	I	1.3	1.3	1.3	1.3
Energy cost	$kJ \cdot d^{-1}$	$J = (1 + I) \times H$	708.4	708.4	708.4	708.4
Dietary energy demand	$kJ \cdot d^{-1}$	$K = J + G$	708.4	834.2	708.4	601.0
Maintenance protein requirement[5]	$g \cdot d^{-1}$	$L = 3.49 \times A^{0.58}$	3.489	3.489	3.489	3.489
Protein cost	$g \cdot d^{-1}$	$M = L + E$	3.5	3.7	3.5	3.3
ME concentration of diet	$kJ \cdot g^{-1}$	N	2.94	2.94	2.94	4.46
Metabolizable protein in diet	$g \cdot g^{-1}$	O	0.114	0.114	0.114	0.017
Food for body energy demand	$g \cdot d^{-1}$	$P = K \div N$	241	284	241	135

Food for body protein demand	$g \cdot d^{-1}$	$Q = M \div O$	31	33	31	196
Egg synthesized	$\cdot d^{-1}$	R	0	0	0.71	0
Egg lipid[6]	$g \cdot d^{-1}$	$S = 6.31 \times R$	0	0	4.51	0
Egg protein[6]	$g \cdot d^{-1}$	$T = 6.53 \times R$	0	0	4.67	0
ME intake for egg synthesis[2,7]	$kJ \cdot d^{-1}$	$U = (S \times 39.3 \div 0.8)+(T \times 23.7 \div 0.5)$	0	0	442.5	0
Food for total energy demand	$g \cdot d^{-1}$	$V = (U + K) \div N$	241	284	391.5	134.7
Food for total protein demand	$g \cdot d^{-1}$	$W = (T + M) \div O$	31	33	71.5	196.1

[1]Body protein coefficient derived from direct measures of body composition of black ducks (Barboza and Jorde 2001).

[2]Energy content of fat and protein (Blaxter 1989).

[3]Metabolizable energy conversions above and below maintenance of body mass (Blaxter 1989).

[4]Basal metabolic rate for roosting non-passerine birds (Aschoff and Pohl 1970).

[5]Maintenance protein requirement of $3.49\,g \cdot kg^{-0.58} \cdot d^{-1}$ (Klasing 1998).

[6]Protein and lipid deposited each day in eggs of black ducks (Barboza and Jorde 2002).

[7]Efficiency of depositing energy as lipid ($0.8\,kJ$ ME $\cdot kJ^{-1}$ NE) and protein ($0.5\,kJ$ ME $\cdot kJ^{-1}$ NE) (Klasing 1998).

and 35.29 kJ • g^{-1} DM for just 4 days whereas horses produce a milk with 12.4% lipid and 20.3 kJ • g^{-1} DM for over 180 days (Bowen et al. 1985; Robbins 1993; National Research Council 2007a). Egg laying is the most demanding reproductive phase for female birds, which typically produce up to 1 egg • d^{-1}. Food supplies and body condition may affect the size and number of eggs produced (Carey 1996). Reproductive costs are generally higher for smaller species because they tend to lay larger eggs as a proportion of body mass. Costs also depend on the number and size of eggs in a clutch, as well as the composition of the egg. Variation in the gross energy content of the dry fraction of eggs in birds (23.8–31.8 kJ • g^{-1}DM) and reptiles (23.8–26.8 kJ • g^{-1} DM) mainly reflects changes in the lipid content of the yolk (Congdon et al. 1982). In contrast to altricial species such as hawks and owls, precocial species like waterfowl tend to have larger yolks to provide for an advanced functional chick at hatching (Vleck and Bucher 1998).

In Table 10.11 the cost of egg production is estimated for a black duck that produces an average of 10 eggs over 14 days or 0.71 eggs • d^{-1}. Egg production increases dietary demands for energy and protein by 62 and 134% respectively above those for the body. The added demand for synthesis of egg protein is easily met on a high-protein diet of amphipods (72 g • d^{-1}), but laying birds may not be able to produce eggs from low-protein foods such as tubers; deposition of 4.7 g protein in the egg would require consumption of 492 g of tubers, which is almost 50% of the body mass of the bird. Many songbirds and upland game birds rely on invertebrate prey for protein deposition during egg laying. However, low-protein foods such as tubers may be subsequently used by ducks during incubation when foraging time is limited and birds mobilize body stores of protein and lipid (Table 10.11).

10.9.2 Capital–Income Continuum

The amount of food required for reproduction depends on how much lipid, protein and other nutrients are derived from body stores. Consequently, reproduction may be supported by a combination of both food intake (income) and body stores (capital) (Fig. 10.10). Income breeders are more limited by the current food supply than capital breeders, and therefore reproduction is timed more closely to meet the pattern of food supplies for income breeders. For example, many desert birds initiate breeding after rainfall because the supply of plants and prey is more likely to coincide with the peak demands for laying and parental care (Wingfield et al. 1992).

In contrast, body stores allow capital breeders to anticipate seasonal changes in food supply to meet the costs of both maintenance and reproduction. Fat and protein stored by reindeer and caribou from food supplied during summer and autumn is used for survival and pregnancy through winter, and also for lactation in the following spring (Parker et al. 2005; Barboza and Parker 2006, 2008). A similar pattern of body storage in Arctic ground squirrels supports survival through winter, but the costs of spring reproduction are mainly derived from food caches and new plant growth in spring (Hock 1960; Long et al. 2005).

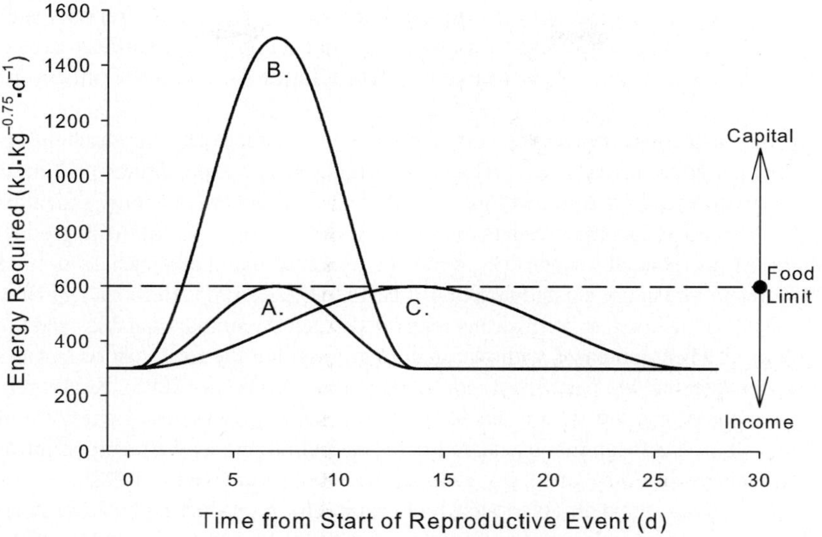

Fig. 10.10 The energy required by an animal during any stage of reproduction is the sum of energy costs for maintenance and production. This model projects a maintenance cost of 300 kJ • kg$^{-0.75}$ • d^{-1} for an animal over 27 days of reproduction. Total cost of reproduction is the product of rate and duration (kJ • d^{-1} × d = kJ), which is the area under the *curve* of energy required against time. *A* An income breeder limits expenditures to the available food supply. *B* A capital breeder uses body stores to invest in reproduction beyond the energy supply from food. *C* Animals may commit the same total investment of energy and nutrients by slowing the rate and extending the duration of the investment; the areas under *curves B* and *C* are the same (36,300 kJ • kg$^{-0.75}$). Animals can use both capital and income for reproduction; body stores of energy may be used briefly by income breeders to meet a sudden increase in environmental demands. Animals may also produce eggs, fetuses or milk with the capital of lean body mass but derive most of their energy needs from dietary income

Shorebirds that migrate to the Arctic in spring use the food available on the nesting grounds for egg production (Klaassen et al. 2001). Body stores allow animals to begin reproduction as early as possible when the window for food availability is narrow, as it is in highly seasonal environments such as the Arctic. Geese that arrive early on nesting grounds may use body stores from distant wintering grounds to initiate egg production, whereas the food available at migratory staging grounds and nesting grounds is important for egg production in birds that arrive later in the season (Klaassen et al. 2006; Schmutz et al. 2006).

Reproductive investments of capital breeders may be more influenced by prior food availability than current availability of food at a nesting or birthing area. The probability of birthing in caribou during spring is related to maternal body mass and thus the stores of lipid and protein at the end of the previous summer foraging window (Cameron and White 1996; Barboza and Parker 2008). Timing of births in primates such as long-tailed macaques follows peak fruit mass by several months, probably because of a similar effect of body stores on reproduction (Brockman and van Schaik 2005). This temporal disconnection between food availability and reproduction may contribute to overutilization of plants by capital breeders such as domestic sheep on islands and by snow geese that winter in agricultural areas but migrate to breed in the Arctic (Ankney 1996; Clutton-Brock and Coulson 2003) (Chapter 2).

The use of capital and income for reproduction is affected by the duration of the investment. Capital breeders can achieve very high energy expenditures and commitments to tissue in a short period of time whereas the same amount of energy and nutrients may be invested by income breeders but is spread over a longer period. Milk production is the most energetically expensive period for placental mammals such as ruminants, with increases in energy expenditure to 4 × BMR (Oftedal 1985; Barboza and Bowyer 2000). Milk production of herbivorous marsupials such as tammar wallabies and koalas is prolonged when compared with placental mammals, but the total cost of lactation is similar between the two patterns of reproduction (Krockenberger 2006). Koalas rely on the low income of a low-quality diet to make the same investment of energy in their young as domestic sheep that use stored body fat and protein as well as food intake to support their lambs (Dove and Cork 1989; Krockenberger and Hume 1998).

Large animals are better able to use body capital for reproduction because stores of lipid may exceed a scalar of 1 (e.g., fat scales to kilograms to the power of 1.145 ($kg^{1.145}$) for mammals), but the energy required for maintenance of the body scales to less than 1 (BMR scales to kilograms to the power of 0.75 ($kg^{0.75}$)). Small animals may therefore rely on body lipids for survival but meet their energy demands for reproduction from their diet (Fig. 10.9). Total energy invested in reproduction decreases with increasing body mass of females at scalars between 0.52 and 0.90 ($kg^{0.52}$ to $kg^{0.90}$). That is, mothers of large species spend proportionately less on reproduction than smaller species. The time required for reproduction declines with maternal body mass; duration of pregnancy in mammals and incubation in birds scale to body mass at kilograms to the power of 0.24 ($kg^{0.24}$) and to the power of 0.17 ($kg^{0.17}$) respectively (Reiss 1989). Mothers of large species therefore have a greater tolerance of environmental perturbations than smaller species because they have potentially greater flexibility in time and body stores to complete the development of their offspring. Large ungulates such as plains bison and Dall's sheep are able to alter the timing of their births when breeding is delayed or when inclement spring weather increases the risks to their offspring (Rachlow and Bowyer 1991; Berger 1992). Conversely, small income breeders are better able to respond to increases in food availability, especially in variable habitats. Small mammals such as ground squirrels use additional food supplies (income) to foster the growth of their pups, presumably through milk production, rather than increasing their body mass (Broussard et al. 2005). However, smaller animals also require smaller absolute increments in food supply to begin reproduction. Small rodents can begin breeding at a much lower threshold of primary plant production than larger mammals, which results in much more rapid recovery of small populations as well as problems of overabundance (Chapter 2) (Davis et al. 2003).

10.10 Growth

Young animals have the lowest rates of survival in a population because they are more vulnerable than adults to environmental exposure, starvation, predation and disease (Chapter 2). Survivorship increases as young animals increase in body size and develop the organ systems to digest and metabolize food, and to respond to infection, as well as the behavioral and locomotor skills to forage and to evade predators. Parental investments of energy and nutrients provide the time and materials to develop their offspring. Parents use different patterns of investment to attain the same goal of independent offspring (Table 10.10). Australian brush turkeys hatch as completely independent chicks long after their parents have left the nest, whereas American kestrel chicks require the constant care of both parents until they molt flight feathers and leave the nest. Brush turkeys invest in provisioning the egg and in building a nest mound that incubates the eggs, whereas kestrels invest less activity in nest building and more time and energy in feeding the nestlings. These different parental investments are returned when offspring become part of the breeding population.

The survival and future reproduction of offspring depend on the quality of core tissues and the size of their body stores (Fig. 10.1), both of which depend on the genes and the environment of the parents. Egg size and neonatal mass are both positively related to maternal body mass, and therefore genetic determinants of adult size affect the mass of dependent young (Reiss 1989). Moose are larger as both adults and calves in Alaska than in Europe; the Alaskan subspecies (*Alces alces gigas*) is a distinct genotype with a larger body size (Hundertmark et al. 2002). Neonatal body mass and function also may be affected by the environment experienced by the mother during pregnancy. Poor maternal body condition in late pregnancy has been associated with reductions in birth mass and with abnormal development in caribou calves (Roffe 1993; Adams 2005). The effect of maternal restriction on the fetus may persist through life in what is called fetal programming because restrictions during fetal development alter the growth and function of core tissues such as liver and muscle (Breier 2000; Wild and Byrne 2004). Cohorts of small infants in primates and ruminants subsequently produce small offspring that may be more vulnerable to environmental extremes and other causes of mortality (Steinheim et al. 2002; Duncan and Scott 2004). These multi-generation effects are more pronounced at high population densities, which reduce food abundance and increase the exposure to pathogens that may make neonates more vulnerable to both disease and predation (Sams et al. 1996).

Young animals are most vulnerable to changes in parental support during the early phases of growth when absolute mass (grams, g) and the daily change in absolute mass (growth rate; grams per day, $g\ d^{-1}$) are small but the proportional change in mass each day is large (relative growth; percent per day, $\%\ d^{-1}$) (Price and White 1985) (Fig. 10.11). Mass-specific rates of metabolism are highest during early growth because most cells are dividing or expanding, with high rates of protein turnover. The component systems of the body are developed in a sequence that sets the minimum duration and maximum rate at which the animal can gain mass and become independent. Bird embryos develop in 38–46 stages that are most variable between species in the latter phases when feathers and external characters are formed (Ricklefs and Starck 1998). Early development is consistently focused on systems that coordinate function: the framework of the skeleton and neuromuscular system, and the circulatory and endocrine systems. Most mass is gained after hatching or birth; early post-natal growth is focused on establishing the nutritive systems of the digestive tract, liver and kidney before the bulk of body mass is gained as muscle and bone (Fig. 10.11).

Consequently, infant age affects the ability to recover from fasting; older infants may be able to resume muscle gains whereas critical systems may be permanently impaired by fasting in the youngest neonates. Storms may interrupt feeding and growth of nestling birds; alpine swifts can resume growth after a storm but fledging is delayed even though relative growth rate of body mass and wings are increased, especially in older nestlings (Bize et al. 2006). Young animals may be able to compensate for interruptions in growth if they can increase food intakes. However, among deer, low-quality diets in seasonal environments may prevent compensation for interruptions in growth, so small calves become small adults (Albon et al. 1987; Gaillard et al. 2003).

The ability to tolerate deficits of energy and nutrients depends on body stores that were deposited in neonates by the mother. More lipids are invested in young that must survive harsh environments or starvation. Muskoxen are born with lipid stored in brown fat that is oxidized to maintain body temperature on ice and snow (Blix et al. 1984). Young phocid seals accumulate large lipid stores before weaning because they must fast until they develop the ability to forage efficiently (Burns et al. 2006). In small species, offspring are provided with minimal stores of lipid because most of the hatchling or neonatal mass is allocated to lean mass with core functions. Large mothers are able to invest in absolutely larger offspring in a wide variety of species (Clutton-Brock 1991). Birth mass varies with maternal age and body size in seals and deer (Clutton-Brock and Albon 1989; Bowen et al. 2001); small, young mothers birth small infants whereas larger females at prime breeding age can produce bigger infants. In sexually dimorphic species such as seals and deer, large infants are more likely to be males (Cameron 2004); this phenomenon is selectively beneficial to the mother because polygynous sons may transfer their mother's genes to many more infants than daughters. Large body size may also confer a greater ability to provide milk and parental care to support the greater daily mass gain of larger offspring (Mellish et al. 1999) (Fig. 10.11). Mothers at peak body condition therefore may provide infants with the greatest chance of survival by beginning their path to independence as early as possible in a foraging season with all the core functions and stores to survive any exigencies.

Parents minimize the amount of energy required for survival of their offspring by providing nests and hatching sites with relatively benign micro-habitats; thermoregulatory and activity costs are minimal for hatchlings in nests, for marsupial joeys in the maternal pouch and for cubs or kits in dens. Parents often select nest and birth sites with relatively low risks of predation, sometimes at the expense of their own condition; female bighorn sheep with young typically use areas with the

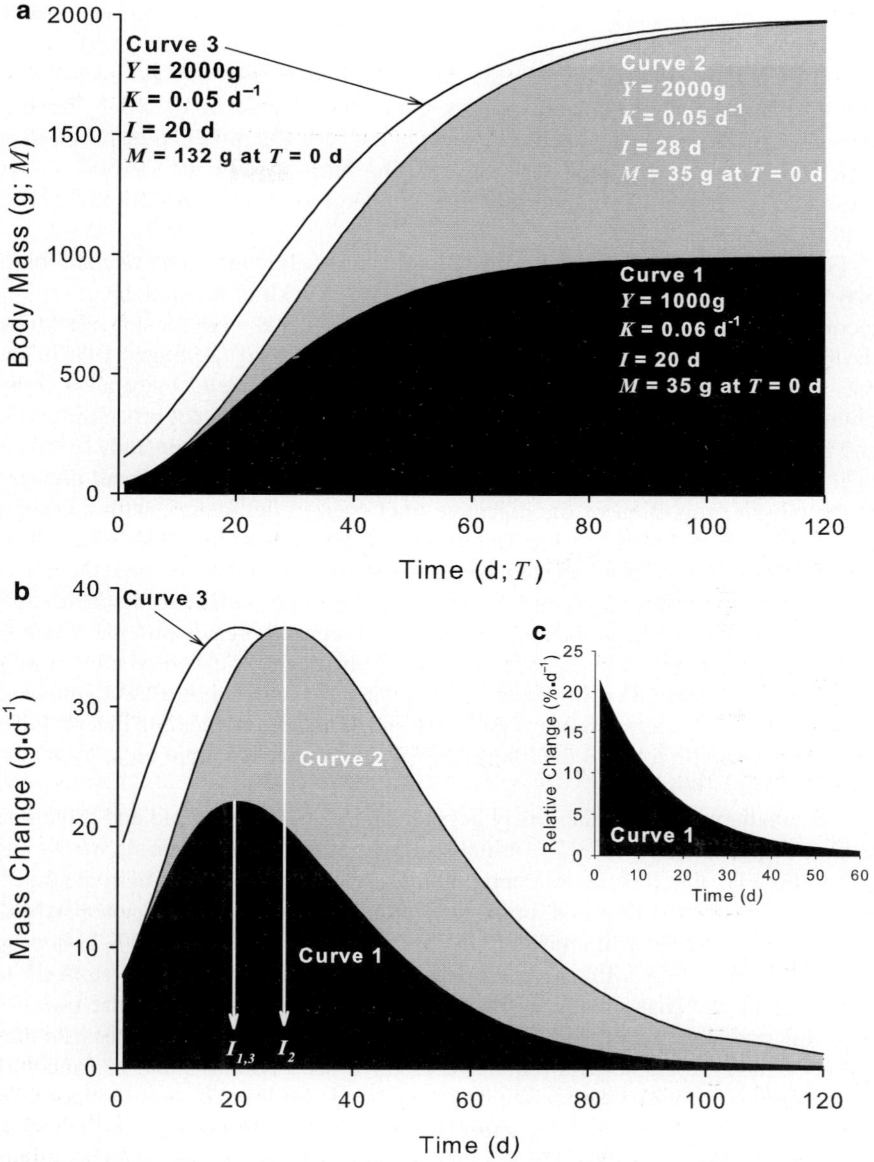

Fig. 10.11 Growth in a mammal or bird can be described by the Gompertz equation:* (Zullinger et al. 1984). **a** Whole body mass (*M*) follows a sigmoid pattern that starts from birth or hatchling mass at 0 days (T) and increases to maximum body mass (*Y*). The same initial mass (35 g) can result in small (*curve 1*; *Y* = 1,000 g) or large (*curve 2*; *Y* = 2,000 g) adults as the duration of growth is increased. Alternatively, the same period of growth may result in large adults (2,000 g) if neonates start life at a larger mass (132 g; *curve 3*). **b** Daily mass gains increase to a maximum at the inflexion point of the growth curve (*I*). Maximum daily gains increase with maximum body mass (*Y*). Large young (*curve 3*) gain more mass each day ($g \bullet d^{-1}$) than small young (*curve 1*) in the same period of time (*I* = 20 d). **c** Relative change in body mass ($\% \bullet d^{-1}$) is highest at day 1 for all curves: 21, 22 and $14\% \bullet d^{-1}$ for *curves 1, 2* and *3* respectively (*curves 2* and *3* are not shown)

$^{*}M = Y \bullet e^{-e^{-K \bullet (T-I)}}$

lowest predation risk even though the quality and abundance of food may be low and insufficient for the demands of milk production (Bleich et al. 1997). Parental protection allows offspring to allocate most of the lipid and protein transferred from their parents to growth; seal pups and nestling birds incorporate 70–80% of the gross energy from milk or food into the net energy of new tissue (Ricklefs et al. 1998; Mellish et al. 1999).

Feeding commences soon after hatching or birth, although some fish and birds may rely on a yolk sac for several hours or days. Cackling Canada geese rapidly incorporate yolk lipids into liver and muscle as an energy reserve for survival during the first few days after hatching when they begin to feed (Sedinger 1986). The masses of the digestive tract, liver and kidney are isometric to body mass (kilograms to the power of 0.9 to 1.0 ($kg^{0.9}$ to $kg^{1.0}$)); organ systems for processing food and metabolites develop at rates that closely match that of the whole body (Ricklefs et al. 1998). Absorptive capacity is well developed in neonates; the small intestine is the largest proportion of the digestive tract even in herbivores, with a relative growth rate of 100 $\%{\bullet}d^{-1}$ in the first few days (Ricklefs et al. 1998; Knott et al. 2004). However, digestive capacity and function are less well developed; the mouth parts may be too soft to mechanically degrade most solids and the stomach has only a modest ability to mix solids and secrete acid. Parents therefore provide neonates with foods that are readily degraded and easily absorbed. Soft-bodied prey or prey parts are fed or partially digested and regurgitated for nestling birds. Penguins and pigeons produce a secretion from lysed cells in the mucosa of the crop that provides their nestlings with a liquid diet that is 65% protein and 35% lipid on a dry matter basis (Klasing 1998).

Mammalian milks are also easily degraded combinations of lipid and protein as well as sugars and minerals. The initial secretion of milk (colostrum) is watery and rich in proteins that include immunoglobulins, which bind to bacteria in the digestive tract and are also absorbed for the developing immune library (Chapter 8). Milk protein content varies from 45% of DM in cottontail rabbits to just 7% in hooded seals (Robbins 1993). Milk proteins denature in the mild acid of the stomach to form a solid clot that slows the flow of the digesta and provides more time for digestion and absorption. Milk proteins have a high biological value because they closely match the amino acid requirements for tissues synthesis and intermediary metabolism (Chapter 8). Milk proteins are also spared from oxidation for energy by the milk sugar lactose, which provides most of the glucogenic requirements (Chapter 6). The transition of neonatal herbivores to adult diets with structural carbohydrates is anticipated by the development of fermentation regions in both the foregut and the hindgut. Ruminal enlargement in muskoxen and reindeer is accompanied by consumption of soil and maternal hair at 30 days of age; the mucosa and the digesta indicate an active fermentation at 60 days of age even though the animals have not been weaned (Knott et al. 2004, 2005). In addition to maintaining the social bond between mother and young, milk provides a supplement to the forage diet for some young herbivores because they may not have the capacity to retain fibrous forages as plants senesce at the end of the season (Chapter 5) (Munn and Dawson 2003, 2006).

The nutritional constituents of the milk contribute to different life-history patterns, even for species living in similar environments. Highly concentrated milk enables rapid summer growth rates of caribou neonates in the Arctic to support fall migration and escape predators by flight, in contrast to the less concentrated milk of lower quality consumed by muskox calves with slower growth rates and more sedentary behavior (Parker et al. 1990).

Lean mass accounts for most of the daily mass gained by young birds and mammals in the first few days after birth or hatching (Fig. 10.12). Consequently, protein supplies are crucial to development and mass gain. In the example in Fig. 10.12, the ratio of metabolizable protein to energy required for growth of a duckling declines from 1.96 to 1.13 $g \cdot 100\,kJ^{-1}$ as relative growth rates decline with age. Fish fry may use diets that are even higher in protein than those of terrestrial neonates because they rely on amino acids for metabolizable energy (Chapter 8). Many herbivorous reptiles begin life as carnivores because of their high demands for protein synthesis and lean mass gain.

Environmental factors may limit the availability of energy and protein for growth. Low ambient temperatures slow growth in ectotherms and divert energy towards thermoregulation and away from growth in endotherms. Most animals therefore slow or cease growing through winter in northern regions or during dry seasons when food abundance is low or environmental demands are high. Growth may not continue in the smooth curves shown in Figures 10.11 and 10.12 but progress in a series of steps or waves as mass is lost each winter and resumed the following spring (Leader-Williams and Ricketts 1981). Large animals therefore may require several seasons to attain a body mass that will allow them to successfully compete for mates or produce eggs. The quality of habitat affects the time required for moose and sheep to reach harvestable body sizes as well as antler and horn dimensions.

Genders, however, may differ in their seasonal mass gain; females may attenuate lean mass gains to establish lipid stores for reproduction whereas males may continue to allocate energy to growth. Sexual size dimorphism in ungulates is the outcome of a longer seasonal window for lean mass gain in males than females (Peltier and Barboza 2003). Increasing windows for plant growth at high latitudes have been related to increasing dimorphism of the sexes in red deer and moose, as females breed earlier and males continue to grow (Post et al. 1999; Garel et al. 2006). The selective benefits of large body size on reproductive output favor some growth throughout life in some species, especially among long-lived reptiles and fish such as crocodiles and sharks. Increasing body size ultimately demands more food and more time for growth that can only be sustained by high food production and large foraging areas. The reproductive benefits of very large size are nonetheless finite; most large herbivores senesce and lose their ability to regain mass each season as they age (Mysterud et al. 2005b; Bowen et al. 2006b).

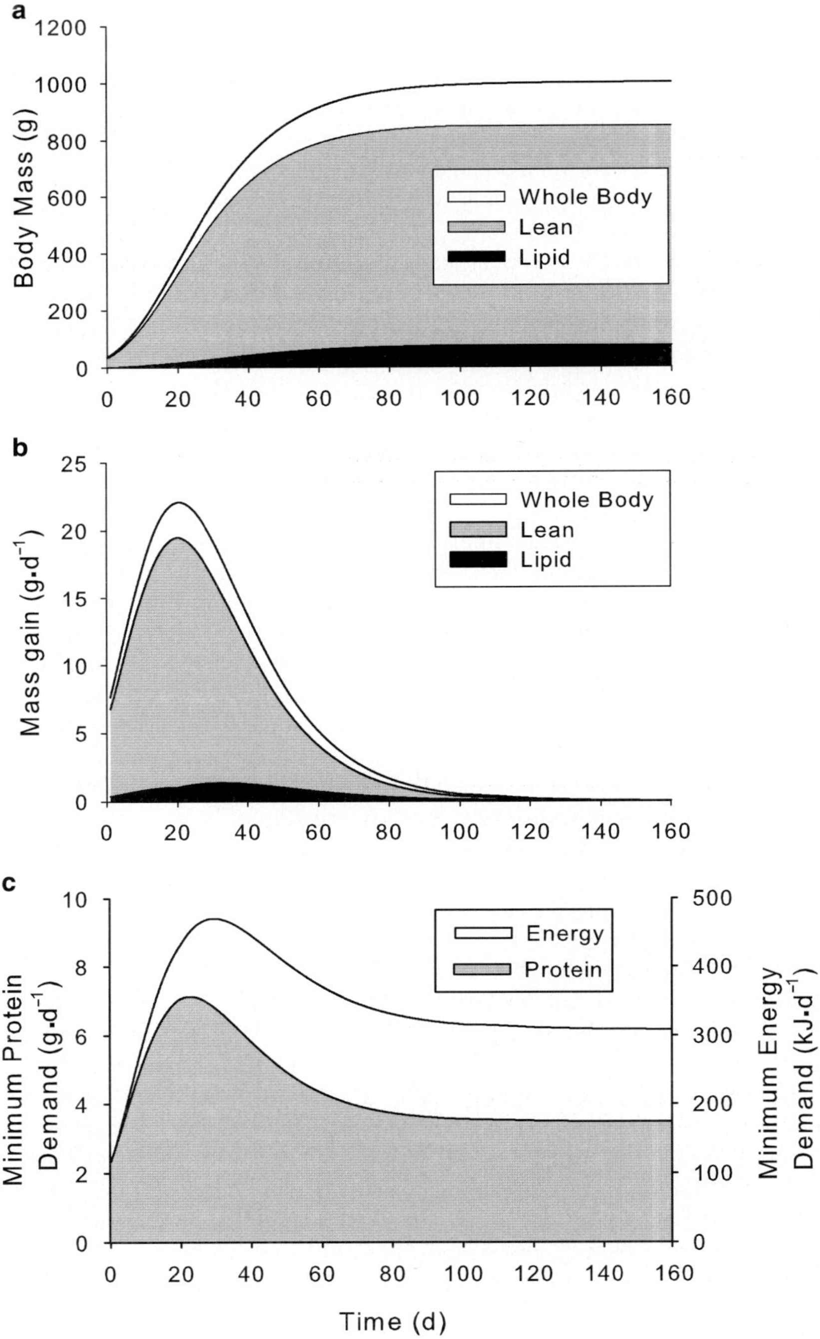
a
Body Mass (g)
Whole Body
Lean
Lipid
b
Mass gain (g•d^{-1})
Whole Body
Lean
Lipid
c
Minimum Protein Demand (g•d^{-1})
Minimum Energy Demand (kJ•d^{-1})
Energy
Protein
Time (d)

Fig. 10.12 A model of growth in a duckling from 36 g at hatching to a subadult of 1000 g at 150 days. **a** Most of the mass gain is associated with lean body mass. **b** Daily gains of body mass increase rapidly during early life to a reach a maximum at the inflexion point (I = 20 days in this example). **c** Rapid gains of protein in lean body mass increase dietary protein demands in early life. The dietary ratio of protein to energy is therefore highest during early growth. Empty body mass was calculated as 93% of whole body mass. Lipid content was calculated on the basis of daily gain in empty body mass at 5% of the daily gain from hatching to 20 days and increased linearly to 38% of the daily gain at 150 days. Lean mass is the difference between empty body mass and lipid. Protein was 26% of lean mass. Net energy contents of lipid and protein were 39.3 and 23.7 $kJ \cdot g^{-1}$, respectively. Net energy was converted to metabolizable energy from the diet with an efficiency of 0.82 for lipid and 0.50 for protein. The minimum requirement for metabolizable energy is the sum of maintenance (308 $kJ \cdot kg^{-0.75}$) and the equivalent daily gain in body energy. The minimum protein requirement is the sum of maintenance (3.489 $g \cdot kg^{-0.58}$) and the daily gain in body protein

10.11 Summary: Energy

The energy of plants and animals is contained in carbohydrates, proteins and lipids. Energy gain and loss by animals can be considered in terms of gross, digestible, metabolizable and net energy. Energy expenditure can be measured using calorimetry with captive animals or with the doubly labeled water technique in free-living animals. Animals vary in their basal metabolic rate (BMR), maintenance requirements (up to 2 × BMR), and the additional requirements for thermoregulation, activity, reproduction and growth. Total energy expended in the field therefore varies from 2 to 7 × BMR. The potential for growth in a population during the year is related to the body condition of individuals within that population. This chapter describes various techniques available to wildlife managers to objectively evaluate body condition of animals. Body condition varies with seasonal changes in food supply and with life history of the species. The time and amount of energy and nutrients required for reproduction and growth vary with allocation of body stores or diet to parental investment, and with the number of offspring and their rate of growth.

Chapter 11
Integrating Nutrient Supply and Demand in Variable Environments

Environmental variation alters the supply of food and water as well as the requirements of animals for these resources (Fig. 11.1). Changes in food intake can affect all components of energy and nutrient balance, which ultimately affects growth and reproduction. The ability to tolerate and resist adverse environmental changes depends on the plasticity of the animal, from its foraging activities (behavioral plasticity) to its production (physiological plasticity). For example, brown bears have some behavioral plasticity to tolerate disturbances from wildlife watchers by altering the times and locations at which they feed on salmon (Rode et al. 2007). Polar bears also can change their foraging behavior by moving onshore when there is insufficient sea-ice cover to allow foraging on seals. The ability to accommodate a change in foraging conditions depends on the duration of the change, the breadth of the animal's dietary niche, the availability of alternative foods and the energy demands of the animal. Decreases in sea-ice cover are associated with poor survival of young polar bears and with increasing conflicts between bears and humans as the bears begin to seek terrestrial foods often near settlements in order to meet their demands for maintenance and growth (Regehr et al. 2007).

Physiological plasticity allows animals to respond to changes in the supply of nutrients. Increased food intakes may require concomitant increases in the capacity of the digestive tract in order to hold more food and absorb more nutrients, and increases in the abilities of the liver and kidney to clear more absorbed metabolites. Seasonal reductions in food abundance may force animals to reduce their food intake and rely instead on their stores of energy and nutrients. Animals can minimize the increased costs of survival under adverse conditions by reducing the energy expended on core functions such as movement and thermoregulation. Bears reduce activity and body temperature when they enter winter dormancy. The ability to respond to changes in food supply or survival costs is determined by the time required to respond and the physiological range over which the response can be sustained. Animals are stressed with potentially negative consequences when they cannot meet the demands of their environment and their life history. Growth and reproduction may be slowed, interrupted or abandoned when stressors cannot be avoided or accommodated by changes in location or food intake (Fig. 11.1). Storms, for example, increase the concentration of stress hormones in breeding

P.S. Barboza et al., *Integrative Wildlife Nutrition*,

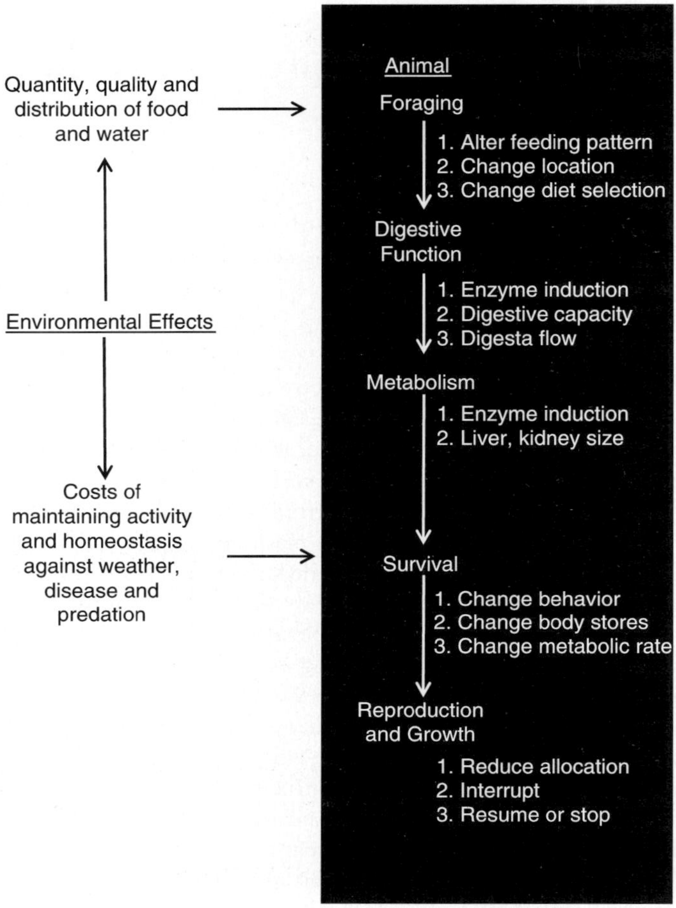

Fig. 11.1 Effects of environmental variation on energy and nutrient flow in animals

songbirds, which alters their allocation of energy from reproduction to survival as they temporarily abandon nesting (Wingfield and Ramenofsky 1999). Similarly, stressful winters with heavy snowfalls are associated with pauses in reproduction of female caribou and their annual production of calves (Adams and Dale 1998).

Behavioral and physiological responses are integrated to maximize the likelihood of survival and reproduction. This integration involves multiple levels of response from genes and metabolic pathways at the cellular level to growth and reproduction of the whole animal. Therefore, responses are integrated over time scales that range from seconds to years as each meal is integrated into the life cycle of the animal. Similarly, responses are integrated over spatial scales ranging from a few square meters at a feeding site (Parker et al. 1996) to the annual range of the animal (Hobbs 2003; Parker 2003).

11.1 Neuro-Endocrine Integration of Food Intake and Metabolism

Animals consume food, and absorb and metabolize nutrients in an attempt to maintain homeostasis throughout the day. Small changes in the balance between supply and demand for energy and nutrients over many days control changes in body mass and composition during each season. Metabolizable carbon (C) is monitored and regulated through the flows of metabolites such as glucose, SCFA and amino acids, which are key components in the pathways of energy metabolism. The flow of metabolizable C fuels core functions such as thermoregulation and cell turnover to maintain the integrity of the body (Chapter 10; Table 11.1). Similarly, water and electrolytes are monitored and regulated closely to maintain the composition of body fluids within a narrow range (Chapter 9). Coordinated increases in the consumption of both metabolizable C and the components of tissues allow animals to gain body mass for growth and reproduction (Chapter 10).

Nutrient flows within the body are controlled through the actions of nerves, hormones and other cellular signals (Fig. 11.2). Nerves are dedicated connections between cells, whereas hormones and cellular signals rely on the circulation of extracellular fluids to reach target cells that are located either within the same organ or in some other part of the body some distance away. Hormones are typically produced by glands whereas individual cells secrete messenger molecules such as the cytokines from cells in the immune system. Neurotransmitters are small molecules such as acetyl choline and epinephrine that are derived from amino acids. Hormones and cellular signals include a wide variety of molecules, from small lipids such as sterols (e.g., cortisol) to large peptides (e.g., insulin). Nerves transmit the sweet taste of sugar from the mouth to the brain, as well as the response from the brain to the mouth, to continue feeding and produce saliva (Chapter 3). Secretions of the pancreatic hormones insulin and glucagon are carried by the blood to the liver, muscles and fat depots to regulate glucose uptake (Chapter 6).

Table 11.1 The role of nutrition in homeostatic functions of animals

Homeostatic function	Nutrient group
Temperature, synthesis and work	Energy substrates as fuels (sugars, fatty acids and amino acids)
Hydration and balance of ions and solutes	Water
	Electrolytes (Na^+, K^+, Cl^-, Ca^{2+}, Mg^{2+}, SO_4^{2-}, PO_4^{2-})
Structural integrity	Amino acids for proteins
	Essential fatty acids for membranes
	Macrominerals (Ca, P, Mg, S) for bones, membranes, proteins
Functional integrity	Antioxidants (vitamins C and E)
	Trace minerals for enzymes
	Vitamins for metabolic pathways

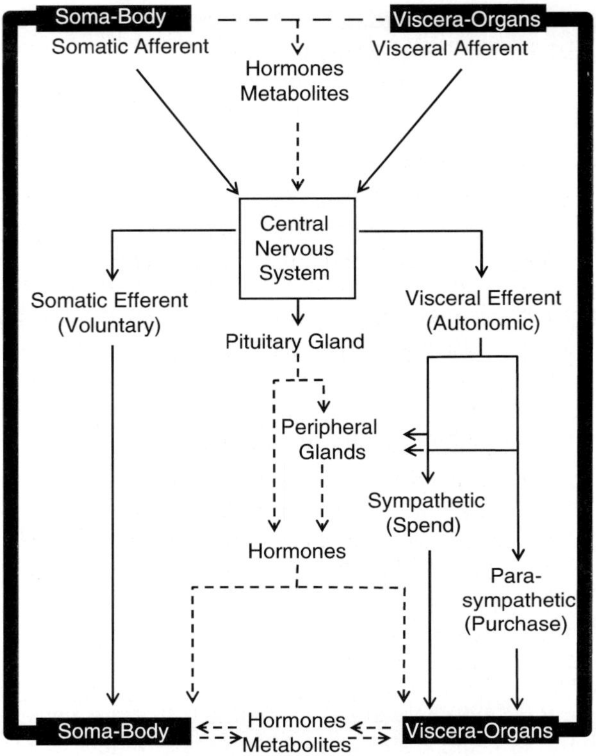

Fig. 11.2 The neuro-endocrine system regulates and integrates food intake and metabolism through signals from nerves (*arrows with solid lines*) and hormones. Extracellular fluids (*arrows with broken lines*) such as blood plasma and interstitial fluids distribute hormones, other cellular signals (e.g., cytokines) and metabolites among cells. The central nervous system (brain and spinal cord) receives signals along afferent nerves from receptors in skeletal muscles and skin at the mouth (soma or body) and from the visceral organs such as the glands and muscles of the gastric stomach. Efferent nerves return signals from the central nervous system to initiate responses at the soma and the viscera

The neural system is divided between the soma and the viscera (Fig. 11.2). The soma consists of the frame and surface of the body, including the skin, skeletal muscles and mouth. The viscera include internal organs such as the liver, kidney, digestive and reproductive tracts. Somatic sensors in the mouth detect the smell, taste, texture and temperature of food. Visceral sensors monitor distension of the stomach wall and acidification of gastric digesta at the start of the meal, as well as the rise in glucose concentrations in the liver during the meal. Afferent nerves bring information to the CNS whereas efferent nerves transmit impulses to the soma and the viscera for a response. Chewing food is a somatic response from the brain to the muscles of the mouth, but mixing food with acid is a visceral response from the brain to the stomach. Somatic responses are under conscious control; the decision to stop feeding and start moving to another feeding patch is a voluntary response that is integrated with other information such as the behavior of conspecifics, the ambient temperature and the presence of predators (Chapter 3). Visceral responses are autonomic and controlled by opposing branches of nerves; parasympathetic nerves stimulate actions that 'purchase' energy by consumption of food (e.g., saliva production, gut motility), whereas the sympathetic nerves stimulate actions that 'spend' energy and suppress food intake (e.g., increasing heart rate). In birds that undergo seasonal changes in body mass, parasympathetic signals predominate over sympathetic signals to increase food intake and thus gain fat and lean mass (Kuenzel et al. 1999).

Sensory feedbacks from the soma and viscera modulate neural responses during a meal. Receptors that sense solutes in the mouth and distension of the foregut are used to control food consumption as well as motility and secretion during a meal. Chickens and red grouse slow food consumption as the crop and gizzard fill with food (Savory 1999). Similarly, sheep and cattle slow food consumption and increase ruminal motility as food fills the forestomach (Forbes 2007). Alterations to the set points for these feedbacks lead to increases in the duration of the feeding bout and thus meal size, and finally daily food intake (Savory 1999). Seasonal mass gains of animals are therefore controlled by changes in the sensitivity to sensory (afferent) inputs as well as the pattern of autonomic (efferent) responses.

Sensory inputs also cause the release of hormones from the pituitary gland, which is intimately connected with the hypothalamus at the base of the brain. An increase in the osmolality of blood flowing to the brain causes the secretion of vasopressin from the pituitary, which stimulates water resorption at the kidney (Chapter 9). The brain also stimulates the pituitary to release hormones that activate peripheral glands such as the thyroid, which releases thyroxine that increases metabolic rate during feeding (Chapters 9 and 10). Peripheral glands also receive neural stimulation; changes in blood glucose concentration stimulate the release of insulin and glucagon from the pancreas, which is further modulated by stimuli from autonomic nerves (Chapter 6). Hormones produced in peripheral tissues stimulate other visceral organs and the soma: cholecystokinin (CCK) from the small intestine slows motility of the stomach and increases release of bile from the liver (Chapter 7); vitamin D produced at the skin is activated in the liver and the kidney to stimulate Ca absorption from the intestine (Chapter 9).

Hormones can change food intake and body mass by directly stimulating the brain, by causing the release of other cytokines that enter the circulation of the brain, or by stimulating organs (e.g., the liver) that return afferent neural signals to the brain (Fig. 11.2) (Rhind et al. 2002). Many signal molecules and cells have been implicated in the stimulation and suppression of food intake during a day. The network of signals is like a stage play; many actors with many lines communicate a common theme such as 'increase food intake and body mass for growth'. One signal (one line in the play) from one organ (one actor) provides only a small portion of the total information required to integrate the response across the network.

The diversity of signals reflects the diversity of supplies from food and the demands of the animal. The supply of metabolizable C must be integrated from foods with different proportions of carbohydrate, protein and lipid to meet the prevailing demands for survival and the anticipated costs of production. For instance, several species of fish, birds and mammals reduce food intake in response to secretion of CCK from the duodenum (Chapters 5 and 7) (De Pedro and Bjornsson 2001; Forbes 2007). Rising concentrations of CCK in the blood indicate to the brain that lipids are being digested. CCK is a good indicator of metabolizable energy supply from most foods because lipids are rich in metabolizable C. CCK is a better indicator of energy intake than the concentrations of fatty acids in the blood because fatty acid concentrations may be slow to increase after a meal and because fatty acids are also mobilized from adipose tissues when food intake is low.

The importance of each signal changes within species with changes in diet and life stage. Sensitivities to insulin and glucagon change with season in hibernating animals; glucagon is more important than insulin during winter because plasma concentrations of glucose are more likely to decline between each bout of torpor (Wang and Lee 1996). Signals have different emphases in different species; evolutionary selection has favored feedbacks between particular organs to ensure homeostasis for survival as well as reproduction. Leptin, a hormone released by adipose cells and liver (Houseknecht and Spurlock 2003), reduces food intake in hibernating rodents such as Arctic ground squirrels and Djungarian hamsters (Ormseth et al. 1996; Klingenspor et al. 2000). High plasma leptin indicates large fat stores in these animals (Florant et al. 2004), but not in all mammals and birds (Speakman et al. 2002); different neuro-endocrine indicators of energy and nutrient status (Rhind et al. 2002) are used in different species to provide similar information for regulating food intake.

Daily patterns of food intake may be regulated by the simple process of filling and emptying the digestive tract, and clearing absorbed nutrients from the blood. This endogenous cycle of feeding is evident in Svalbard reindeer during the continuous night of the polar winter and the continuous day of the polar summer (van Oort et al. 2007). Endogenous cycles of food intake are modulated by internal cues for growth and reproduction. Pituitary hormones that stimulate production include growth hormone that controls development and mass gain, as well as luteinizing hormone (LH) and follicle stimulating hormone (FSH) that promote the production of eggs and sperm. Feeding patterns are also modulated by environmental cues that allow animals to anticipate periods that are favorable for foraging, that is, when risks of predation or thermal stress are lowest and the net gains of nutrients are most likely to be high (Chapter 3). Fish may modify their feeding patterns in relation to regular tides and currents that change the temperature and concentrations of dissolved O_2 and CO_2 in water (Madrid et al. 2001). Daylight is used to schedule feeding in animals that are nocturnal, diurnal or crepuscular (active at dusk and dawn). Changes in the duration of day (photoperiod) and night (scotoperiod) may result in seasonal changes in food availability and environmental demands. Daily (circadian) and seasonal (circannual) responses are stimulated by melatonin, a hormone released from the pineal gland. Melatonin is secreted in the dark and suppressed by light exposure; plasma melatonin there-

Fig. 11.3 Pelage change is one of the metabolic systems affected by photoperiod. The pelage of this snowshoe hare is changing from white to brown as day length increases in spring. The hare will turn back to white again as day length shortens in autumn. Exposure to light suppresses the secretion of the hormone melatonin from the pineal gland. Melatonin regulates the secretion of melanotropin from the pituitary gland, which in turn regulates the production of melanin at the skin and the hair follicle. Decreasing day length in autumn increases melatonin, which suppresses melanotropin and stops melanin synthesis as animals molt into their white pelage for winter (Norris 1997)

fore signals day length (Norris 1997). Northern ruminants such as red deer increase food intake and gain body mass in the long days of spring and summer (low plasma melatonin) and decrease food intakes in the short days of winter (high plasma melatonin) (Rhind et al. 2002). Day length and melatonin also are used to schedule molt and regrowth of fur and feathers, as well as growth and reproduction in mammals and birds (Fig. 11.3).

11.2 Stressors

Any challenge to the homeostasis of an animal may be considered a stressor that requires an integrated response from the neuro-endocrine system (Woebeser 2006). Environmental changes usually present multiple challenges to homeostasis. Droughts challenge animals to regulate their body fluids against declining plasma volume and increasing plasma concentration of Na. Vasopressin and atrial naturetic hormone (ANH) secretions stimulate responses from the brain to search for water, minimize water loss and eliminate excess Na by changing renal function (Chapter 9). Droughts also cause heat loads that stimulate voluntary responses such as a shift in foraging pattern to minimize exposure to the sun, autonomic responses such as sweating or panting, and the production of protective agents such as melanin and heat-shock proteins (Chapter 10). Water-stressed animals may reduce food intake, which then limits the supply of energy and nutrients for maintaining the structural and functional integrity of the body (Table 11.1).

Each stressor can be considered as the difference between supply and demand of energy or any nutrient (e.g., water, Na or N) at three different levels: eustress, neutral stress and distress (Breazile 1987) (Fig. 11.4). Eustress is the condition of beneficial stress when supply exceeds demand and the animal is in positive or zero balance. The metabolic load of absorbed nutrients that must be cleared after a meal

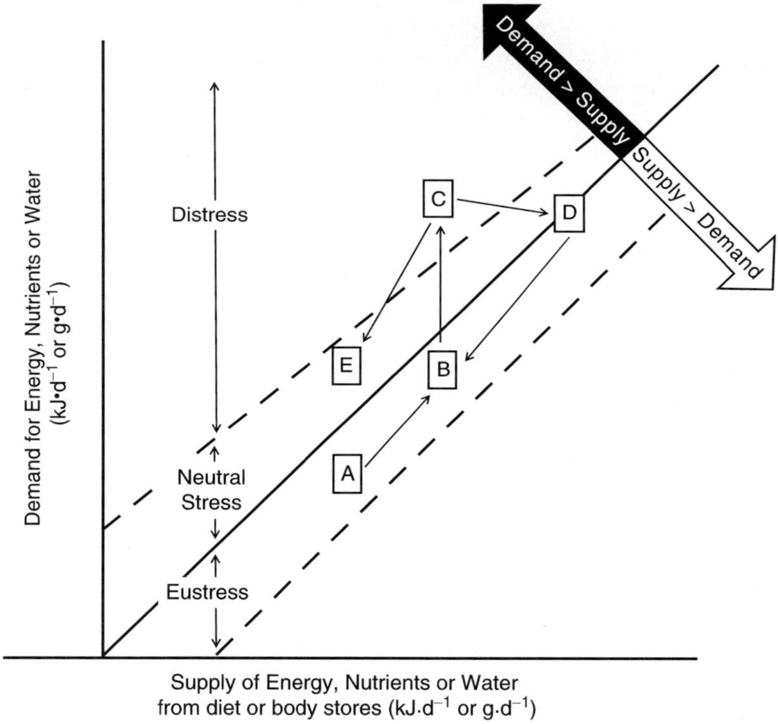

Fig. 11.4 Stress may be considered as positive (eustress), neutral or negative (distress). Stress levels are related to the difference between supply and demand of any nutrient required for homeostasis or integrity of the animal. The *solid line* indicates a complete match between supply and demand. Eustress is a moderate excess of supply over demand (below *solid line*), whereas neutral stress is incurred when animals contend with a moderate deficit of supply (above *solid line*). Animals are distressed as the deficit in supply increases. An animal at maintenance (*A*) increases its demands as it begins to reproduce (*B*). A stressor such as a storm or disturbance increases the demands of the animal without changing supply and the animal becomes distressed (*C*). The animal may alleviate its distress by moving to an area with more food or shelter (*D*) or by reducing its demands by interrupting breeding (*E*)

is an example of eustress that maintains the normal range of vascular, liver and kidney functions. Selection of plants that contain plant secondary metabolites (PSMs) is a form of eustress that maintains a complement of detoxification enzymes, which defend the herbivore against toxins; this phenomenon of inoculation or induction by low exposure to stressors is called 'hormesis' (Lindsay 2005).

Neutral stress is incurred when demand begins to exceed supply and animals enter negative balances for some nutrients but can still tolerate the deficit by using body stores (Fig. 11.4). Animals can consume a meal with little N or vitamin A because they can use labile proteins and vitamin A that are stored in the liver. Dietary imbalances in Ca and P are neutral stressors that can be tolerated if stores of Ca and P from bone are available, and if the animal is able to excrete the excess mineral (Chapter 9).

Animals are distressed when demands exceed supplies and the deficit of energy or nutrients exceeds the tolerance threshold provided by body stores (neutral stress). Unpredictable events such as storms force songbirds to mobilize their body stores to either fly away or stay and meet the added demands for thermoregulation when foraging is limited (Fig. 11.4) (Wingfield and Ramenofsky 1999). Animals are more likely to be distressed by these additional demands when their requirements are already elevated for growth or reproduction. Consequently, distressed animals may reduce their demands by slowing or interrupting growth or reproduction so that body stores can be allocated to behavioral and physiological responses that favor survival (Figs. 11.1 and 11.4) (Wingfield 2005). Frequent distresses may prevent reproduction and ultimately debilitate the animal, but reproduction may continue under neutral stress when smaller deficits are met by mobilizing body stores (capital breeders; Chapter 10).

Distress responses are integrated at the brain by both hormonal and neural pathways. Sympathetic autonomic nerves activate the distress response from organs and fat depots as well as the adrenal gland, which releases epinephrine to the blood (Fig. 11.2). The neural path is fast acting and part of the 'fight or flight response' that increases blood flow and O_2 delivery to the brain and muscle, and suppresses the blood supply and activity of the digestive tract, kidney and skin (Breazile 1987). The CNS also initiates the hormonal pathway for distress by secreting hormones from the hypothalamus and pituitary gland that stimulate the adrenal gland to release the glucocorticoid (GCORT) hormones cortisol or corticosterone. GCORTs are longer acting than epinephrine and more likely to be detected in blood and hair and as excretory conjugates in urine and feces (Parker et al. 1993a; Wasser et al. 2000; Washburn et al. 2002; Davenport et al. 2006). GCORTs increase the concentration of glucose in plasma by suppressing insulin responses, increasing gluconeogenesis in the liver and mobilizing protein and lipid from muscle and fat. GCORTs mobilize stored energy to prepare the animal for the sudden demands of intense activity that may be required to survive an external threat to the body or to flee the area to alleviate a challenge to homeostasis.

Food intake is both stimulated and suppressed by GCORTs. Mild distress at low concentrations of GCORTs stimulates food intake and increases energy supply to meet the anticipated demand; chronic low elevations of GCORTs may be associated with overfeeding and obesity in some mammals, including humans (Wingfield 2005). Small rises in plasma GCORT may serve as a signal to resume feeding in fasting animals that also use body stores for reproduction. In lactating Antarctic fur seals, for example, plasma GCORT increases before they return to sea to forage (Guinet et al. 2004). High concentrations of GCORTs suppress food intake as well as immune responses and wound healing, which diverts energy away from maintenance of the body. High concentrations of cortisol in breeding males impair renal function, and increase the incidence of mortality from wounds and infection among reindeer, Arctic ground squirrels and the semelparous marsupial brown antechinus (Barboza et al. 2004; Boonstra 2005; Naylor et al. 2008). Interactions between GCORTs and other hormones may alleviate some adverse effects. Testosterone reduces the effect of cortisol on renal function in brown antechinus, so that the dominant males with the highest concentrations of testosterone survive and breed for the longest period, but still eventually die from the consequences of impaired body maintenance (Naylor et al. 2008).

High concentrations of GCORTs also impair development and may suppress reproduction (Sapolsky 2001). The adverse effects of GCORTs are ameliorated by binding the hormones to proteins (corticosteroid binding globulins) in the plasma, which reduces the concentration of unbound hormone (Reeder and Kramer 2005). Alternatively, the release of GCORT may be suppressed by increasing the threshold for distress. Cortisol levels, for example, are not related to changes in body fat or food intake of pregnant females among Sitka black-tailed deer or reindeer during winter (Parker et al. 1993a). The beneficial effects of GCORTs are only realized when the animal has sufficient stores to escape the stressor or when increased activity can increase the supply of energy (Wingfield and Ramenofsky 1999). Plasma corticosterone in small, food-deprived Magellanic penguin chicks increases in response to the stress of handling during early development, but not when birds are handled at fledging. Small chicks therefore avoid the adverse effects of corticosterone when body stores must be conserved at fledging (Walker et al. 2005). Adult king penguins and common eiders do not use corticosterone but rather other hormones such as prolactin from the pituitary to stimulate resumption of feeding and mass regain (Cherel et al. 1994b; Criscuolo et al. 2006).

Distress responses are not always due to deficits in energy (Walsberg 2003). Secretion of GCORT is stimulated by behaviors that perceive immediate threats such as pursuit by a predator, and perceived threats such as the sight, smell or sound of a predator. Plasma concentrations of cortisol in wild snowshoe hares increase with predation risk (Boonstra 2005). These reactions may be altered by experience; flight responses of caribou are reduced after one generation following the elimination of predators (Berger 2007a). GCORT concentrations are affected further by dominance interactions among social animals such as baboons and wolves. That is, social rank and breeding status affect the apparent distress of animals within a group or population (Sapolsky 2001; Creel 2005).

Population cycles are correlated with GCORT concentrations in some species. Cyclical declines in a population of water voles, for example, are accompanied by elevated fecal metabolites of cortisol that indicate distress as animals respond to overcrowding and to shortages of food or water (Charbonnel et al. 2008). Concentrations of GCORTs vary with season in many species of amphibians, reptiles, birds and mammals. These seasonal changes in the distress response may be associated with changes in the sensitivity of animals to energy demands (e.g., locomotor activity and field metabolic rate (FMR)) and behavioral cues (e.g. predation, mating) that ultimately prepare the animal for each stage of its life history (e.g., growth, pregnancy) (Romero 2002).

Distress responses therefore may indicate the net outcome of multiple stressors on a population and must be considered in the context of the species and season. The procedures for handling wild or domestic animals often include behavioral stressors of unfamiliar sounds, smells and sights in confined spaces close to other animals, and physiological stressors of fasting, dehydration and increased exposure to infection (Spraker 1993; Hogan et al. 2007). Consequently, a true evaluation of stressors may require multiple indices combined with measures of body condition. Indicators that integrate responses over longer time frames should be more useful for evaluating neutral stress as well as the frequency of distress. Extraction of GCORTs from excreta rather than blood samples provides a longer integration of distress responses and can be collected with little or no disturbance to the animal (Berger et al. 1999; Wasser et al. 2000). Extraction of N metabolites from urine and feces for isotope measures of nutritional stress can also provide a 'hands off' estimate of body protein loss in wintering reindeer (Barboza and Parker 2006). Antibodies and cytokines such as tumor necrosis factors, interleukins and haptoglobins can provide additional indices of exposure to stressors such as toxins or disease that may ultimately lead to more specific indicators for the species and population under study (Ditchkoff et al. 2001b; Bowyer et al. 2003; Woebeser 2006).

11.3 Plasticity of Food Intake and Production

The patterns of feeding in animals are ultimately determined by variation in their environments. In terrestrial animals, seasonal variations in light, temperature and water affect the abundance and distribution of food, and circadian variations in light and temperature affect the animal's ability to consume that food in a foraging window with minimum risk (Chapter 3). Consequently, environmental variations in food abundance and quality often affect seasonal food intakes of some herbivores more than the endogenous cues for feeding (Iason et al. 2000). The ability to tolerate variation in food supply depends on the physiological plasticity of the species. Fluctuating food supplies require integrated responses of digestion, intermediary metabolism, body mass regulation and reproduction.

Variation in food supplies and thus food intakes can be considered as an oscillating pattern of energy intake based on calculations in Chapter 10. The pattern of metabolizable energy (ME) intake oscillates around a mean (ordinate) with varying amplitude and frequency. In Fig. 11.5a, seasonal food intakes are modeled on a frequency of

365 days for one annual cycle. Large herbivores such as Arctic muskoxen increase food intakes by 28% above the mean in summer when plant biomass is greatest, and decrease intakes to 56% of the peak intake in winter at the nadir of food abundance (Fig. 11.5a) (Peltier et al. 2003). The model includes two costs of maintenance: energy demands at rest (resting metabolic rate, RMR) and the energy lost when metabolizing food (diet-induced thermogenesis, DIT). These costs follow the pattern of food intake because muskoxen and other northern herbivores decrease RMR from summer to winter (Lawler and White 1997) and because DIT is proportional to ME intake (Blaxter 1989; Lawler and White 2006).

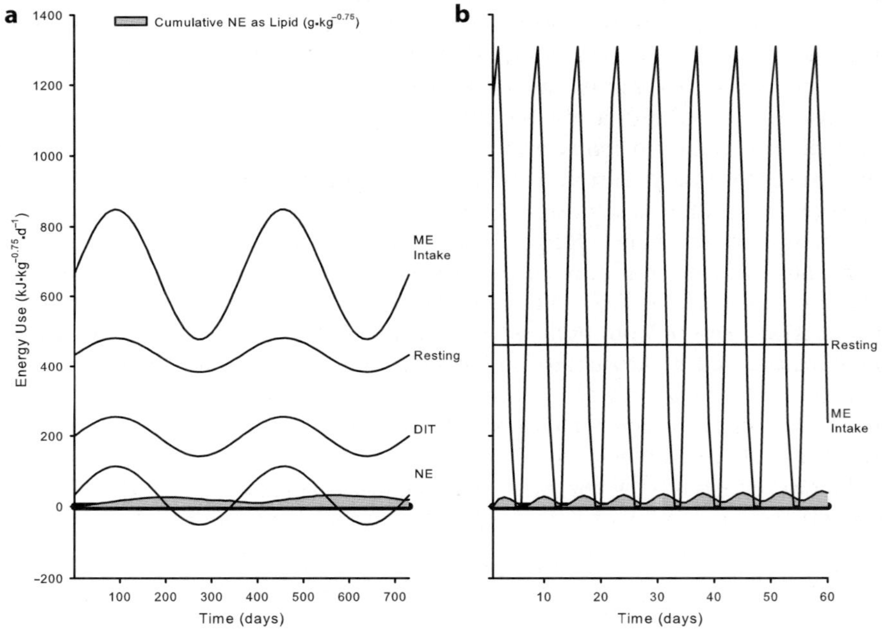

Fig. 11.5 Two models of variable food intake. **a** Annual variation in food intake of a northern ungulate such as muskoxen. **b** Weekly variation in food intake of a carnivore (e.g., mountain lion) or migratory bird on wintering grounds (e.g., black duck). Oscillations in metabolizable energy (ME) intake (*Y*) were modeled with the equation $Y = Y_0 + a \bullet Sin\left(\frac{2\ X}{b}\right)$, where Y_0 is the ordinate of the oscillation, X is time in days, a is the amplitude of the oscillation and b is frequency (Barboza and Hume 2006). The seasonal intake of muskoxen for model **a** was calculated as: $Y_0 = 662.4\,kJ{\cdot}kg^{-0.75}{\cdot}d^{-1}$; $a = 186.5\,kJ{\cdot}kg^{-0.75}{\cdot}d^{-1}$; $b = 365\,d$ (Peltier et al. 2003). Feeding responses of intermittently fasted black ducks for model **b** were calculated as: $Y_0 = 568.5\,kJ{\cdot}kg^{-0.75}{\cdot}d^{-1}$; $a = 758.6\,kJ{\cdot}kg^{-0.75}{\cdot}d^{-1}$; $b = 7\,d$ (Barboza and Jorde 2001). Diet-induced thermogenesis (*DIT*) was calculated at 0.3·ME intake in model **a** and at 0.15·ME intake in model **b** (Blaxter 1989). Seasonal resting metabolic rate (RMR) for model **a** was estimated as: $Y_0 = 432.0\,kJ{\cdot}kg^{-0.75}{\cdot}d^{-1}$; $a = 48.0\,kJ{\cdot}kg^{-0.75}{\cdot}d^{-1}$; $b = 365\,d$ (Lawler and White 2006). RMR was fixed at $460.8\,kJ{\cdot}kg^{-0.75}{\cdot}d^{-1}$ for model **b** (Kaseloo and Lovvorn 2003). Net energy (*NE*) is the energy remaining from ME intake after deducting RMR and DIT. The available NE was converted to a change in body lipid (*shaded areas*) with the calculations described in Table 10.1 and summed over the duration of each model

The net energy (NE) available for production of fat increases in autumn but declines during winter when ME intakes are insufficient to meet the combined costs of RMR and DIT (Fig. 11.5a).

Slow oscillations of food intake and mass gain in seasonal herbivores do not include regular periods of fasting (Fig. 11.5a). Many animals, however, feed and fast intermittently in much shorter cycles with greater amplitudes (Fig. 11.5b). Intermittent feeders include predators such as lions and pythons that fast between large kills, and shorebirds or waterfowl that are restricted to foraging between tides or migratory flights (Barboza and Hume 2006). Mean energy intake and the frequency of feeding increase with metabolic demands of endotherms and ectotherms. Endothermic lions feed every few days, whereas ectothermic pythons may feed at intervals of several weeks, depending on ambient temperatures (Chapter 10). Black ducks can tolerate fasting for 4–5 days and feeding for 2–3 days in a weekly cycle that mimics frequent interruptions of feeding by winter storms and tides (Barboza and Jorde 2001). These birds resume feeding at 233% of their mean intake to compensate for each fast (Fig. 11.5b). Other birds respond in a similar manner to increased energy demands during exposure to cold temperatures (McWilliams and Karasov 1998).

Each oscillation in the pattern of food intake switches between hyperphagia and hypophagia. Hyperphagia presents a load on both digestive and metabolic systems because intakes of ME and other nutrients exceed requirements. Hypophagia is a deficit that must be met from body stores because intakes are insufficient for requirements. Large changes in the abundance of foods select for hyperphagic responses in animals because the subsequent ability to tolerate food shortages is dependent on prior accumulation of nutrients. Highly variable environments that frequently alter food abundance for animals select for rapid adjustments to both hyperphagia and hypophagia. These responses are quickly integrated by the neuroendocrine system, but more time and nutrients may be required to alter the activities of tissues and metabolic pathways.

The time to respond to a sudden change in energy demands and food intake is met by 'spare capacity', which is a safety margin for increasing supply in response to a sudden demand. Spare capacities can be measured as the increase in daily digestible intake of animals switched from low to high demands for thermoregulation or reproduction. Spare capacity varies widely from 9–50% in small endotherms such as yellow-rumped warblers, white-crowned sparrows and prairie voles that were switched from mild to cold temperatures (Karasov and McWilliams 2005). Enzyme induction is the first response to a shift in diet and food intake (Fig. 11.1). Endogenous enzymes are induced within a day by synthesis at the surface of the intestinal mucosa and the pancreas in response to changes in carbohydrate and protein intake (Karasov and Hume 1997). Dietary induction of exogenous enzymes that are produced by fermentative microbes may require more than a week, depending on the composition and turnover rate of the microbial community (Chapter 6). Similarly, enzymes and transporters that oxidize and eliminate PSMs may also require longer periods of induction, depending on the frequency of exposure to the toxin (Chapter 3).

Structural changes in organs follow enzymatic changes as dietary shifts and intakes are sustained (Fig. 11.1). Morphological changes in the digestive tract and liver are observed within 1–2 days of a change in diet for birds such as Japanese quail, red grouse and red knots (Battley and Piersma 2005; Starck 2005). Food composition affects the sequence of changes by stimulating different organs: abrasive diets stimulate mass gains of the gizzard, high lipid loads promote increases in liver mass, and solute loads promote enlargement of salt glands and kidneys (Chapters 5 and 9).

The time available to process increasing amounts of food depends on rates of digesta flow and digestion and the volume of the digestive tract (Chapter 5) (Fig. 11.1). Foods that are quickly degraded by endogenous enzymes can be processed in a tubular gut system (midgut plug flow reactor, PFR). This allows migratory shorebirds, eagles and bears to attain high food intakes and utilize brief periods of food abundance, such as spawning salmon and horseshoe crabs. Slowly digested substrates such as lipids are most affected by high flow rates and high food intakes. In harbor seals, high intakes of lipid in herring reduced lipid digestibility from 90% to 50%, but protein digestibility was maintained over a wide range of protein intakes (Trumble et al. 2003). The maximum capacity to process lipids, proteins and carbohydrates varies with the composition of the diet and with the secretory and absorptive capacity of the species. Carnivores such as seals and cats have very high capacities to process proteins, whereas the ability to handle sugars is greatest among nectarivores and frugivores (Chapters 5, 6 and 8).

Slow flow rates and large digestive volumes are required to utilize structural carbohydrates. Fermentation is a slow, multi-step process that is best suited to large mixing chambers in either the foregut (e.g., rumen continuous-flow, stirred-tank reactor, CSTR) or the hindgut (e.g., colonic modified plug flow reactor, MPFR; Chapter 5). High food intakes and flow rates may limit fiber digestion to the hemicellulose fraction of the cell wall in waterfowl (Dawson et al. 2000). Intermittent feeding by black ducks decreases the digestibility of cellulose but does not affect hemicellulose digestion (Barboza and Jorde 2001). Herbivores such as ruminants and equids that utilize cellulose are poorly suited to rapid increases in food intake and sudden changes in diet composition because those changes may exceed the capacity to maintain homeostasis. Captive ruminants that are quickly switched to diets with high concentrations of soluble carbohydrates are vulnerable to bloat and acidosis because one portion of the microbial community can dominate the fermentation (Owens et al. 1998). Conversely, alligators consume large meals (10% of body mass), which are mixed in a muscular stomach that produces large amounts of acid through adaptations of the vascular system and the pathways for acid–base balance (Wang et al. 2005; Farmer et al. 2008). The large digestive capacities of herbivores allow animals to increase food intake, especially as food quality declines (Chapter 5), but those increases in food intake are usually achieved over several days or weeks (Bunnell and Gillingham 1985; Barboza and Parker 2008). Muskoxen increase total digesta fill by 58% when food intakes increase by 73% over several weeks between spring and autumn (Barboza et al. 2006).

Declines in food intake are associated with a reduction in the mass of intestine, liver and kidney. Herbivores such as alpine marmots reduce the digestive tract as food intakes decline before hibernation but quickly restore those tissues when feeding resumes in spring (Hume et al. 2002). Reductions in the mass of nutritional organs are associated with reduction in the basal metabolic rate (BMR) of migratory shorebirds. These reductions reduce their wing loading and increase their flight range (Battley and Piersma 2005). Conversely, migratory garden warblers restore digestive function as food intake increases during the first 3 days after arriving at their destination (Hume and Biebach 1996). Frequent cycles of feeding and fasting, however, may not allow animals to reduce their digestive tissues and their BMR during a fast because they must maintain the capacity for digestion when feeding resumes. The high BMR of seals and dolphins is related to the high costs of maintaining a long intestine (Williams et al. 2001, 2004a). The energy required for maintenance of the body may therefore increase with the frequency of intermittent feeding and fasting.

Frequent oscillations between deposition and mobilization of energy stores may further increase food intakes because energy is lost in converting body tissue to metabolic demand (e.g., 80%) and restoring that body tissue with food (e.g., 50%; Chapter 10) (Blaxter 1989). For example, if an animal requires 1 $MJ \cdot d^{-1}$ of ME for maintenance, it would use 2 $MJ \cdot d^{-1}$ from the diet (50% efficiency) or 1.25 $MJ \cdot d^{-1}$ from body tissue (80% efficiency). For an animal meeting its maintenance energy requirement from body stores, it must replenish the lost body energy by consuming an additional 2.5 $MJ \cdot d^{-1}$ (1.25 $MJ \cdot d^{-1}$ at 50% efficiency) during refeeding. That is, each day of fasting increases the subsequent food intake from 4.0 to 4.5 $MJ \cdot d^{-1}$ (2.0 for current maintenance + 2.5 for regain). Frequent fluctuations in food abundance can increase energy costs for animals, which may reduce the number of animals that can use the same amount of food, and increase the time required to regain mass. Daily interruptions to feeding may extend the time required for waterfowl to regain mass at migratory stopover sites and reduce the number of birds that can be sustained on a wintering ground.

Energy expenditures of small mammals and birds during a fast may be reduced by entering torpor, especially in winter when reductions in food abundance are combined with increased thermoregulatory demands (Chapter 10). Torpor, however, interrupts the development of gametes and embryos in mammals (Wang and Lee 1996; Naylor et al. 2008). Male Arctic ground squirrels arouse from hibernation earlier than females and consume cached food to maintain body temperatures that are needed for sperm production. Reproduction can therefore continue during hypophagia only if body stores are used to maintain parental homeostasis and to synthesize gametes or new tissues of the offspring. Female bears can give birth and suckle cubs during winter dormancy because the mother uses body fat and protein to sustain the cubs and maintain body temperatures at only a few degrees below the active body temperature in summer (Farley and Robbins 1995). Similarly, female reindeer can give birth and commence lactation before peak food abundance because they use body stores of fat and protein deposited in autumn for the production of the fetus as well as the synthesis of milk in the first 4 weeks after birth in spring (Barboza and Parker 2008). Increasing body size may be associated with a greater tolerance of hypophagia during reproduction. In comparison with small species, large animals have proportionately larger fat stores (scalar >1) but smaller BMRs (scalar <1) that allow them to survive on the same proportion of body fat for longer periods of time (Chapter 10). Large species also require a longer period to develop, which may allow offspring more opportunities to compensate for brief interruptions in growth (Chapter 10) (Barboza and Hume 2006).

Some animals may be unable to dampen dietary fluctuations in all nutrients, especially when demands for synthesis are high and body stores are small. Thus intermittent feeding impairs the synthesis of red colors in male house finches and three-spined sticklebacks because the skin cells require a constant supply of carotenoids, energy and amino acids for these changes in pelage and skin (Chapter 9) (Frischknecht 1993; Hill 2000). Fluctuations in food supply may also alter the response to stressors such as infections, parasites, disease and toxins (Krasnov et al. 2005; Smith et al. 2005). Lower food availability reduces immune function in deer mice even though body mass is not affected (Martin et al. 2008). Consequently, intermittent fasts may increase the requirements of some nutrients even though animals may be able to achieve the same ME intake during refeeding. Intermittent fasts may be a neutral stress that could be alleviated by altering food selection to enhance the intake of some nutrients (Fig. 11.4). Migratory songbirds rely on a diversity of fruits and insects at staging grounds to meet demands for energy, protein and trace nutrients depleted during hypophagia (Smith et al. 2007).

The timing of productive events that increase demands for energy and nutrients depends on environmental conditions that regularly provide adequate supplies. The transition from low food abundance in winter to high food abundance in spring also may be accompanied by a shift from high to low variance in weather conditions and thus environmental demands. Environmental variation in feeding cycles affects the rate at which animals can gain mass for production. High variation in feeding conditions increases the variation in body condition among animals and the time at which they attain sufficient stores to begin reproduction (Fig. 11.6). Timing of body stores may be one of many factors that affect the reproduction of migratory ducks such as lesser scaup (Anteau and Afton 2004). Changes in the availability and species of invertebrate prey affect foraging times for scaup (Poulton et al. 2002) and the body composition of birds as they migrate to their breeding grounds in the prairies and boreal forests (Vest et al. 2006). Variation in the quality and abundance of foods in breeding areas may influence reproduction of ducks by affecting the supplies of energy and nutrients for replenishing body stores (capital) as well as those nutrients that are required from the diet each day (income) for production of eggs and feathers. Stressors associated with variable foraging conditions may therefore impair reproductive development and behavior with regard to nesting and laying eggs. Black ducks that were intermittently fasted did not complete egg laying until food was available each day even though birds had sufficient body mass and lipid stores to complete the clutch of eggs when food was provided intermittently (Barboza and Jorde 2002). Diversity of foods and feeding areas may minimize variation between sites for reproducing females. Birthing windows for red deer are

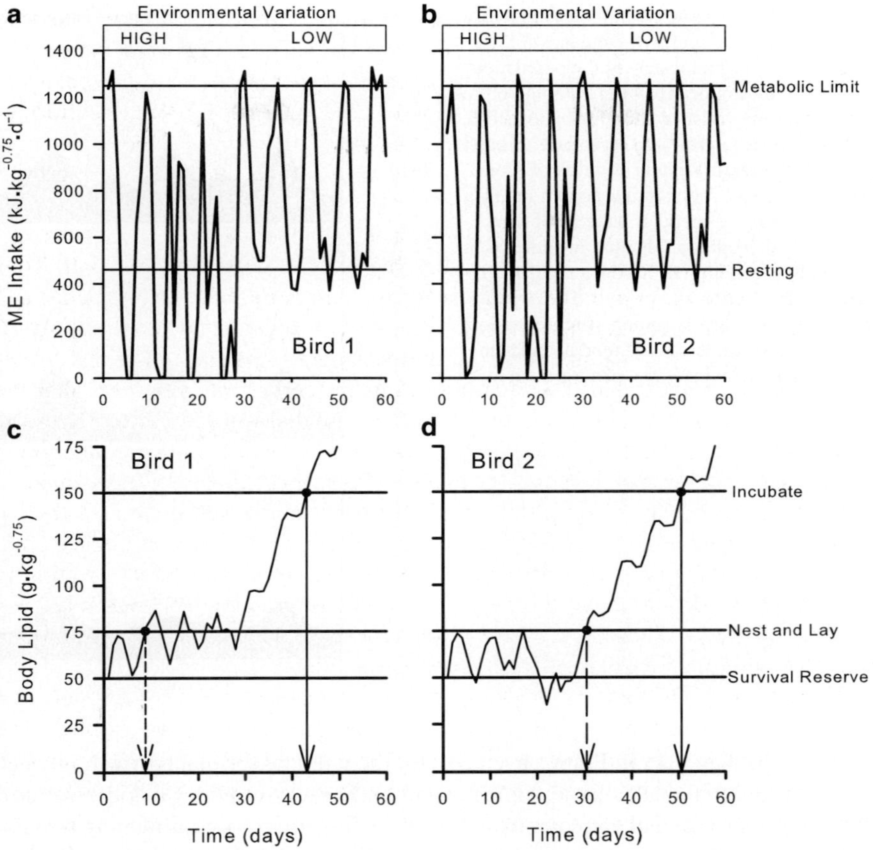

Fig. 11.6 Environmental variations that affect food intake and survival costs ultimately affect timing of productive events such as mass regain for migration, reproduction and growth in waterfowl. The model for black ducks from Fig. 11.5**b** was used to simulate high variation in late winter and lower variation in spring. Two model runs (*Bird 1* and *Bird 2*) are plotted for ME intake and the corresponding cumulative change in body lipid. Birds consistently gain body lipid when variation in food intakes decline from winter to spring. High environmental variation in winter increases the variation in condition between birds during both periods. Bird 1 experienced slightly better conditions during the period of high environmental variation. Consequently, it could begin nesting and laying earlier than Bird 2; body fat first exceeded the threshold for nesting in Bird 1 at 9 days whereas Bird 2 only initiated nesting at 31 days (*broken lines*). Bird 1 also met the incubation threshold for body fat 8 days earlier than Bird 2 (43 vs. 51days) (*solid lines*). Model parameters for high variation: $Y_0 = 568.5\,kJ \cdot kg^{-0.75} \cdot d^{-1}$; $a = 758.6\,kJ \cdot kg^{-0.75} \cdot d^{-1}$; $b = 7\,d$, with 10% variation on all parameters. Model parameters for low variation: $Y_0 = 852.7\,kJ \cdot kg^{-0.75} \cdot d^{-1}$; $a = 474.38\,kJ \cdot kg^{-0.75} \cdot d^{-1}$; $b = 7\,d$, with 2% variation on all parameters. RMR was fixed at $460.8\,kJ \cdot kg^{-0.75} \cdot d^{-1}$. An upper limit for ME intake was set at $1{,}250\,kJ \cdot kg^{-0.75} \cdot d^{-1}$, so no NE for production was available when ME intake exceeded this limit

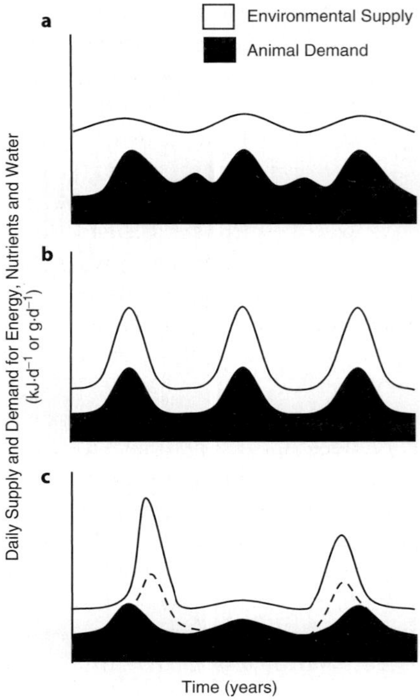

Fig. 11.7 Annual patterns of food availability for a population of animals. Demands for energy, nutrients and water for the population are highest during periods of growth and reproduction. **a** Low variation in food abundance allows animals to grow and reproduce through most of the year. **b** Regular patterns of food abundance select for a similar pattern in animal production. **c** Irregular patterns of food abundance prevent reproduction in some years, especially when animals follow a regular pattern of reproduction and growth. The *dotted line* indicates a shift in timing of reproduction to follow immediate cues of food abundance such as rainfall and primary plant growth in deserts

wider in Norway than in France even though the window for plant growth is much narrower at the higher latitude; alpine terrain in Norway provides a wider window of plant emergence and a greater diversity of feeding areas for reproducing females (Loe et al. 2005).

Timing of reproduction and growth in a population may be aseasonal when plants or prey are abundant throughout the year, as they are in some tropical habitats and in captivity (Fig. 11.7a). Pulses of births in primates from these largely aseasonal environments are mostly due to the variation among individuals in the time required to grow and establish sufficient condition to reproduce (Brockman and van Schaik 2005). Synchrony of births in herbivores is also affected by risks of predation and thus the density of predators. In African ungulates, species that form large herds (e.g., wildebeest) birth in narrow windows that overwhelm the predator response, whereas species that form small herds (e.g., impala) have a wider window of births that presumably reduces the attraction of predators (Sinclair et al. 2000).

Seasonal plant production favors some form of synchrony between environmental supply and animal demands (Fig. 11.7b). Births of African ungulates are timed in relation to the emergence of new plant growth and peak protein availability or to the subsequent peak in plant biomass (Sinclair et al. 2000). The correspondence between food production and animal reproduction is closest for income breeders. Capital breeders such as sheep are less sensitive to the timing of plant production because body stores are used for reproductive demands (Durant et al. 2005). Food abundance affects the supply of nutrients for current reproduction of income breeders as well as stores for the subsequent commitments of capital breeders. Thus high metabolic demands make populations of endotherms (e.g., Atlantic puffins) more sensitive to seasonal food abundance than those of ectotherms (e.g., Atlantic cod) (Durant et al. 2005).

The cues for timing reproduction may be immediate indicators of food availability (e.g., number of meals, delays between meals) or correlative cues such as day length, water temperature and lunar cycles. Animals that use correlative cues to anticipate seasonal food availability for reproduction may mismatch supply and demand (trophic mismatch) when they change habitat or when the habitat changes around them (Fig. 11.7c). Blue tits that breed long before the peak abundance of insect prey expend more energy to feed their offspring than birds that breed close to the peak in prey (Thomas et al. 2001). Trophic mismatches may increase as animals move further between seasonal ranges, as the seasonal window for food production narrows, and as species rely on food rather than body stores for reproduction (income vs. capital breeding). Birds that migrate long distances may be more vulnerable to trophic mismatches than resident birds because environmental cues for departing the wintering grounds may not reflect the changes in food at their destination hundreds of kilometers away. For example, photoperiodic changes for songbirds in Mexico may not reflect changes in air temperature and insect abundance at their breeding grounds in Canada. Narrow windows of plant and insect abundance at high latitudes increase the likelihood of trophic mismatches for Arctic and Antarctic animals, many of which are also long-distance migrants. Small, income breeders such as Arctic-nesting shore birds (Klaassen et al. 2001) are probably more vulnerable to trophic mismatches than large-bodied birds such as trumpeter swans or snow geese that can rely on body capital to breed in the Arctic (Ankney and MacInnes 1978).

Correlative cues such as photoperiod become less informative as food availability becomes less seasonal or more unpredictable. Consequently, birds from less predictable habitats (e.g., hot deserts) and species that breed opportunistically when adequate conditions arise use immediate indicators of food availability to time breeding (Wingfield et al. 1992; Hahn et al. 1997). Zebra finches initiate breeding when rains break a drought in their desert habitat. Commitments to reproduction are further modulated by the emergence of grass seeds and insects that are required for feeding nestlings in the weeks that follow rainfall (Hahn et al. 1997). The immediate cues of food abundance and food quality (e.g., metabolizable energy or protein content) also stimulate ovulation in domestic sheep, pigs and chickens (Torrell et al. 1972; Hocking 1987; Downing and Scaramuzzi 1991; Johnson 2000).

11.4 Global Climate Change

Climate is the description of weather patterns at large scales of space (national, continental, global) and time (decadal, millennial, epochal) that affect food supplies and the environmental demands for wildlife at multiple levels, from habitats to ecosystems. The longest scale of changes in global climate are associated with Milankovitch cycles of solar radiation that reflect changes in the orbit of the earth around the sun in periods of 23,000–100,000 years, which correspond with changes in the size of the polar ice caps and continental glaciers (Chapin et al. 2002; van Schaik and Brockman 2005). These changes in ice cover altered sea levels that opened and subsequently closed land bridges, which allowed wildlife to disperse between continents. Low sea levels allowed moose to spread from Europe to Alaska across the Bering land bridge whereas high seas isolated kangaroos in Australia and kiwis in New Zealand.

Interactions between the atmosphere, ocean, land and ice drive currents of air and water. Albatross use the strong winds of the Southern Ocean to forage across large areas where upwellings of cold water bring nutrients to the surface for their prey of invertebrates and fish (Weimerskirch et al. 2005). Atmospheric interactions vary widely between decades and result in contrasting weather across oceans and regions. The El Nino Southern Oscillation (ENSO) intensifies droughts and fires in Australia but enhances rainfall on the other side of the Pacific Ocean in Peru. However, El Nino also slows the cold, nutrient-rich Humboldt current that supports schools of anchovy and other prey for seals and seabirds in South America (Chapin et al. 2002). The North Atlantic Oscillation (NAO) causes warm winters that result in rain on the coast and snow in the mountains of Norway when the NAO index is high. Norwegian red deer use coastal habitat in winter and mountain pastures in summer where longer periods of snow melt increase the window for plant growth. A series of winters with high NAO indices allow female red deer to produce proportionately more male calves that are larger than female calves at birth and continue to grow more than young females in the subsequent summers (Post et al. 1999; Mysterud et al. 2000).

Small changes in temperature have profound effects on air and water currents that affect the distribution of rainfall and nutrients in both terrestrial and aquatic systems. The global climate has warmed by 0.6°C over the last 100 years (Root et al. 2003; Parmesan 2007) and is projected to increase in the northern hemisphere by another 1–4°C over the next 100 years (Berteaux et al. 2004). Warming has increased changes in precipitation over many areas of the globe, resulting in drier sub-tropical regions in Africa and wetter north-temperate regions of North America and Central Asia (Walther et al. 2002). Temperature also affects the metabolic rates of bacteria, plants and ectothermic animals such as invertebrates, fish, amphibians and reptiles; a rise of 10°C can double energy expenditures in these organisms and enhance their rates of growth and reproduction ($Q_{10} = 2$; Chapter 10). Consequently, global warming has advanced the seasonal window for production of plants and ectotherms in several areas (Parmesan 2006). In the northern hemisphere, the window for plant growth has advanced at an average rate of 2.8 days·decade^{-1} and is now increasing by up to 4.9 days·decade^{-1} (Parmesan 2007). Some animals such as yellow-bellied marmots are advancing their annual cycles of feeding by emerging earlier from hibernation (Inouye et al. 2000). Many birds that migrate short distances in Europe have likewise advanced their arrival and nesting dates, presumably to match the emergence of insect prey (Walther et al. 2002; Both et al. 2004).

Global warming is also shifting the distribution of many plants and animals to higher latitudes at an average rate of 6.1 km·decade^{-1}, and to higher altitudes at an average rate of 6.1 m·decade^{-1} (Parmesan and Yohe 2003). Expansion of shrubs and trees into Arctic tundra enhances the rate of snow melt and warming of the land, which favors a progression to habitats that are more suited to browsing moose than to grazing muskoxen (Chapin et al. 2006). Conversely, global warming is shrinking mountain cloud forests used by frogs and neotropical rainforests used by migrant songbirds (Norris et al. 2004; Parmesan 2006).

The effects of climate warming vary widely because animals and their habitats respond differently to changes in temperature and precipitation. Warm temperatures that reduce snowfall may improve mobility for moose but increase the exposure of voles to cold temperatures and predators by reducing the sub-nivean space (Guthrie 1990). The shift from snow to rain is not always beneficial for large mammals; freezing rain increases thermoregulatory costs of reindeer by reducing insulation of the pelage (Chapter 10) and impedes feeding by covering the tundra with ice (Tyler et al. 2007). Warming may also alter the spread of diseases in wildlife. Mild winters increase the survival of ectotherms that include both external parasites (e.g., ticks) and internal parasites (e.g., nematodes) and their intermediate hosts (e.g., snails and insects). Warm summers allow these parasites to develop more rapidly and to increase rates of infection of their endothermic hosts (reindeer, muskoxen, moose and elk) (Kutz et al. 2005). Furthermore, changes in the distribution of a host will alter the dispersal of the disease it transmits to other populations and other species (Crowl et al. 2008). The effect of climate warming on a population is therefore the outcome of several responses and interactions among species to changes in temperature and the availability of food and water.

Climatic changes that rapidly alter habitats are likely to affect many species. Warming of the Antarctic affects the entire trophic chain, from zooplankton to penguins, because reductions in sea ice are accompanied by declines in production of phytoplankton and algae that are the base of the food web (Parmesan 2006). Consequently, extinctions of populations have been correlated with climate warming, especially in communities that are restricted to a narrow range of conditions, such as the shallow coastal areas colonized by corals and also the species that live in the reef. Populations may avoid extinction if animals have enough time to change their behavior or physiology in response to habitat changes. Female red squirrels advanced the timing of births by 6 days each generation as their diet of white spruce cones increased in abundance with increasing spring temperatures and decreasing precipitation over 10 years. However, most of the variation in timing of birth was attributed to phenotypic plasticity rather than evolution by natural selection (Berteaux et al. 2004). Long-term changes in the morphology of mammals probably reflect both phenotypic and evolutionary changes. Bushy-tailed woodrats declined in body size as climate warmed over a period of 25,000 years, which indicates that either growth was limited by food and water supply or natural selection favored smaller body sizes and absolute requirements for energy and water in dry climates (Smith et al. 1995) (Chapters 1, 9 and 10). Evolutionary changes in response to recent climate change have been observed in insects because those animals usually have higher intrinsic growth rates and shorter generation times than mammals and birds (Parmesan 2006). However, phenotypic plasticity may be the principal response to recent changes in climate for large, long-lived species of wildlife. Pacific black brant use the same physiological plasticity in body lipid stores to migrate to Mexico in autumn or to overwinter in Alaska (Mason et al. 2007). The growing population of brant that overwinter in Alaska probably reflects abundant supplies of their principal food (sea grass) as well as higher winter temperatures that allow the geese to survive to spring and start reproduction with minimum delay and cost for migration (Mason et al. 2006).

11.5 Resilience and Wildlife

The concept of resilience is an approach to understanding the sustainability of human societies in ecosystems by considering multiple levels of responses to environmental change (Walker and Salt 2006). Our societies value wildlife for a wide variety of consumptive and aesthetic purposes, from wildlife watching to subsistence harvest and trophy hunting. Populations of fish and wildlife are therefore an important source of food and commerce for our society. Societies have long observed animals for indications of change because wildlife populations integrate environmental stressors. Movements of wildlife have marked our seasons of food supply for centuries; the arrival of migrant songbirds indicates the start of our season for plant cultivation and their departure indicates the times for harvest and hunting (Leopold 1949). We now look to arrivals of migrant songbirds as one indicator of a changing climate (Walther et al. 2002). Furthermore, wildlife populations are being monitored for diseases, toxins and variation in productivity because they integrate many of the stressors that challenge human societies (Crowl et al. 2008). Measures of radioactive isotopes in the food web of lichen, reindeer and wolves, for example, have been used to monitor the fallout from nuclear tests and accidents and to indicate the risks for human consumption of reindeer in northern Europe (Staaland et al. 1991). Similarly, selenium and mercury are monitored in salmon, seals and bowhead whales in order to evaluate the health of the animals and the quality of their tissues for subsistence hunters (Dehn et al. 2006).

Wildlife populations are resilient to environmental change if they can recover from the stressors that accompany the change. Stressors cause a deficit in the available energy, nutrients, space or time required to complete the life cycles of individuals in the population. A population is resilient to the change when enough individuals can complete their life cycle for the population to persist (Chapter 2). Resistance of the stressor may lead to distress, which may prompt individuals to avoid the stressor by moving to find more food and space (e.g., change in distribution). Natural selection for individuals that tolerate the stressor expands the region of neutral stress and may lead to an adaptive change in the species (e.g., changes in kidney function for water resorption). Resilient populations have multiple routes of response to avoid and tolerate stressors. Behavioral plasticity, physiological plasticity, genetic diversity and demography all contribute to the resilience of a population.

Boreal forests are disturbed by fire, logging and roads which present a variety of stressors to wildlife. Moose are often resilient to these changes because they are large enough to avoid the initial disturbance by moving out of the area (Fig. 11.8)

Fig. 11.8 Moose populations often are favored by early stages of plant succession in boreal forests. Fires provide abundant food for moose within 25 years. Land management practices that limit the cover of large trees also increase forage for moose. Consequently, moose are often seen foraging in clearings adjacent to roads. Vehicle collisions and other conflicts with moose increase as the moose population grows and continues to exploit areas that provide abundant food near human developments

(Osko et al. 2004). Large body size also allows moose to tolerate the initial deficit in food supply by using their body stores (Schwartz 1992). Moose are able to utilize a wide variety of plants throughout the year and quickly adapt to consuming new species that emerge after fires and along roads (Franzmann and Schwartz 1998). Fires improve forage quality for moose as plants colonize the area; plants consumed by moose are higher in N content at 10 years than 30 years after fire (Regelin et al. 1987). In addition, moose have relatively high rates of intrinsic growth for ungulates, which allow populations to recover quickly from a decline or disturbance. Males can begin breeding as yearlings; females can begin reproduction at 2 years of age and are able to produce two calves per year when body fat stores are high (Schwartz et al. 1982; Franzmann and Schwartz 1985; Keech et al. 2000). However, no species is resilient to all stressors or their combined effects. Persistent declines in a population of moose in north-central USA have been related to multiple stressors associated with poor nutrition, disease and climate warming (Murray et al. 2006).

The interactions between populations may result in positive, neutral or negative effects on one or more species; that is, one population can become a stressor for another population. The microbial community in the digestive tract includes all three of these effects among microbes as well as between microbes and host; the microbial community includes potentially pathogenic bacteria, protozoa that have little effect on the host, and bacteria that degrade cellulose or provide vitamins to the host (Chapter 5). Resilient balances are those that allow populations to persist by both tolerating and resisting stressors. Mice use their immune system to resist an infection by destroying a parasite or tolerate the infection through replacement of infected cells. Energy and nutrients are expended by the mouse in both responses, but the parasite is exposed to different processes that select for either greater ability to evade detection during resistance, or faster replication to persist against tolerance. Responses of tolerance and resistance may have different and unexpected outcomes as each species evolves in response to changes in the prevailing environment (Raberg et al. 2007). Gradual changes in a habitat result in the replacement of species as one population is replaced or displaced. The succession of plant communities after a fire provides food for moose in the first 25 years, but as the forest develops and matures resources become available over a longer period for flying squirrels, owls and caribou. The ecosystem may be most resilient when the largest number of species and trophic levels are supported; this provides the whole system with the greatest number of options to respond to changes (Chapter 2).

Resilience of wildlife is further influenced by numbers of individuals. As populations of wildlife grow, they can have large impacts on the structure of their habitat. Large species such as elephants can alter woodlands by breaking trees that are used by a wide variety of other animals (Owen-Smith 1988). Similarly, growing populations of moose can alter their habitat and reduce the abundance of migrant songbirds (Berger et al. 2001a). Populations of white-tailed deer attain high densities when acorns and other foods are abundant and when deaths from winter exposure and predation are low. Dense populations of animals include distressed individuals that are vulnerable to disease. White-tailed deer are hosts for a brain-worm (*Paraelaphostrongylus tenuis*) that can spread with the population and quickly infect other ungulates including moose, elk, mule deer, caribou, bighorn sheep and pronghorns (Woebeser 2006). Acorn production also affects the number and movements of white-tailed deer that carry ticks infected with *Borrelia burgdorferi*, which causes Lyme disease in humans. A large acorn production increases the risk of human Lyme disease within 2 years because it increases the production and transfer of the pathogens from ticks that live on both white-tailed deer and deer mice (Woebeser 2006).

Harvest and predation can have both positive and negative effects on populations of herbivores and thus their resilience to change. Hunting can quickly reduce a population from a high density and improve the body condition of individuals by alleviating competition for food (Swihart et al. 1998; Boertje et al. 2007). Predation may have a greater impact on animal condition than hunting because weaker individuals are preferred as prey. The average age of female elk selected by gray wolves in the Yellowstone ecosystem (13.9 years) was much greater than that selected by hunters (6.5 years) (Wright et al. 2006). Low levels of wolf predation therefore had less effect on the potential production of the elk herds because wolves removed females that were at or near the end of their reproductive lives.

Many predators select young animals because they are easier to catch than adults. Predation of neonatal and growing animals directly reduces the annual production of a population. Wolves, bears, wolverines and mountain lions prey on the calves of various ungulates including caribou, moose, bighorn sheep and mule deer. Consequently, female ungulates may select birth sites with the lowest predation risks for their young even though food abundance may be inadequate for the demands of lactation (Bleich et al. 1997; Gustine et al. 2006; Poole et al. 2007). Behavioral plasticity allows females to choose birthing sites and subsequent foraging areas that balance the risks of predation with other stressors such as impaired calf growth and exhaustion of maternal body stores. In the Yellowstone ecosystem, female moose selected birth sites closer to roads because grizzly bears avoided these disturbed areas (Berger 2007b).

High rates of predation can limit populations and even extirpate small populations. Predators with high intrinsic growth rates such as wolves are able to respond quickly to changes in the populations of their prey. Predators that use multiple prey items can have a dramatic effect on one prey population, especially when they select young animals in a small population of prey. Grizzly bears consume both prey and plants during the year (Fig. 11.9). Grizzly bear predation on moose calves during the spring can have a large effect on the moose population. However, the

Fig. 11.9 Grizzly bears in Alaska prey on the calves of moose, caribou and muskoxen as well as on adult ground squirrels and salmon. The effects of a population of bears on any one of these prey species depend on the relative availability of each prey and on the growth rates of the populations of both predator and prey. Bears have low intrinsic growth rates in comparison with other predators such as wolves; that is, bears cannot reproduce and recruit young into the population as quickly as can wolves. Populations of bears and wolves often switch among prey items. The persistence of large populations of bears and wolves can maintain populations of caribou and moose at levels below the limit of available forage. Populations of prey with low intrinsic growth rates are more vulnerable to high densities of predators. Moose are more fecund than caribou, and therefore the same number of predated calves has less of an effect on a moose population than on a caribou population. A population of moose may sustain enough bears and wolves in an area to keep a caribou population from increasing because some caribou are predated when predator numbers are high. Complete removal of bears and wolves from the area could allow caribou and moose populations to increase. However, high populations of these herbivores affect the populations of plants they consume and can eventually alter the habitats they share with other species

bear population is less affected by changes in the availability of moose calves because other sources of energy and nutrients contribute to the maintenance of female bears and the growth of their cubs. Populations of bears and wolves must be suppressed over several years to increase the production of moose in an area (National Research Council 1997) because these populations of predators are resilient and have sufficient behavioral and physiological plasticity to recover from translocations and hunting.

Enormous efforts also are required to remove introduced predators (e.g., red foxes) and rodents (e.g., brown rats) from islands where they prey on seabirds and their nests. Predator reductions often attempt to displace the predator so that human society can improve their harvest of the prey (e.g., moose) or prevent local extirpation of the prey (e.g., woodland caribou). These reductions are usually based on the assumption that the population of the native predator is resilient and is able to recover by reproduction and immigration from other areas. Small populations of predators that are isolated by barriers such as mountains or seas, however, may be vulnerable to outbreaks of disease and genetic bottlenecks. Canine distemper virus from large populations of coyotes, skunks and dogs has caused large mortalities in small populations of Ethiopian wolf, African wild dogs, African lions, Baikal seals and American black-footed ferrets (Woebeser 2006). The point at which a predator becomes vulnerable to extirpation is difficult to judge and often only recognized when efforts to suppress the predator are no longer required because the population is no longer resilient to change. Californian grizzly bears and Tasmanian tigers were both abundant populations of predators that were suppressed and failed to recover.

Extirpation of a natural predator may reduce the resilience of an ecosystem because predators reduce the adverse effects of high herbivore density such as habitat degradation and disease transmission. Alternatively, extirpation of an introduced predator or herbivore may allow an ecosystem to return to a more stable state that is more resilient to stressors. Introductions of predators such as Indian mongoose to the Hawaiian Islands and herbivores such as domestic goats to the Galapagos Islands threaten to extirpate a diverse range of native animals. Control and restoration of native predators or introduced species are extremely controversial because of the enormous amounts of public money and time that may be required to address the problem and because each sector of society may have a different value for the predator, prey or habitat in question (Lessard et al. 2005). Approaches that attempt to maximize the number of trophic levels and species for the financial and human resources available may be the most sustainable and thus most resilient to unexpected changes (Walker and Salt 2006).

Wildlife populations are part of complex systems with properties that result from interactions among species and between each species and the environment at multiple levels (Fig. 11.10). For example, conditions associated with global climate change in Australia, such as increased atmospheric CO_2 concentrations and average temperature, could lead to increases in the concentration of PSMs such as phenolics and a reduction in the concentration of N in the leaves of *Eucalyptus* spp. These reductions in leaf nutritional quality have been shown to adversely affect the growth of an invertebrate herbivore (Christmas beetle). It is possible that they may also

Fig. 11.10 Species that rely on foods in complex habitats include the giant panda and migratory shorebirds. **a** Bamboo, which comprises most of the diet of pandas, flowers in cycles of decades. These flowering events cause sudden reductions in food abundance for giant pandas, which must then use alternative patches of bamboo over large areas. **b** Intertidal marshes and mangroves are sustained by complex patterns of water flow that are affected by sediments from nearby rivers as well as ocean currents and tides. Translocation of nutrients from the ocean to freshwater and intertidal communities supports production of many invertebrates, fish, mammals and birds such as the black-winged stilt. The eggs of spawning horseshoe crabs in Chesapeake Bay in northeastern USA feed a variety of shorebirds, including red knots that migrate from the southernmost parts of South America to the Canadian Arctic

have a negative impact on the intake of energy and nutrients by folivorous marsupials such as koalas. As a result, a large proportion of forests presently suitable for folivorous marsupials could become unsuitable. The potential for substantial reductions in population densities and ranges is such that local extinctions may well be widespread. The effects of projected increases in atmospheric CO_2 concentrations over the next century will be most pronounced in forests growing on poor soils; it is these forests that dominate conservation reserves in Australia (Hume 1999).

We cannot anticipate the effects on our society of perturbations such as these to such complex communities. We can, however, look closely at wildlife populations for the cumulative effects of changes in the environment. Body condition, health and productivity of wildlife are measures of the health of the ecosystems that sustain both their populations and our society.

11.6 Conclusion

Wildlife nutrition informs management of the condition of habitats and populations because changes in supplies and demands for one population may affect the entire food web and the structure of animal and plant communities. This book began with two questions that confront wildlife biologists:

- What does a population need?
- Will that population grow or decline, and why?

The approaches used in the book emphasize the importance of the energy and nutrient requirements of a species to complete its life cycle, and therefore of a population of individuals from each class of age and sex of this species to survive. An understanding of nutritional requirements and linkages is essential to understand the likely range of responses that will sustain a population or increase its resilience to anticipated and unexpected environmental changes. Wildlife biologists are therefore tasked with answering a third question for societies:

- What is the risk of changes to a population and its ecosystem?

The answers to this question will improve our stewardship of wildlife populations and the ecosystems we share with them.

References

Adams LG (2005) Effects of maternal characteristics and climatic variation on birth masses of Alaskan caribou. Journal of Mammalogy 86:506–513

Adams LG and Dale BW (1998) Reproductive performance of female Alaskan caribou. Journal of Wildlife Management 62:1184–1195

Adams NJ, Pinshow B, and Gannes LZ (1997) Water influx and efflux in free-flying pigeons. Journal of Comparative Physiology B 167:444–450

Adams SH and Costa DP (1993) Water conservation and protein metabolism in northern elephant seal pups during the postweaning fast. Journal of Comparative Physiology B 163:367–373

Agre P, King LS, Yasui M, Guggino WB, Ottersen OP, Fujiyoshi Y, Engel A, and Nielsen S (2002) Aquaporin water channels – from atomic structure to clinical medicine. Journal of Physiology London 542:3–16

Agricultural Research Service (2007) USDA National Nutrient Database for Standard Reference, Release 20. US Department of Agriculture, Beltsville, MD

Al-kahtani MA, Zuleta C, Caviedes-Vidal E, and Garland T Jr (2004) Kidney mass and relative medullary thickness of rodents in relation to habitat, body size, and phylogeny. Physiological and Biochemical Zoology 77:346–365

Albon SD, Clutton-Brock TH, and Guinness FE (1987) Early development and population dynamics in red deer. II. Density-independent effects and cohort variation. Journal of Animal Ecology 56:69–81

Albon SD, Stien A, Irvine RJ, Langvatn R, Ropstad E, and Halvorsen O (2002) The role of parasites in the dynamics of a reindeer population. Proceedings of the Royal Society London B 269:1625–1632

Alisauskas RT (1998) Winter range expansion and relationships between landscape and morphometrics of midcontinent lesser snow geese. Auk 115:851–862

Alisauskas RT, Ankney CD, and Klaas EE (1988) Winter diets and nutrition of midcontinental lesser snow geese. Journal of Wildlife Management 52:403–414

Allaye-Chan AC (1991) Physiological and ecological determinants of nutrient partitioning in caribou and reindeer. PhD Thesis, University of Alaska, Fairbanks, AK

Allaye Chan-McLeod AC, White RG, and Holleman DF (1994) Effects of protein and energy intake, body condition, and season on nutrient partitioning and milk production in caribou and reindeer. Canadian Journal of Zoology 72:938–947

Allaye Chan-McLeod AC, White RG, and Russell DE (1999) Comparative body composition strategies of breeding and nonbreeding female caribou. Canadian Journal of Zoology 77:1901–1907

Allen LR and Hume ID (1997) The importance of green seed in the nitrogen nutrition of the zebra finch *Taeniopygia guttata*. Australian Journal of Ecology 22:412–418

Allen ME, Oftedal OT, and Baer DJ (1996) The feeding and nutrition of carnivores. In: Kleiman DG, Allen ME, Thompson KV, Lumpkin S, and Harris H (eds) Wild mammals in captivity. University of Chicago Press, Chicago, IL, pp 139–147

Allison MJ, Mayberry WR, McSweney CS, and Stahl DA (1992) *Synergistes jonesii* gen. nov., sp. nov.: a rumen bacterium that degrades toxic pyridine diols. Systematic and Applied Microbiology 15:522–529

Altmann SA (1998) Foraging for survival. Yearling baboons in Africa. University of Chicago Press, Chicago, IL

Ancel A, Petter L, and Groscolas R (1998) Changes in egg and body temperature indicate triggering of egg desertion at a body mass threshold in fasting incubating blue petrels (*Halobaena caerulea*). Journal of Comparative Physiology B 168:533–539

Ankney CD (1996) An embarassment of riches: too many geese. Journal of Wildlife Management 60:217–223

Ankney CD and MacInnes CD (1978) Nutrient reserves and reproductive performance of female lesser snow geese. Auk 95:459–471

Anteau MJ and Afton AD (2004) Nutrient reserves of lesser scaup (*Aythya affinis*) during spring migration in the Mississippi flyway: a test of the spring condition hypothesis. Auk 121:917–929

Argo CM, Smith JS, and Kay RNB (1999) Seasonal changes of metabolism and appetite in Soay rams. Animal Science 69:191–202

Arnold W, Ruf T, Reimoser S, Tartaruch F, Onderscheka K, and Schober F (2004) Nocturnal hypometabolism as an overwintering strategy of red deer (*Cervus elaphus*). American Journal of Physiology. Regulatory, Integrative and Comparative Physiology 286:R174–R181

Asch A and Roby DD (1995) Some factors affecting precision of the total body electrical conductivity technique for measuring body composition in live birds. Wison Bulletin 107:306–316

Aschoff J and Pohl H (1970a) Der ruheumsatz von vogeln als funktion der tageszeit und der korpergrosse. Journal of Ornithology 111:38–47

Aschoff J and Pohl H (1970b) Rhythmic variations in energy metabolism. Federation Proceedings 29:1541–1552

Asleson MA, Hellgren EC, and Varner LW (1996) Nitrogen requirements for antler growth and maintenance in white-tailed deer. Journal of Wildlife Management 60:744–752

Atwater AO and Bryant AP (1900) Availability and fuel values of food materials. Connecticut Agricultural Experimental Research Station Annual Report 12:73–100

Ayotte JB, Parker KL, Arocena JM, and Gillingham MP (2006) Chemical composition of lick soils: functions of soil ingestion by four ungulate species. Journal of Mammalogy 87: 878–888

Bakken GS (1980) The use of standard operative temperature in the study of the thermal energetics of birds. Physiological Zoology 53:108–119

Balcells J, Ganura JM, Perez JF, Martin-Orue SM, and Ronquillo MG (1998) Urinary excretion of purine derivatives as an index of microbial-nitrogen intake in growing rabbits. British Journal of Nutrition 79:373–380

Ballard BM, Thompson JE, Petrie MJ, Chekett M, and Hewitt DG (2004) Diet and nutrition of northern pintails wintering along the southern coast of Texas. Journal of Wildlife Management 68:371–382

Barboza PS (1993a) Digestive strategies of the wombats: feed intake, fiber digestion and digesta passage in two grazing marsupials with hindgut fermentation. Physiological Zoology 66:983–999

Barboza PS (1993b) Effects of restricted water intake on digestion, urea recycling and renal function in wombats (Marsupialia: Vombatidae) from contrasting habitats. Australian Journal of Zoology 41:527–536

Barboza PS (1995) Digesta passage and functional anatomy of the digestive tract in the desert tortoise (*Xerobates agassizii*). Journal of Comparative Physiology B 165:193–202

Barboza PS (1996) Nutrient balances and maintenance requirements for energy and nitrogen in desert tortoises (*Xerobates agassizii*) consuming forages. Comparative Biochemistry and Physiology A 112:537–545

Barboza PS and Blake JE (2001) Ceruloplasmin as an indicator of copper reserves in wild ruminants at high latitudes. Journal of Wildlife Diseases 37:324–331

Barboza PS and Bowyer RT (2000) Sexual segregation in dimorphic deer: a new gastrocentric hypothesis. Journal of Mammalogy 81:473–489.

Barboza PS and Hume ID (1992) Digestive tract morphology and digestion in the wombats (Marsupialia: Vombatidae). Journal of Comparative Physiology B 162:552–560

Barboza PS and Hume ID (2006) Physiology of intermittent feeding: integrating responses of vertebrates to nutritional deficit and excess. Physiological and Biochemical Zoology 79:250–264

Barboza PS and Jorde DG (2001) Intermittent feeding in a migratory omnivore: digestion and body composition of American black duck during autumn. Physiological and Biochemical Zoology 74:307–317

Barboza PS and Jorde DG (2002) Intermittent feeding during winter and spring affects body composition and reproduction of a migratory duck. Journal of Comparative Physiology B 172:419–434

Barboza PS and Parker KL (2006) Body protein stores and isotopic indicators of N balance in female reindeer (*Rangifer tarandus*) during winter. Physiological and Biochemical Zoology 79:628–644

Barboza PS and Parker KL (2008) Allocating protein to reproduction in arctic reindeer and caribou. Physiological and Biochemical Zoology 81:835–855

Barboza PS and Reynolds PE (2004) Monitoring nutrition of a large grazer: muskoxen on the Arctic Refuge. International Congress Series 1275C:327–333

Barboza PS and Vanselow BA (1990) Copper toxicity in captive wombats (Marsupialia: Vombatidae). Proceedings of the American Association of Zoo Veterinarians, 21—26 October, South Padre Island, Texas, pp. 204–206

Barboza PS, Allen ME, Rodden M, and Pojeta K (1994) Feed intake and digestion in the maned wolf (*Chrysocyon brachyurus*): consequences for dietary management. Zoo Biology 13:375–389

Barboza PS, Farley SD, and Robbins CT (1997) Whole-body urea cycling and protein turnover during hyperphagia and dormancy in growing bears (*Ursus americanus* and *U. arctos*). Canadian Journal of Zoology 75:2129–2136.

Barboza PS, Rombach EP, Blake JE, and Nagy JA (2003) Copper status of muskoxen: a comparison of wild and captive populations. Journal of Wildlife Diseases 39:610–619

Barboza PS, Hartbauer DW, Hauer WE, and Blake JE (2004) Polygynous mating impairs body condition and homeostasis in male reindeer (*Rangifer tarandus tarandus*). Journal of Comparative Physiology B 174:309–317

Barboza PS, Peltier TC, and Forster RJ (2006) Ruminal fermentation and fill change with season in an arctic grazer: responses to hyperphagia and hypophagia in muskoxen (*Ovibos moschatus*). Physiological and Biochemical Zoology 79:497–513

Barker S (1961a) Copper, molybdenum and inorganic sulphate levels in rottnest plants. Journal of the Royal Society of Western Australia 44:49–52

Barker S (1961b) Studies on marsupial nutrition. Australian Journal of Biological Sciences 14:646–658

Barry TN and McNabb WC (1999) The implications of condensed tannins on the nutritive value of temperate forages fed to ruminants. British Journal of Nutrition 81:263–272

Baskin L and Danell K (2003) Ecology of ungulates: a handbook of species in eastern europe and northern and central Asia. Springer, Berlin

Bateson M and Kacelnik A (1998) Risk-sensitive foraging: decision making in variable environments. In: Dukas R (ed) Cognitive ecology. The evolutionary ecology of information processing and decision making. University of Chicago Press, Chicago, IL, pp 297–342

Battley PF and Piersma T (2005) Adaptive interplay between feeding ecology and features of the digestive tract in birds. In: Starck JM and Wang T (eds) Physiological and ecological adaptations to feeding in vertebrates. Science Publishers, Enfield, NH, pp 201–228

Battley PF, Dekinga A, Dietz MW, Piersma T, Tang S, and Hulsman K (2001) Basal metabolic rate declines during long-distance migratory flight. Condor 103:838–845

Batzli GO (1986) Nutritional ecology of the California vole: effects of food quality on reproduction. Ecology 67:406–412

Bauman DE, Lock AL, Cori BA, Salter AM, and Parodi PW (2006) Milk fatty acids and human health: potential role of conjugated linoleic acid and trans fatty acids. In: Sejrsen K, Hvelpund T, and Nielsen MO (eds) Ruminant physiology. Wageningen Academic Publishers, Wageningen, pp 529–561

Bayliss P and Choquenot D (2003) The numerical response: rate of increase and food limitation in herbivores and predators. In: Sibley RM, Hone J, and Clutton-Brock TH (eds) Wildlife population growth rates. Cambridge University Press, Cambridge, pp 148–179

Belitz H-D, Grosch W, and Schieberle P (2004) Food chemistry, 3rd edn. Springer, Berlin

Bell RHV (1971) A grazing ecosystem in the Serengeti. Scientific American 225:86–93

Bellrose FC (1976) Ducks, geese and swans of North America. Stackpole Books, Harrisburg, PA

Belovsky GE, Fryxell J, and Schmitz OJ (1999) Natural selection and herbivore nutrition: optimal foraging theory and what it tells us about the structure of ecological communities. In: Jung H-JG and Fahey GC Jr (eds) Nutritional ecology of herbivores. American Society of Animal Science, Savoy, IL, pp 1–70

Bennett AF (1982) The energetics of reptilian activity. In: Gans C and Pough FH (eds) Biology of the Reptilia. Academic Press, New York, pp 155–199

Bennett AF (1994) Exercise performance of reptiles. Advances in Veterinary Science and Comparative Medicine 38A:113–138

Bennett DC, Gray DA, and Hughes MR (2003) Effect of saline intake on water flux and osmotic homeostasis in Pekin ducks (*Anas platyrhynchos*). Journal of Comparative Physiology B 173:27–36

Berger J (1992) Facilitation of reproductive synchrony by gestation adjustment in gregarious mammals: a new hypothesis. Ecology 73:323–329

Berger J (2007a) Carnivore repatriation and holarctic prey: narrowing the deficit in ecological effectiveness. Conservation Biology 21:1105–1116

Berger J (2007b) Fear, human shields and the redistribution of prey and predators in protected areas. Biology Letters 3:620–623

Berger J, Testa JW, Roffe T, and Monfort SL (1999) Conservation endocrinology: a noninvasive tool to understand relationships between carnivore colonization and ecological carrying capacity. Conservation Biology 13:980–989

Berger J, Stacey PB, Bellis L, and Johnson MP (2001a) A mammalian predator–prey imbalance: grizzly bear and wolf extinction affect neotropical migrants. Ecological Applications 11:947–960

Berger J, Swenson JE, and Persson I-L (2001b) Recolonizing carnivores and naive prey: conservation lessons from Pleistocene extinctions. Science 291:1036–1039

Berteaux D (2000) Energetic cost of heating ingested food in mammalian herbivores. Journal of Mammalogy 81:683–690

Berteaux D, Reale D, McAdam AG, and Boutin S (2004) Keeping pace with fast climate change: can arctic life count on evolution? Integrative and Comparative Biology 44:140–151

Beuchat CA and Chong CR (1998) Hyperglycemia in hummingbirds and its consequences for hemoglobin glycation. Comparative Biochemistry and Physiology A 120:409–416

Bidwell MT and Dawson RD (2005) Calcium availability limits reproductive output of tree swallows (*Tachycineta bicolor*) in a non-acidified landscape. Auk 122:246–254

Bieber C (1998) Population dynamics, sexual activity, and reproduction failure in the fat dormouse (*Myoxus glis*). Journal of Zoology London 244:223–229

Bize P, Metcalfe NB, and Roulin A (2006) Catch-up growth strategies differ between body structures: interactions between age and structure-specific growth in wild nestling alpine swifts. Functional Ecology 20:857–864

Bjorndal KA (1989) Flexibility of digestive responses in two generalist herbivores, the tortoises *Geochelone carbonaria* and *Geochelone denticulata*. Oecologia 78:317–321

Bjorndal KA (1997) Foraging ecology and nutrition of sea turtles. In: Lutz PL and Musick JA (eds) The biology of sea turtles. CRC Press, Boca Raton, Florida, pp 199–231

Bjorndal KA and Bolten AB (1990) Digestive processing in a herbivorous freshwater turtle: consequences of small-intestine fermentation. Physiological Zoology 63:1232–1247

Blanchard P, Festa-Bianchet M, Gaillard JM, and Jorgenson JT (2003) A test of long-term fecal nitrogen monitoring to evaluate nutritional status in bighorn sheep. Journal of Wildlife Management 67:477–484

Blaxter KL (1989) Energy metabolism in animals and man. Cambridge University Press, Cambridge

Bleich VC, Bowyer RT, and Wehausen JD (1997) Sexual segregation in mountain sheep: resources or predation? Wildlife Monographs 134:1–50

Blix AS (2005) Arctic animals and their adaptations to life on the edge. Tapir Academic Press, Trondheim

Blix AS, Grav HJ, Markussen KA, and White RG (1984) Modes of thermal protection in newborn muskoxen. Biological Papers of University of Alaska Special Report 4: 207

Block BA (2005) Physiological ecology in the 21st century: advancements in biologging science. Integrative and Comparative Biology 45:305–320

Blood DC and Studdert VP (1988) Bailliere's comprehensive veterinary dictionary. W.B. Saunders, London

Boertje RD, Kellie KA, Seaton CT, Keech MA, Young DD, Dale BW, Adams LG, and Aderman AR (2007) Ranking Alaska moose nutrition: signals to begin liberal antlerless harvests. Journal of Wildlife Management 71: 1494–1506

Boness DJ, Clapham PJ, and Mesnick SL (2002) Life history and reproductive strategies. In: Hoelzel R (ed) Marine mammal biology: an evolutionary approach. Blackwell Science, Oxford, pp 278–324

Bonham M, O'Connor JM, Hannigan BM, and Strain JJ (2002) The immune system as a physiological indicator of marginal copper status? British Journal of Nutrition 87:393–403

Boonstra R (2005) Equipped for life: the adaptive role of the stress axis in male mammals. Journal of Mammalogy 86:236–247

Bordel R and Haase E (2000) Influence of flight on protein catabolism, especially myofilament breakdown, in homing pigeons. Journal of Comparative Physiology B 170:51–58

Both C, Artemyev AV, Blaauw B, Cowie RJ, Dekhuijzen AJ, Eeva T, Enemar A, Gustafsson L, Ivankina EV, Jarvinen A, Metcalfe NB, Nyholm NEI, Potti J, Ravussin P-A, Sanz JJ, Silverin B, Slater FM, Sokolov LV, Torok J, Winkel W, Wright J, Zang H, and Visser ME (2004) Large-scale geographical variation confirms that climate change causes birds to lay earlier. Proceedings of the Royal Society London B 271:1657–1662

Bothma JP and Coertze RJ (2004) Motherhood increases hunting success in southern Kalahari leopards. Journal of Mammalogy 85:756–760

Boveé KC, Bush M, Dietz J, Jezyk P, and Segal S (1981) Cystinuria in the maned wolf of South America. Science 212:919–920

Bowen WD and Iverson SJ (1998) Estimation of total body water in pinnipeds using hydrogen-isotope dilution. Physiological Zoology 71:329–332

Bowen WD, Oftedal OT, and Boness DJ (1985) Birth to weaning in 4 days: remarkable growth in the hooded seal, *Cystophora cristata*. Canadian Journal of Zoology 71:2841–2846

Bowen WD, Ellis SL, Iverson SJ, and Boness DJ (2001) Maternal effects on offspring growth rate and weaning mass in harbour seals. Canadian Journal of Zoology 79:1088–1101

Bowen WD, Beck CA, Iverson SJ, Austin D, and McMillan JI (2006a) Linking predator foraging behaviour and diet with variability in continental shelf ecosystems: grey seals of eastern Canada. In: Boyd IL, Wanless S, and Campuysen CJ (eds) Top predators in marine ecosystems. Cambridge University Press, Cambridge, pp 63–81

Bowen WD, Iverson SJ, McMillan JI, and Boness DJ (2006b) Reproductive performance in grey seals: age-related improvement and senescence in a capital breeder. Journal of Animal Ecology 75:1340–1351

Bowyer RT (1991) Timing of parturition and lactation in southern mule deer. Journal of Mammalogy 72:138–145

Bowyer RT and Kie JG (2006) Effects of scale on interpreting life-history characteristics of ungulates and carnivores. Diversity and Distributions 12:244–257

Bowyer RT, Stewart KM, Kie JG, and Gasaway WC (2001) Fluctuating asymmetry in antlers of Alaskan moose: size matters. Journal of Mammalogy 82:814–824

Bowyer RT, Blundell GM, Ben-David M, Jewett SC, Dean TA, and Duffy LK (2003) Effects of the *Exxon Valdez* oil spill on river otters: injury and recovery of a sentinel species. Wildlife Monographs 153:1–53

Bradshaw FJ and Bradshaw SD (2001) Maintenance nitrogen requirement of an obligate nectarivore, the honey possum, *Tarsipes rostratus*. Journal of Comparative Physiology B 171:59–67

Bradshaw SD (1992) Ecophysiology of desert reptiles. In: Adler K (ed) Herpetology: current research on the biology of amphibians and reptiles. Society for the Study of Amphibians and Reptiles, Oxford, pp 121–139

Bradshaw SD (2003) Vertebrate ecophysiology. An introduction to its principles and applications. Cambridge University Press, Cambridge

Braun EJ (1999) Integration of renal and gastrointestinal function. Journal of Experimental Zoology 283:495–499

Breazile JE (1987) Physiologic basis and consequences of distress in animals. Journal of the American Veterinary Medical Association 10:1212

Breier BH (2000) Prenatal nutrition, fetal programming and opportunities for farm animal research. In: Cronjé PB (ed) Ruminant physiology. Digestion, metabolism and impact of nutrition on gene expression, immunology and stress. CAB International, Wallingford, pp 347–361

Bremner I and Beattie JH (1990) Metallothionein and the trace minerals. Annual Review of Nutrition 10:63–83

Breves G and Wolffram S (2006) Transport systems in the epithelia of the small and large intestines. In: Sejrsen K, Hvelpund T, and Nielsen MO (eds) Ruminant physiology. Digestion, metabolism and impact of nutrition on gene expression, immunology and stress. Wageningen Academic Publishers, Wageningen, pp 139–154

Brice AT and Grau CR (1991) Protein requirements of Costa's hummingbirds *Calypte costae*. Physiological Zoology 64:611–626

Bridge ES (2004) The effects of intense wing molt on diving in alcids and potential influences on the evolution of molt patterns. Journal of Experimental Biology 207:3003–3014

Brockman DK and van Schaik CP (2005) Seasonality and reproductive function. In: Brockman DK and van Schaik CP (eds) Seasonality in primates: studies of living and extinct human and non-human primates. Cambridge University Press, Cambridge, pp 3–20

Brody T (1999) Nutritional biochemistry, 2nd edn. Academic Press, San Diego, CA

Brooker B and Withers P (1994) Kidney structure and renal indices of dasyurid marsupials. Australian Journal of Zoology 42:163–176

Brosh A, Sneh B, and Shkolnik A (1983) Effect of severe dehydration and rapid rehydration on the activity of the rumen microbial population of black Bedouin goats. Journal of Agricultural Science Cambridge 100:413–421

Broussard DR, Dobson FS, and Murie JO (2005) The effects of capital on an income breeder: evidence from female Columbian ground squirrels. Canadian Journal of Zoology 83:546–552

Brown JS (1999) Vigilance, patch use and habitat selection: foraging under predation risk. Evolutionary Ecology Research 1:49–71

Brown RD, Hellgren EC, Abbott M, Ruthven DC, and Bingham RL (1995) Effects of dietary energy and protein restriction on nutritional indices of female white-tailed deer. Journal of Wildlife Management 59:595–609

Brugger KE (1992) Repellency of sucrose to American robins (*Turdus migratorius*). Journal of Wildlife Management 56:793–798

Bryant DM and Tatner P (1991) Intraspecies variation in avian energy expenditure: correlates and constraints. Ibis 133:236–245

Bryant JP and Reichardt PB (1992) Controls over secondary metabolite production by arctic woody plants. In: Chapin FSI, Jeffries RL, Reynolds JF, Shaver GR, Svoboda J, and Chu EW (eds) Arctic ecosystems in a changing climate. An ecophysiological perspective. Academic Press, San Diego, CA, pp 377–390

Buck CL and Barnes BM (2000) Effects of ambient temperature on metabolic rate, respiratory quotient, and torpor in an arctic hibernator. American Journal of Physiology 279:R255–R262

Buffenstein R (2008) Negligible senescence in the longest living rodent, the naked mole-rat: insights from a successfully aging species. Journal of Comparative Physiology B 178:439–445

Buffenstein R, Sergeev IN, and Pettifor JM (1993) Vitamin D hydroxylases and their regulation in a naturally vitamin D-deficient subterranean mammal, the naked mole rat (*Heterocephalus glaber*). Journal of Endocrinology 138:59–64

Bugalho MN, Milne JA, Mayes RW, and Rego FC (2005) Plant-wax alkanes as seasonal markers of red deer dietary components. Canadian Journal of Zoology 83:465–473

Bunnell FL and Gillingham MP (1985) Foraging behavior: dynamics of dining out. In: Hudson RJ and White RG (eds) Bioenergetics of wild herbivores. CRC Press, Boca Raton, FL, pp 53–80

Bureau DP, Kaushik SJ, and Cho CY (2002) Bioenergetics. In: Halver JE and Hardy RW (eds) Fish nutrition, 3rd edn. Academic Press, Amsterdam, pp 1–59

Burns JM, Clark CA, and Richmond JP (2004) The impact of lactation strategy on physiological development of juvenile marine mammals: implications for the transition to independent foraging. International Congress Series 1275:341–350

Burns JM, Williams TM, Secor SM, Owen-Smith N, Bargmann NA, and Castellini MA (2006) New insights into the physiology of natural foraging. Physiological and Biochemical Zoology 79:242–249

Butler PJ, Green JA, Boyd IL, and Speakman JR (2004) Measuring metabolic rate in the field: the pros and cons of the doubly labelled water and heart rate methods. Functional Ecology 18:168–183

Buttemer WA and Dawson TJ (1989) Body temperature, water flux and estimated energy expenditure of incubating emus (*Dromaius novaehollandiae*). Comparative Biochemistry and Physiology 94A:21–24

Calder WAI and Braun EJ (1983) Scaling of osmotic regulation in mammals and birds. American Journal of Physiology 244:R601–R606

Cameron EZ (2004) Facultative adjustment of mammalian sex ratios in support of the Trivers-Willard hypothesis: evidence for a mechanism. Proceedings of the Royal Society London B 271:1723–1728

Cameron RD and White RG (1996) Importance of summer weight gain to the reproductive success of caribou in arctic Alaska. Rangifer 9:397

Campbell JW (1994) Excretory nitrogen metabolism and gluconeogenesis during development of higher vertebrates. Israel Journal of Zoology 40:363–381

Carey C (1996) Female reproductive energetics. In: Carey C (ed) Avian energetics and nutritional ecology. Chapman and Hall, New York, pp 324–374

Carey HV (2005) Gastrointestinal responses to fasting in mammals: lessons from hibernators. In: Starck JM and Wang T (eds) Physiological and ecological adaptations to feeding in vertebrates. Science Publishers, Enfield, NH, pp 229–254

Caro T (2005) Antipredator defenses in birds and mammals. University of Chicago Press, Chicago, IL

Cash HL, Whitham CV, Behrendt CL, and Hooper LV (2006) Symbiotic bacteria direct expression of an intestinal bactericidal lectin. Science 313:1126–1130

Castellini MA, Kooyman GL, and Ponganis PJ (1992) Metabolic rates of freely diving Weddell seals: correlations with oxygen stores, swim velocity and diving duration. Journal of Experimental Biology 165:181–194

Castro G, Myers JP, and Ricklefs RE (1992) Ecology and energetics of sanderlings migrating to four latitudes. Ecology 73:833–844

Chang MH, Chediak JG, Caviedes-Vidal E, and Karasov WH (2004) L-glucose absoprtion in house sparrows (*Paser domesticus*) is nonmediated. Journal of Comparative Physiology B 174:181–188

Chapin FS III, Matson PA, and Mooney HA (2002) Principles of ecosystem ecology. Springer, New York

Chapin FS III, Sturm M, Serreze MC, McFadden JP, Key JR, Lloyd AH, McGuire AD, Rupp TS, Lynch AH, Schimel JP, Beringer J, Chapman WL, Epstein HE, Euskirchen ES, Hinzman LD, Jia G, Tape KD, Thompson CDC, Walker DA, and Welker JM (2006) Role of land-surface changes in arctic summer warming. Science 310:657–660

Charbonnel N, Chaval Y, Berthier K, Deter J, Morand S, Palme R, and Cosson J-F (2008) Stress and demographic decline: a potential effect mediated by impairment of reproduction and

immune function in cyclic vole populations. Physiological and Biochemical Zoology 81:63–73
Charnov EL (1976) Optimal foraging: the marginal value theorem. Theoretical Population Biology 9:129–136
Chase JM and Leibold MA (2003) Ecological niches. Linking classical and contemporary approaches. University of Chicago Press, Chicago, IL
Chen X, Dickman CR, and Thompson MB (2004) Selective consumption by predators of different body regions of prey: is rate of energy intake important? Journal of Zoology London 264:189-196
Cherel Y, Gilles J, Handrich Y, and Le Maho Y (1994a) Nutrient reserve dynamics and energetics during long-term fasting in the king penguin (*Aptenodytes patagonicus*). Journal of Zoology London 234:1–12
Cherel Y, Mauget R, Lacroix A, and Gilles J (1994b) Seasonal and fasting-related changes in circulating gonadal steroids and prolactin in king penguins, *Aptenodytes patagonicus*. Physiological Zoology 67:1154–1173
Chilcott ME and Hume ID (1984) Nitrogen and urea metabolism and nitrogen requirements of the common ringtail possum (*Pseudocherius peregrinus*) fed *Eucalyptus andrewsii* foliage. Australian Journal of Zoology 32:615–622
Childs-Sanford SE and Angel R (2006) Transit time and digestibility of two experimental diets in the maned wolf (*Chrysocyon brachyurus*) and domestic dog (*Canis lupus*). Zoo Biology 25:369–381
Choshniak I, Wittenberg C, Rosenfeld J, and Shkolnik A (1984) Rapid rehydration and kidney function in the black Bedouin goat. Physiological Zoology 57:573–579
Choubert G (2001) Carotenoids and pigmentation. In: Guillaume J, Kaushik S, Bergot P, and Métailler R (eds) Nutrition and feeding of fish and crustaceans. Springer–Praxis, Chichester, pp 183–196
Clauss M, Lechner-Doll M, and Streich WJ (2003) Ruminant diversification as an adaptation to the physiochemical characteristics of forage. A reevaluation of an old debate and a new hypothesis. Oikos 102:253–262
Clauss M, Streich WJ, Schwarm A, Ortmann S, and Hummel J (2007) The relationship of food intake and ingesta passage predicts feeding ecology in two different megaherbivore groups. Oikos 116:209–216
Clements KD and Raubenheimer D (2006) Feeding and nutrition. In: Evans DH and Claiborne JB (eds) The physiology of fishes, 3rd edn. CRC Press, Boca Raton, FL, pp 47–82
Clunies M, Etches RJ, Fair C, and Leeson S (1993) Blood, intestinal and skeletal calcium dynamics during egg formation. Canadian Journal of Animal Science 73:517–532
Clutton-Brock TH (1991) The evolution of parental care. Princeton University Press, Princeton, NJ
Clutton-Brock TH and Albon SD (1989) Red deer in the Highlands. BSP Professional Books, Oxford
Clutton-Brock TH and Coulson T (2003) Comparative ungulate dynamics: the devil is in the detail. In: Sibley RM, Hone J, and Clutton-Brock TH (eds) Wildlife population growth rates. Cambridge University Press, Cambridge, pp 249–268
Collins ED and Norman AW (1991) Vitamin D. In: Machlin LJ (ed) Handbook of vitamins, 2nd edn. Marcel Dekker, New York, pp 59–98
Congdon JD, Dunham AE, and Tinkle DW (1982) Energy budgets and life histories of reptiles. In: Gans C and Pough FH (eds) Biology of the Reptilia. Academic Press, New York, pp 155–199
Conover MR (2002) Resolving human–wildlife conflicts. The science of wildlife damage management. Lewis, Boca Raton, FL
Conway PL (1997) Development of intestinal microbiota. In: Mackie RI, White BA, and Isaacson RE (eds) Gastrointestinal microbiology. Chapman and Hall, New York, pp 3–38
Cook JG (2002) Nutrition and food. In: Toweill DE and Thomas JW (eds) North American elk: ecology and management. Smithsonian Institution Press, Washington, DC, pp 259–349
Cook RC, Cook JG, Murray DL, Zager P, Johnson BK, and Gratson MW (2001a) Development of predictive models of nutritional condition for Rocky Mountain elk. Journal of Wildlife Management 65:973–987

Cook RC, Cook JG, Murray DL, Zager P, Johnson BK, and Gratson MW (2001b) Nutritional condition models for elk: which are the most sensitive, accurate, and precise? Journal of Wildlife Management 65:988–997

Cook RC, Cook JG, and Mech LD (2004) Nutritional condition of northern Yellowstone elk. Journal of Mammalogy 85:714–722

Cook RC, Stephenson TR, Myers WL, Cook JG, and Shipley LA (2007) Validating predictive models of nutritional condition for mule deer. Journal of Wildlife Management 71:1934–1943

Cook WE, Williams ES, and Dubay SA (2004) Disappearance of bovine fetuses in northwestern Wyoming. Wildlife Society Bulletin 32:254–259

Cooper MH, Iverson SJ, and Heras H (2005) Dynamics of blood chylomicron fatty acids in a marine carnivore: implications for lipid metabolism and quantitative estimation of predator diets. Journal of Comparative Physiology B 175:133–145

Cork SJ, Hume ID, and Faichney GJ (1999) Role of the hindgut in nonruminant herbivores. In: Jung H-JG and Fahey GC Jr (eds) Nutritional ecology of herbivores. American Society of Animal Science, Savoy, IL, pp 210–260

Corraze G (2001) Lipid nutrition. In: Guillaume J, Kaushik S, Bergot P, and Métailler R (eds) Nutrition and feeding of fish and crustaceans. Springer–Praxis, Chichester, pp 111–130

Cox MK and Hartman KJ (2004) Nonlethal estimation of proximate composition in fish. Canadian Journal of Fisheries and Aquatic Sciences 62:269–275

Crater AR and Barboza PS (2007) The rumen in winter: cold shocks in naturally feeding muskoxen (*Ovibos moschatus*). Journal of Mammalogy 88:625–631

Crater AR, Barboza PS, and Forster RJ (2007) Regulation of fermentation during seasonal fluctuations in food intake of muskoxen. Comparative Biochemistry and Physiology A 146:233–241

Creel S (2005) Dominance, aggression, and glucocorticoid levels in social carnivores. Journal of Mammalogy 86:255–264

Creel S, Winnie J, Maxwell B, Hamlin K, and Creel M (2005) Elk alter habitat selection as an antipredator response to wolves. Ecology 86:3387–3397

Crête M, Huot J, Nault R, and Patenaude R (1993) Reproduction, growth and body composition of Rivière George caribou in captivity. Arctic 46:189–196

Criscuolo F, Bertile F, Durant JM, Raclot T, Gabrielsen GW, Massemin S, and Chastel O (2006) Body mass and clutch size may modulate prolactin and corticosterone levels in eiders. Physiological and Biochemical Zoology 79:514–521

Crowl TA, Crist TO, Parmenter RR, Belovsky G, and Lugo AE (2008) The spread of invasive species and infectious disease as drivers of ecosystem change. Frontiers in Ecology and the Environment 6:238–246

Dailey TV and Hobbs NT (1989) Travel in alpine terrain: energy expenditures for locomotion by mountain goats and bighorn sheep. Canadian Journal of Zoology 67:2368–2375

Dauphine TCJ (1975) Kidney weight fluctuations affecting the kidney fat index in caribou. Journal of Wildlife Management 39:379–386

Davenport MD, Tiefenbacher S, Lutz CK, Novak MA, and Meyer JS (2006) Analysis of endogenous cortisol concentrations in the hair of rhesus macaques. General and Comparative Endocrinology 147:255–261

Davis SA, Pech RP, and Catchpole EA (2003) Populations in variable environments: the effect of variability in a species' primary resource. In: Sibley RM, Hone J, and Clutton-Brock TH (eds) Wildlife population growth rates. Cambridge University Press, Cambridge, pp 180–197

Dawson RD and Bidwell MT (2005) Dietary calcium limits size and growth of nestling tree swallows (*Tachycineta bicolor*) in a non-acidified landscape. Journal of Avian Biology 36:127–134

Dawson TJ (1983) Monotremes and marsupials: the other mammals. Edward Arnold, London

Dawson TJ and Hulbert AJ (1970) Standard metabolism, body temperature, and surface areas of Australian marsupials. American Journal of Physiology 218:1233–1238

Dawson TJ, Johns AB, and Beal AM (1989) Digestion in the Australian wood duck (*Chenonetta jubata*): a small avian herbivore showing selective digestion of the hemicellulose component of fiber. Physiological Zoology 62:522–540

Dawson TJ, Maloney SK, and Skadhauge E (1991) The role of the kidney in electrolyte and nitrogen excretion in a large flightless bird, the emu, during different osmotic regimes, including dehydration and nesting. Journal of Comparative Physiology B 161:165–171

Dawson TJ, Whitehead PJ, McLean A, Fanning FD, and Dawson WR (2000) Digestive function in Australian magpie geese (*Anseranas semipalmata*). Australian Journal of Zoology 48:265–279

De Pedro N and Bjornsson BT (2001) Regulation of food intake by neuropeptides and hormones. In: Houlihan D, Boujard T, and Jobling M (eds) Food intake in fish. Blackwell Science, Oxford, pp 269–296

Deaton KE, Bishop CM, and Butler PJ (1997) The effect of thyroid hormones on the aerobic development of locomotor and cardiac muscles in the barnacle goose. Journal of Comparative Physiology B 167:319–327

Dehn L-A, Follmann EH, Rose C, Duffy LK, Thomas DL, Bratton GR, Taylor RJ, and O'Hara TM (2006) Stable isotope and trace element status of subsistence-hunted bowhead and beluga whales in Alaska and gray whales in Chukotka. Marine Pollution Bulletin 52:301–319

Dehority BA (2003) Rumen microbiology. Nottingham University Press, Nottingham

DelGiudice GD, Seal US, and Mech LD (1988) Response of urinary hydroxyproline to dietary protein and fasting in white-tailed deer. Journal of Wildlife Diseases 24:75–79

DelGiudice GD, Mech LD, and Seal US (1989) Physiological assessment of deer populations by analysis of urine in snow. Journal of Wildlife Management 53:284–291

DelGiudice GD, Asleson MA, Varner L, Hellgren EC, and Riggs MR (1996) Creatinine ratios in random sampled and 24-hour urines of white-tailed deer. Journal of Wildlife Management 60:381–387

DelGiudice GD, Kerry KD, Mech LD, Riggs MR, and Seal US (1998) Urinary 3-methylhistidine and progressive winter undernutrition in white-tailed deer. Canadian Journal of Zoology 76:2090–2095

DelGiudice GD, Moen RA, Singer FJ, and Riggs MR (2001) Winter nutritional restriction and simulated body condition of Yellowstone elk and bison before and after the fires of 1988. Wildlife Monographs 147:1–60

Derickson WK (1976) Lipid storage and utilization in repiles. American Zoologist 16:711–723

Diana JS (2004) Biology and ecology of fishes, 2nd edn. Cooper Publishing Group, Traverse City, MI

Dickman CR and Braithwaite RW (1992) Postmating mortality of males in the dasyurid marsupials, Dasyurus and Parantechinus. Journal of Mammalogy 73:143–147

Dierenfeld ES (1994) Vitamin E in exotics: effects, evaluation and ecology. Journal of Nutrition 124:2579S–2581S

Ditchkoff SS, Lochmiller RL, and Masters RE (2001a) Major-histocompatability -complex-associated variation in secondary sexual traits for white-tailed deer (*Odocoileus virginianus*): evidence for good-genes advertisement. Evolution 55:616–625

Ditchkoff SS, Sams MG, Lochmiller RL, and Leslie DM Jr (2001b) Utility of tumor necrosis factor-α and interleukin-6 as predictors of neonatal mortality in white-tailed deer. Journal of Mammalogy 82:239–245

Doherty PF Jr, Williams JB, and Grubb TC Jr (2001) Field metabolism and water flux of Carolina chickadees during breeding and nonbreeding seasons: a test of the 'peak-demand' and 'reallocation' hypotheses. Condor 103:370–375

Dove H (1988) Estimation of the intake of milk by lambs, from the turnover of deuterium- or tritium-labelled water. British Journal of Nutrition 60:375–387

Dove H and Cork SJ (1989) Lactation in the tammar wallaby (*Macropus eugenii*). I. Milk consumption and the algebraic description of the lactation curve. Journal of Zoology London 219:385–397

Dove H and Mayes RW (2003) Wild and domestic herbivore diet characterization. Technical manual for satellite meeting. Proceedings of the 6th International Symposium on the Nutrition of Herbivores, 19—24 October, Merida, Mexico. Universidad Autonoma de Yucatan, pp. 1–88

Downing JA and Scaramuzzi RJ (1991) Nutrient effects on ovulation rate, ovarian function and the secretion of gonadotrophic and metabolic hormones in sheep. Journal of Reproduction and Fertility 43:209–227

Drent RH and Daan S (1980) The prudent parent: energetic adjustments in avian breeding. Ardea 68:225–252

Drew KL, Osborne PG, Frerichs KU, Hu Y, Koren RE, Hallenbeck JM, and Rice ME (1999) Ascorbate and glutathione regulation in hibernating ground squirrels. Brain Research 851:1–8

Duncan CJ and Scott S (2004) The key role of nutrition in controlling human population dynamics. Nutrition Research Reviews 17:163–175

Durant JM, Hjermann DØ, Anker-Nillsen T, Beaugrand G, Mysterud A, Pettorelli N, and Stenseth NC (2005) Timing and abundance as key mechanisms affecting trophic interactions in variable environments. Ecology Letters 8:952–958

Eberhardt LL (2002) A paradigm for population analysis of long-lived vertebrates. Ecology 83:2841–2854

Eeva T and Lehikoinen E (2004) Rich calcium availability diminishes heavy metal toxicity in pied flycatcher. Functional Ecology 18:548–553

Ehleringer JR and Rundel PW (1988) Stable isotopes: history, units, and instrumentation. In: Rundel PW, Ehleringer JR, and Nagy KA (eds) Stable isotopes in ecological research. Springer, New York, pp 1–15

Ehrhardt N, Heldmaier G, and Exner C (2005) Adaptive mechanisms during food restriction in *Acomys russatus*: the use of torpor for desert survival. Journal of Comparative Physiology B 175:193–200

Eiben B, Scharla St, Fischer K, and Schmidt-Gayk H (1984) Seasonal variations of serum 1,25-dihydroxyvitamin D_3 and alkaline phosphatase in relation to the antler formation in the fallow deer (*Dama dama* L). Acta Endocrinologica 107:141–144

Elix JA (1996) Biochemistry and secondary metabolites. In: Nash THI (ed) Lichen biology. Cambridge University Press, Cambridge, pp 154–180

Ellis WAH, Melzer A, Green B, Newgrain K, Hindell MA, and Carrick FN (1995) Seasonal variation in water flux, field metabolic rate and food consumption of free-ranging koalas (*Phascolarctos cinereus*). Australian Journal of Zoology 43:59–68

Fahey GCJ and Jung HG (1983) Lignin as a marker in digestion studies: a review. Journal of Animal Science 57:220–225

Faichney GJ (1975) The use of markers to partition digestion within the gastro-intestinal tract of ruminants. In: McDonald IW and Warner ACI (eds) Digestion and metabolism in the ruminant. University of New England Press, Armidale, NSW, pp 277–291

Fancy SG and White RG (1985a) Energy expenditures by caribou while cratering in snow. Journal of Wildlife Management 49:987–993

Fancy SG and White RG (1985b) Incremental cost of activity. In: Hudson RJ and White RG (eds) Bioenergetics of wild herbivores. CRC Press, Boca Raton, FL, pp 143–160

Fancy SG, Blanchard JM, Holleman DF, Kokjer KJ, and White RG (1986) Validation of doubly labeled water method using a ruminant. American Journal of Physiology 251:R143–R149

Fantle MS, Dittel AI, Schwalm SM, Epifanio CE, and Fogel ML (1999) A food web analysis of the juvenile blue crab, *Callinectes sapidus*, using stable isotopes in whole animals and individual amino acids. Oecologia 120:416–426

Farley SD and Robbins CT (1994) Development of two methods to estimate body composition of bears. Canadian Journal of Zoology 72:220–226

Farley SD and Robbins CT (1995) Lactation, hibernation, and mass dynamics of American black bears and grizzly bears. Canadian Journal of Zoology 73:2216–2222

Farmer CJ, Person DK, and Bowyer RT (2006) Risk factors and mortality of black-tailed deer in a managed forest landscape. Journal of Wildlife Management 70:1403–1415

Farmer CJ, Uriona TJ, Olsen DB, Stenbik M, and Sanders K (2008) The right-to-left shunt of crocodilians serves digestion. Physiological and Biochemical Zoology 81:125–137

Felicetti LA, Robbins CT, and Shipley LA (2003a) Dietary protein content alters energy expenditure and composition of the mass gain in grizzly bears (*Ursus arctos horribilis*). Physiological and Biochemical Zoology 76:256–261

Felicetti LA, Schwartz CC, Rye RO, Haroldson MA, Gunther KA, Philllips DL, and Robbins CT (2003b) Use of sulfur and nitrogen stable isotopes to determine the importance of whitebark pine nuts to Yellowstone grizzly. Canadian Journal of Zoology 81:763–770

Feret M, Bety J, Gauthier G, Biroux J-F, and Picard G (2005) Are abdominal profiles useful to assess body condition of spring staging greater snow geese? Condor 107:694–702

Ferguson GW, Gehrmann WH, Karsten KB, Landwer AJ, Carman EN, Chen TC, and Holick MF (2005) Ultraviolet exposure and vitamin D synthesis in a sun-dwelling and a shade-dwelling species of *Anolis*: are there adaptations for lower ultraviolet B and dietary vitamin D_3 availability in the shade? Physiological and Biochemical Zoology 78:193–200

Festa-Bianchet M and Côté SD (2008) Mountain goats: ecology, behavior, and conservation of an alpine ungulate. Island Press, Washington, DC

Florant GL, Porst H, Peiffer A, Hudachek SF, Pittman C, Summers SA, Rajala MW, and Scherer PE (2004) Fat-cell mass, serum leptin and adiponectin changes during weight gain and loss in yellow-bellied marmots (*Marmota flaviventris*). Journal of Comparative Physiology B 174:633–639

Floyd T, Nelson RA, and Wynne GF (1990) Calcium and bone metabolic homeostasis in active and denning black bears (*Ursus americanus*). Clinical Orthopaedics and Related Research 225:301-309

Flueck WT (1994) Effect of trace elements on population dynamics: selenium deficiency in free-ranging black-tailed deer. Ecology 75:807–812

Flynn A, Franzmann AW, Arneson PD, and Oldemeyer JL (1977) Indications of copper deficiency in a subpopulation of Alaskan moose. Journal of Nutrition 107:1182–1189

Foley WJ, Iason GR, and McArthur C (1999) Role of secondary metabolites in the nutritional ecology of mammalian herbivores: how far have we come in 25 years? In: Jung H-JG and Fahey GC Jr (eds) Nutritional ecology of herbivores. American Society of Animal Science, Savoy, IL, pp 130–209

Foose TJ (1982) Trophic strategies of ruminant versus nonruminant ungulates. PhD Thesis, University of Chicago, Chicago, IL

Forbes GB (1999) Body composition: overview. Journal of Nutrition 129:270S–272S

Forbes JM (2007) Voluntary food intake and diet selection of farm animals, 2nd edn. CABI, Wallingford, Oxon

Forsberg CW, Forano E, and Chesson A (2000) Microbial adherence to the plant cell wall and enzymatic hydrolysis. In: Cronje PB (ed) Ruminant physiology. Digestion, metabolism, growth and reproduction. CABI, New York, pp 79–97

Forsyth DM, Duncan RP, Tustin KG, and Gaillard JM (2005) A substantial energetic cost to male reproduction in a sexually dimorphic ungulate. Ecology 86:2154–2163

Fortin D, Morales JM, and Boyce MS (2005) Elk winter foraging at fine scale in Yellowstone National Park. Oecologia 145:335–343

Fowler ME (1986) Poisoning in wild animals. In: Fowler ME (ed) Zoo and wild animal medicine, 2nd edn. W.B. Saunders, Denver, CO, pp 91–96

Frank A, Galgan V, and Petersson LR (1994) Secondary copper deficiency, chromium deficiency and trace element imbalance in the moose (*Alces alces* L): effect of anthropogenic activity. Ambio 23:315–317

Frank A, Sell DR, Danielsson R, Fogarty JF, and Monnier VM (2000) A syndrome of molybdenosis, copper deficiency, and type 2 diabetes in the moose population of south-west Sweden. Science of the Total Environment 249: 123–131

Frank CL, Hood WR, and Donnelly MC (2004) The role of linolenic acid (18:3) in mammalian torpor. In: Barnes BM and Carey HV (eds) Life in the cold: evolution, mechanisms, adaptation, and application. Proceedings of the 12th International Hibernation Symposium, University of Alaska Fairbanks, Fairbanks, pp 71–80

Franson JC, Hoffman DJ, and Schmutz JA (2002) Blood selenium concentrations and enzyme activities related to glutathione metabolism in wild emperor geese. Environmental Toxicology and Chemistry 21:2179–2184

Franzmann AW and Schwartz CC (1985) Moose twinning rates: a possible population condition assessment. Journal of Wildlife Management 49:394–396

Franzmann AW and Schwartz, CC (1998) Ecology and management of the North American moose. Smithsonian Institution Press, Washington, DC

Frappell PB, Baudinette RV, MacFarlane PR, and Wiggins PR (2002) Ventilation and metabolism in a large semifossorial marsupial: the effect of graded hypoxia and hypercapnia. Physiological and Biochemical Zoology 45:77–82

Freudenberger DO (1992) Gut capacity, functional allocation of gut volume and size distributions of digesta particles in two macropodid marsupials (*Macropus robustus erubescens* and *M. r. robustus*) and the feral goat (*Capra hircus*). Australian Journal of Zoology 40:551–561

Freudenberger DO and Hume ID (1993) Effects of water restriction on digestive function in two macropodid marsupials from divergent habitats and the feral goat. Journal of Comparative Physiology B 163:247–257

Frischknecht M (1993) The breeding coloration of male three-spined sticklebacks (*Gasterosteus aculeatus*) as an indicator of energy investment in vigour. Evolutionary Ecology 7:439–450

Fryxell JM and Sinclair ARE (2000) A dynamic view of population regulation. In: Demarais S and Krausman PR (eds) Ecology and management of large mammals in North America. Prentice Hall, Upper Saddle River, NJ, pp 156–174

Fuglesteg BN, Haga OE, Folkow LP, Fuglei E, and Blix AS (2006) Seasonal variations in basal metabolic rate, lower critcial temperature and responses to temporary starvation in the arctic fox (*Alopex lagopus*) from Svalbard. Polar Biology 29:308–319

Furnival EP, Corbett JL, and Inskip MW (1990a) Evaluation of controlled release devices for administration of chromium sesquioxide using fistulated grazing sheep. I. Variation in marker concentration in faeces. Australian Journal of Agricultural Research 41:969–975

Furnival EP, Ellis KJ, and Pickering FS (1990b) Evaluation of controlled release devices for administration of chromium sesquioxide using fistulated grazing sheep. II. Variation in rate of release from the device. Australian Journal of Agricultural Research 41:977–986

Gabel G and Aschenbach JR (2000) Ruminal SCFA absorption: channeling acids without harm. In: Cronje PB (ed) Ruminant physiology. Digestion, metabolism, growth and reproduction. CABI, New York, pp 173–195

Gaillard J-M, Loison A, Toigo C, Delorme D, and Van Laere G (2003) Cohort effects and deer population dynamics. Ecoscience 10:412–420

Gannes LZ (2001) Comparative fuel use of migrating passerines: effects of fat stores, migration distance, and diet. Auk 118:665–677

Garel M, Solberg EJ, Saether B-E, Herfindal I, and Hogda K-A (2006) The length of growing season and adult sex ratio affect sexual size dimorphism in moose. Ecology 87:745–758

Garmo TH (1986) Chemical composition and in vivo digestibility of lichens. Rangifer 6:8–13

Garrott RA, White PJ, Vagnoni DB, and Heisey DM (1996) Purine derivatives in snow-urine as a dietary index for free-ranging elk. Journal of Wildlife Management 60:735–743

Garrott RA, Eberhardt LL, Otton JK, White PJ, and Chaffee MA (2002) A geothermal trophic cascade in Yellowstone's geothermal environments. Ecosystems 5:659–666

Garroway CJ and Broders HG (2005) The quantitative effects of population density and winter weather on the body condition of white-tailed deer (*Odocoileus virginianus*) in Nova Scotia, Canada. Canadian Journal of Zoology 83:1246–1256

Gaskins HR (1997) Immunological aspects of host/microbiota interactions at the intestinal epithelium. In: Mackie RI, White BA, and Isaacson RE (eds) Gastrointestinal microbiology. Chapman and Hall, New York, pp 537–587

Gass CL, Romich MT, and Suarez RK (1999) Energetics of hummingbird foraging at low ambient temperature. Canadian Journal of Zoology 77:314–320

Gaughwin MD, Judson GJ, Macfarlane WV, and Siebert BD (1984) Effect of drought on the health of wild hairy-nosed wombats, *Lasiorhinus latifrons*. Australian Wildlife Research 11:455–463

Gaunt AS, Hikida RS, Jehl JR Jr, and Fenbert L (1990) Rapid atrophy and hypertrophy of an avian flight muscle. Auk 107:649–659

Gedir JV and Hudson RJ (2000a) Seasonal foraging behavioural compensation in reproductive wapiti hinds (*Cervus elaphus canadensis*). Applied Animal Behaviour Science 67:137–150

Gedir JV and Hudson RJ (2000b) Seasonal intake determination in reproductive wapiti hinds (*Cervus elaphus canadensis*) using n-alkane markers. Canadian Journal of Animal Science 80:137–144

Geiser F and Pavey CR (2007) Use of basking and torpor in a rock-dwelling desert marsupial: survival strategies in a resource-poor environment. Journal of Comparative Physiology B 177:885–892

Geiser F, McAllan BM, and Kenagy GJ (1994) The degree of dietary fatty acid unsaturation affects torpor patterns and lipid composition of a hibernator. Journal of Comparative Physiology B 164:299–305

Gerhart KL, White RG, Cameron RD, and Russell DE (1996) Estimating fat content of caribou from body condition scores. Journal of Wildlife Management 60:713–718

Gill RE Jr, Piersma T, Hufford G, Servranckx R, and Riegen A (2005) Crossing the ultimate ecological barrier: evidence for an 11 000-km-long nonstop flight from Alaska to New Zealand and eastern Australia by bar-tailed godwits. Condor 107:1–20

Gillingham MP, Parker KL, and Hanley TA (1997) Forage intake by large herbivores in a natural environment: bout dynamics. Canadian Journal of Zoology 75:1118–1128

Gogan PJP, Jessup DA, and Akeson M (1989) Copper deficiency in tule elk at Point Reyes, California. Journal of Range Management 42:233–238

Grand JB, Franson JC, Flint PL, and Petersen MR (2002) Concentrations of trace elements in eggs and blood of spectacled and common eiders of the Yukon-Kuskokwim delta, Alaska, USA. Environmental Toxicology and Chemistry 21:1673–1678

Grand TI (1997) How muscle mass is part of the fabric of behavioral ecology in East African bovids (*Madoqua, Gazella, Damaliscus, Hippotragus*). Anatomy and Embryology 195: 375–386

Grand TI and Barboza PS (2001) Anatomy and development of the koala, *Phascolarctos cinereus*: an evolutionary perspective on the superfamily Vombatoidea. Anatomy and Embryology 203:211–223

Grant PR (1999) Ecology and evolution of Darwin's finches. Princeton University Press, Princeton, NJ

Grasman BT and Hellgren EC (1993) Phosphorus nutrition in white-tailed deer: nutrient balance, physiological responses, and antler growth. Ecology 74:2279–2296

Graveland J (1996) Avian eggshell formation in calcium-rich and calcium-poor habitats: importance of snail shells and anthropogenic calcium sources. Canadian Journal of Zoology 74:1035–1044

Green B and Merchant JC (1988) The composition of marsupial milk. In: Tyndale-Biscoe CH and Janssens PA (eds) The developing marsupial. Springer, Berlin, pp 41–54

Groff JL, Gropper SS, and Hunt SM (1995) Advanced nutrition and human metabolism, 2nd edn. West Publishing Company, St. Paul, MN

Gugliemo CG, Cersale DJ, and Eldermire C (2005) A field validation of plasma metabolite profiling to assess refueling performance of migratory birds. Physiological and Biochemical Zoology 78:116–125

Guillaume J and Choubert G (2001) Digestive physiology and nutrient digestibility in fishes. In: Guillaume J, Kaushik S, Bergot P, and Métailler R (eds) Nutrition and feeding of fish and crustaceans. Springer–Praxis, Chichester, pp 27–58

Guinet C, Servera N, Mangin S, Georges J-Y, and Lacroix A (2004) Change in plasma cortisol and metabolites during the attendance period ashore in fasting lactating subantarctic fur seals. Comparative Biochemistry and Physiology A 137:523–531

Guppy M (1986) The hibernating bear: why is it so hot, and why does it cycle urea through the gut? Trends in Biochemical Science 11:274–276

Gustine DD, Parker KL, Lay RA, Gillingham MP, and Heard DC (2006) Calf survival of woodland caribou in a multi-predator ecosystem. Wildlife Monographs 165:1–32

Gustine DD, Parker KL, and Heard DC (2007) Using ultrasound measurements of rump fat to assess nutritional condition of woodland caribou in northern British Columbia, Canada. Rangifer 17:249–256

Guthery FS, Rybak AR, Fuhlendorf SD, Hiller TL, Smith SG, Puckett WH Jr, and Baker RA (2005) Aspects of the thermal ecology of bobwhites in north Texas. Wildlife Monographs 159:1–36

Guthrie RD (1990) Frozen fauna of the mammoth steppe. The story of Blue Babe. University of Chicago Press, Chicago, IL

Gutowska MA, Drazen JC, and Robison BH (2004) Digestive chitinolytic activity in marine fishes of Monterey Bay, California. Comparative Biochemistry and Physiology A 139:351–358

Hagerman AE and Robbins CT (1993) Specificity of tannin-binding salivary proteins relative to diet selection by mammals. Canadian Journal of Zoology 71:628–633

Hahn TP, Boswell T, Wingfield JC, and Ball GF (1997) Temporal flexibility in avian reproduction. Patterns and mechanisms. Current Ornithology 14:39–79

Halver JE (2002) The vitamins. In: Halver JE and Hardy RW (eds) Fish nutrition, 3rd edn. Academic Press, Amsterdam, pp 61–141

Hammond KA (1993) Seasonal changes in gut size of the wild prairie vole (*Microtus ochrogaster*). Canadian Journal of Zoology 71:820–827

Hammond KA (1998) The match between load and capacity during lactation: where is the limit to energy expenditure? In: Weibel ER, Taylor CR, and Bolis L (eds) Principles of animal design: the optimization and symmorphosis debate. Cambridge University Press, Cambridge, pp 205–211

Hammond KA and Wunder BA (1991) The role of diet quality and energy need in the nutritional ecology of a small herbivore, *Microtus ochrogaster*. Physiological Zoology 64:541–567

Hammond KA and Wunder BA (1995) Effect of cold temperatures on the morphology of gastrointestinal tracts of two microtine rodents. Journal of Mammalogy 76:232–239

Hanley TA, Robbins CT, Hagerman AE, and McArthur C (1992) Predicting digestible protein and digestible dry matter in tannin-containing forages consumed by ruminants. Ecology 73:537–541

Hansson L (1991) Bark consumption by voles in relation to mineral contents. Journal of Chemical Ecology 17:735–743

Harborne JB (1993) Introdction to ecological biochemistry. Academic Press, San Diego, CA

Harlow H (1981) Effect of fasting on rate of food passage and assimilation efficiency in badgers. Journal of Mammalogy 62:173–177

Harlow HJ and Frank CL (2001) The role of dietary fatty acids in the evolution of spontaneous and facultative hibernation in prairie dogs. Journal of Comparative Physiology B 171:77–84

Harlow HJ, Lohuis T, Beck TDI, and Iaizzo PA (2001) Muscle strength in overwintering bears. Nature 409:997

Hayes JP and Shonkwiller JS (2001) Morphometric indicators of body condition: worthwhile or wishful thinking? In: Speakman JR (ed) Body composition analysis of animals. A handbook of non-destructive methods. Cambridge University Press, Cambridge, pp 8–38

Hayssen V and Lacy RC (1985) Basal metabolic rates in mammals: taxonomic differences in the allometry of BMR and body mass. Comparative Biochemistry and Physiology A 81:741–754

Heitmeyer ME (1987) The prebasic moult and basic plumage of female mallards (*Anas platyrhynchos*). Canadian Journal of Zoology 65:2248–2261

Hellgren EC and Pitts WJ (1997) Sodium economy in white-tailed deer (*Odocoileus virginianus*). Physiological Zoology 70:547–555

Hendriks WH, Moughan PJ, and Tarttelin MF (1997) Urinary excretion of endogenous nitrogen metabolites in adult domestic cats using a protein-free diet and the regression technique. Journal of Nutrition 127:623–629

Henen BT, Peterson CC, Wallis IR, Berry KH, and Nagy KA (1998) Effects of climatic variation on field metabolism and water relations of desert tortoises. Oecologia 117:365–373

Henke SE and Demarais S (1990) Effect of diet on condition indices in black-tailed jackrabbits. Journal of Wildlife Diseases 26:28–33

Herd RM and Dawson TJ (1984) Fiber digestion in the emu, *Dromaius novaehollandiae*, a large bird with a simple gut and high rates of passage. Physiological Zoology 57:70–84

Hershey JD, Robbins CT, Nelson OL, and Lin DC (2008) Minimal seasonal alterations in the skeletal muscle of captive brown bears. Physiological and Biochemical Zoology 18:138–147

Hespell RB, Aiken DE, and Dehority BA (1997) Bacteria, protozoa and fungi of the rumen. In: Mackie RI, White BA, and Isaacson RE (eds) Gastrointestinal microbiology. Chapman and Hall, New York, pp 59–141

Hill GE (2000) Energetic constraints on expression of carotenoid-based plumage coloration. Journal of Avian Biology 31:559–566

Hill GE (2006) Environmental regulation of ornamental coloration. In: Hill GE and McGraw KJ (eds) Bird coloration. Harvard University Press, Cambridge, MA, pp 507–560

Hill GE, Doucet SM, and Buchholz R (2005) The effect of coccidial infection on irridescent plumage coloration in wild turkeys. Animal Behaviour 69:387–394

Hillman JR, Davis RW, and Abdelbaki YZ (1973) Cyclic bone remodeling in deer. Calcified Tissue Research 12:323–330

Hobbs NT (2003) Challenges and opportunities in integrating ecological knowledge across scales. Forest Ecology and Management 181:223–238

Hochachka PW and Somero GN (2002) Biochemical adaptation: mechanism and process in physiological evolution. Oxford University Press, New York

Hock RJ (1960) Seasonal variations in physiologic functions in arctic ground squirrels and black bears. In: Lyman CP and Dawe AR (eds) Mammalian hibernation. Proceedings of the 1st International Symposium on Natural Mammalian Hibernation (Bulletin of the Museum of Comparative Zoology, v. 124). The Museum at Harvard College, Cambridge, Massachusetts, pp 155–171

Hocking PM (1987) Nutritional interactions with reproduction in birds. Proceedings of the Nutrition Society 46:217–225

Hoddle MS (2004) Biological control in support of conservation: friend or foe? In: Gordon M and Bartol SM (eds) Experimental approaches to conservation biology. University of California Press, Berkeley, CA, pp 202–237

Hofmann RR (1989) Evolutionary steps of ecophysiological adaptation and diversification of ruminants: a comparative view of their digestive system. Oecologia 78:443–457

Hogan JP, Petherick JC, and Phillips CJC (2007) The physiological and metabolic impacts on sheep and cattle of feed and water deprivation before and during transport. Nutrition Research Reviews 20:17–28

Holick MF (1989) Phylogenetic and evolutionary aspects of vitamin D from phytoplankton to humans. In: Pang PKT and Schreibman MP (eds) Vertebrate endocrinology: fundamentals and implications. Academic Press, San Diego, pp 7–43

Holleman DF and White RG (1989) Determination of digesta fill and passage rate from nonabsorbed particulate phase markers using the single dosing method. Canadian Journal of Zoology 67:488–494

Holleman DF, Luick JR, and White RG (1979) Lichen intake estimates for reindeer and caribou during winter. Journal of Wildlife Management 43:192–201

Holmgren A and Holmberg A (2005) Control of gut motility and secretion in fasting and fed nonmammalian vertebrates. In: Starck JM and Wang T (eds) Physiological and ecological adaptations to feeding in vertebrates. Science Publishers, Enfield, NH, pp 229–254

Horn MH and Messer KS (1992) Fish guts as chemical reactors: a model of the alimentary canals of marine herbivorous fishes. Marine Biology 113:527–535

Houseknecht KL and Spurlock ME (2003) Leptin regulation of lipid homeostasis: dietary and metabolic implications. Nutrition Research Reviews 16:83–96

Houser DS, Champagne CD, and Crocker DE (2007) Lipolysis and glycerol gluconeogenesis in simultaneously fasting and lactating northern elephant seals. American Journal of Physiology. Regulatory, Integrative and Comparative Physiology 293:R1129–R1137

Houston AI and McNamara JM (1999) Models of adaptive behaviour. An approach based on state. Cambridge University Press, Cambridge

How O-J and Nordoy ES (2007) Seawater drinking restores water balance in dehydrated harp seals. Journal of Comparative Physiology B 177:535–542

Hudson PJ, Dobson AP, Cattadori IM, Newborn D, Haydon DT, Shaw DJ, Benton TG, and Grenfell BT (2003) Trophic interactions and population growth rates: describing patterns and identifying mechanisms. In: Sibley RM, Hone J, and Clutton-Brock TH (eds) Wildlife population growth rates. Cambridge University Press, Cambridge, pp 198–224

Hudson RJ and Christopherson RJ (1985) Maintenance metabolism. In: Hudson RJ and White RG (eds) Bioenergetics of wild herbivores. CRC Press, Boca Raton, FL, pp 143–160

Hudson RJ and Jeon B-T (2003) Are nutritional adaptations of wild deer relevant to commercial venison production? Ecoscience 10:462–471

Hughes MR, Bennett DC, Sullivan TM, and Hwang H (1999) Retrograde movement of urine into the gut of salt water acclimated mallards (*Anas platyrhynchos*). Canadian Journal of Zoology 77:342–346

Huhtanen P, Ahenjärvi S, Weisberg MR, and Nřrgaard P (2006) Digestion and passage of fibre in ruminants. In: Sejrsen K, Hvelpund T, and Nielsen MO (eds) Ruminant physiology. Wageningen Academic Publishers, Wageningen, pp 87–135

Hulbert AJ and Else PL (2000) Mechanisms underlying the cost of living in mammals. Annual Review of Physiology 62:207–235

Hulbert AJ, Pamplona R, Buffenstein R, and Buttemer WA (2007) Life and death: metabolic rate, membrane composition, and life span of animals. Physiological Reviews 87:1175–1213

Hulbert IAR, Iason GR, and Mayes RW (2001) The flexibility of an intermediate feeder: dietary selection by mountain hares measured using faecal n-alkanes. Oecologia 129:197–205

Hume ID (1987) Native and introduced herbivores in Australia. In: Hacker JB and Ternouth JH (eds) The nutrition of herbivores. Academic Press, Sydney, pp 1–22

Hume ID (1989) Optimal digestive strategies in mammalian herbivores. Physiological Zoology 62:1145–1163

Hume ID (1997) Fermentation in the hindgut of mammals. In: Mackie RI and White BA (eds) Gastrointestinal microbiology. Chapman and Hall, New York, pp 84–115

Hume ID (1999) Marsupial nutrition. Cambridge University Press, Cambridge

Hume ID (2005) Concepts of digestive efficiency. In: Starck JM and Wang T (eds) Physiological and ecological adaptations to feeding in vertebrates. Science Publishers, Enfield, NH, pp 43–58

Hume ID (2006) Nutrition and digestion. In: Armati PJ, Dickman CR, and Hume ID (eds) Cambridge University Press, Cambridge, pp 137–158

Hume ID and Biebach H (1996) Digestive tract function in the long-distance migratory garden warbler, *Sylvia borin*. Journal of Comparative Physiology B 166:388–395

Hume ID, Carlisle CH, Reynolds K, and Pass MA (1988) Effects of fasting and sedation on gastrointestinal function in two potoroine marsupials. Australian Journal of Zoology 36:411–420

Hume ID, Karasov WH, and Darken BW (1993) Acetate, butyrate and proline uptake in the caecum and colon of prairie voles (*Microtus orchogaster*). Journal of Experimental Biology 176:285–297

Hume ID, Beiglböck C, Ruf T, Frey-Roos F, Bruns U, and Arnold W (2002) Seasonal changes in morphology and function of the gastrointestinal tract of free-living alpine marmots (*Marmota marmota*). Journal of Comparative Physiology B 172:197–207

Hume ID, Gibson LA, and Lapidge SJ (2004) Conservation of Australian arid-zone marsupials. In: Gordon MS and Bartol SM (eds) Experimental approaches to conservation biology. University of California Press, Berkeley, CA, pp 122–131

Humphreys WF (1979) Production and respiration in animal populations. Journal of Animal Ecology 48:427–453

Hundertmark KJ, Shields GF, Bowyer RT, and Schwartz CC (2002) Genetic relationships deduced from cytochrome-b sequences among moose. Alces 38:113–122

Huot J, Poulle M-L, and Crete M (1995) Evaluation of several indices for assessment of coyote (*Canis latrans*) body composition. Canadian Journal of Zoology 73:1620–1624

Hupp JW, White RG, Sedinger JS, and Robertson DG (1996) Forage digestibility and intake by lesser snow geese: effects of dominance and resource heterogeneity. Oecologia 108:232–240

Hyvarinen H and Nygren T (1993) Accumulation of copper in the liver of moose in Finland. Journal of Wildlife Management 57:469–474

Iason GR, Sim DA, and Gordon IJ (2000) Do endogenous seasonal cycles of food intake influence foraging behaviour and intake by grazing sheep? Functional Ecology 14:614–622

Illius AW and Gordon IJ (1999) The physiological ecology of mammalian herbivory. In: Jung H-JG and Fahey GC Jr (eds) Nutritional ecology of herbivores. American Society of Animal Science, Savoy, IL, pp 71–96

Inouye DW, Barr B, and Armitage KB (2000) Climate change is affecting altitudinal migrants and hibernating species. Proceedings National Academy of Sciences USA 97:1630–1633

Irvine RJ, Stien A, Dallas JF, Halvorsen O, Langvatn R, and Albon SD (2001) Contrasting regulation of fecundity in two abomasal nematodes of Svalbard reindeer (*Rangifer tarandus platyrhynchus*). Parasitology 122:673–681

Irving L (1972) Arctic life of birds and mammals including man. Springer, Berlin

Iverson SJ, McDonald JE, and Smith LH (2001) Changes in diet of free-ranging black bears in years of contrasting food availability revealed through milk fatty acids. Canadian Journal of Zoology 79:2268–2279

Iverson SJ, Field C, Bowen WD, and Blanchard W (2004) Quantitative fatty acid signature analysis: a new method of estimating predator diets. Ecological Monographs 74:211–235

Jackson S, Place AR, and Seiderer LJ (1992) Chitin digestion and assimilation by seabirds. Auk 109:758–770

Jeffries RL (1999) Herbivores, nutrients and trophic cascades in terrestrial environments. In: Olff H, Brown VK, and Drent RH (eds) Herbivores: between plants and predators. Blackwell Scientific, Oxford, pp 301–332

Jenkins KJ and Barten NL (2005) Demography and decline of the Mentasta caribou herd in Alaska. Canadian Journal of Zoology 83:1174–1188

Jenks JA and Leslie DM Jr (1988) Effect of lichen and in vitro methodology on digestibility of winter deer diets in Maine. Canadian Field-Naturalist 102:216–220

Jensen PG, Pekins PJ, and Holter JB (1999) Compensatory effect of the heat increment of feeding on thermoregulation costs of white-tailed deer fawns in winter. Canadian Journal of Zoology 77:1474–1485

Jobling M (1994) Fish bioenergetics. Chapman and Hall, London

Jobling M, Coves D, Damsgard B, Kristiansen HR, Koskela J, Petursdottir TE, Kadri S, and Gudmundsson O (2001) Techniques for measuring feed intake. In: Houlihan D, Boujard T, and Jobling M (eds) Food intake in fish. Blackwell Science, Oxford, pp 49–87

Jodice PGR, Roby DD, Suryan RM, Irons DB, Kaufman AM, Turco KR, and Visser GH (2003) Variation in energy expenditure among black-legged kittiwakes: effects of activity-specific rates and activity budgets. Physiological and Biochemical Zoology 76:375–388

Johnsgard PA (1983) Cranes of the world. Indiana University Press, Bloomington, IN

Johnson AL (2000) Reproduction in the female. In: Whittow GC (ed) Sturkie's avian physiology. Academic Press, San Diego, CA, pp 569–596

Joly DO and Messier F (2004a) The distribution of *Echinococcus granulosus* in moose: evidence for parasite-induced vulnerability to predation by wolves? Oecologia 140:586–590

Joly DO and Messier F (2004b) Testing hypotheses of bison population decline (1970–1999) in Wood Buffalo National Park: synergism between exotic disease and predation. Canadian Journal of Zoology 82:1165–1176

Jordano P (1992) Fruits and frugivory. In: Fenner M (ed) Seeds. The ecology of regeneration in plant communities. CABI, Wallingford, Oxon, pp 105–156

Jorde DG and Owen RB Jr (1988) Efficiency of nutrient use by American black ducks wintering in Maine. Journal of Wildlife Management 52:209–214

Jorde DG and Owen RB (1990) Foods of black ducks, *Anas rubripes*, wintering in marine habitats of Maine. Canadian Field-Naturalist 104:300–302

Jorde DG, Krapu GL, Crawford RD, and Hay MA (1984) Effects of weather on habitat selection and behavior of mallards wintering in Nebraska. Condor 86:258–265

Jorde DG, Haramis GM, Bunck CM, and Pendleton GW (1995) Effects of diet on rate of body mass gain by wintering canvasbacks. Journal of Wildlife Management 59:31–39

Karasov WH (1990) Digestion in birds: chemical and physiological determinants and ecological implications. Journal of Avian Biology 13:391–415

Karasov WH and Cork SJ (1994) Glucose absorption by a nectarivorous bird: the passive pathway is paramount. American Journal of Physiology 267:G18–G26

Karasov WH and Cork SJ (1996) Test of a reactor-based digestion optimization model for nectar-eating rainbow lorikeets. Physiological Zoology 69:117–138

Karasov WH and Hume ID (1997) Vertebrate gastrointestinal system. In: Dantzler WH (ed) Handbook of physiology. American Physiological Society, Washington, DC, pp 409–480

Karasov WH and McWilliams SR (2005) Digestive constraints in mammalian and avian ecology. In: Starck JM and Wang T (eds) Physiological and ecological adaptations to feeding in vertebrates. Science Publishers, Enfield, NH, pp 88–112

Karasov WH, Phan D, Diamond JM, and Carpenter FL (1986) Food passage and intestinal nutrient absorption in hummingbirds. Auk 103:453–464

Kaseloo PA and Lovvorn JR (2003) Heat increment of feeding and thermal substitution in mallard ducks feeding voluntarily on grain. Journal of Comparative Physiology B 173:207–213

Kaushik S (2001) Mineral nutrition. In: Guillaume J, Kaushik S, Bergot P, and Métailler R (eds) Nutrition and feeding of fish and crustaceans. Springer–Praxis, Chichester, pp 169–181

Keech MA, Bowyer RT, Ver Hoef JM, Boertje RD, Dale BW, and Stephenson TR (2000) Life-history consequences of maternal condition in Alaskan moose. Journal of Wildlife Management 64:450–462

Kelly JF (2000) Stable isotopes of carbon and nitrogen in the study of avian and mammalian trophic ecology. Canadian Journal of Zoology 78:1–27

Kennedy BP, Chamberlain CP, Blum JD, Nislow KH, and Folt CL (2005) Comparing naturally occurring stable isotopes of nitrogen, carbon, and strontium as markers for the rearing locations of Atlantic salmon (*Salmo salar*). Canadian Journal of Fisheries and Aquatic Sciences 62:48–57

Kennedy PM, Boniface AN, Liang ZJ, Muller D, and Murray RM (1992) Intake and digestion in swamp buffaloes and cattle. 2. The comparative response to urea supplements in animals fed tropical grasses. Journal of Agricultural Science Cambridge 119:243–254

Kestemont P and Baras E (2001) Environmental factors and feed intake: mechanisms and interactions. In: Houlihan D, Boujard T, and Jobling M (eds) Food intake in fish. Blackwell Science, Oxford, pp 131–156

Kie JG (1999) Optimal foraging and risk of predation: effects on behavior and social structure in ungulates. Journal of Mammalogy 80:1114–1129

Kie JG, Bowyer RT, and Stewart KM (2003) Ungulates in western forests: habitat requirements, population dynamics, and ecosytstem processes. In: Zabel CJ and Anthony RG (eds) Mammal community dynamics: management and conservation in the coniferous forests of western North America. Cambridge University Press, New York, pp 296–340

Kilgore DL and Armitage KB (1978) Energetics of yellow-bellied marmot populations. Ecology 59:78–88

Kinnear JE, Purohit KG, and Main AR (1968) The ability of the tammar wallaby (*Macropus eugenii*, Marsupialia) to drink sea water. Comparative Biochemistry and Physiology 25:761–782

Kirschner LB (1991) Water and ions. In: Prosser CL (ed) Environmental and metabolic animal physiology, 4th edn. Wiley-Liss, New York, pp 13–107

Kjellander P, Gaillard J-M, and Hewison AJM (2006) Density-dependent responses of fawn cohort body mass in two contrasting roe deer populations. Oecologia 146:521–530

Klaassen M (1995) Water and energy limitations on flight range. Auk 112:260–262

Klaassen M, Lindström Å, Meltofte H, and Piersma T (2001) Arctic waders are not capital breeders. Nature 413:794

Klaassen M, Abraham KF, Jefferies RL, and Vrtiska M (2006) Factors affecting the site of investment, and the reliance on savings for arctic breeders: the capital–income dichotomy revisited. Ardea 94:371–384

Klasing KC (1998) Comparative avian nutrition. CAB International, New York

Klasing KC (2005) Interplay between diet, microbes, and immune defenses of the gastrointestinal tract. In: Starck JM and Wang T (eds) Physiological and ecological adaptations to feeding in vertebrates. Science Publishers, Enfield, NH, pp 229–254

Kleiber M (1947) Body size and metabolic rate. Physiological Reviews 27:511–541

Klein DR (1987) Vegetation recovery patterns following overgrazing by reindeer on St. Matthew Island. Journal of Range Management 40:336–338

Klein DR and Bay C (1994) Resource partitioning by mammalian herbivores in the high Arctic. Oecologia 97:439–450

Klein DR and Thing H (1989) Chemical elements in mineral licks and associated muskoxen feces in Jameson Land, northeast Greenland. Canadian Journal of Zoology 67:1092–1095

Klingenspor M, Niggeman H, and Heldmaier G (2000) Modulation of leptin sensitvity by short photoperiod acclimation in Djugarian hamster, *Phodopus sungorus*. Journal of Comparative Physiology B 170:37–43

Knott KK, Barboza PS, Bowyer RT, and Blake JE (2004) Nutritional development of feeding strategies in arctic ruminants: digestive morphometry of reindeer, *Rangifer tarandus*, and muskoxen, *Ovibos moschatus*. Zoology 107:315–333

Knott KK, Barboza PS, and Bowyer RT (2005) Postnatal development and organ maturation in *Rangifer tarandus* and *Ovibos moschatus*. Journal of Mammalogy 86:121–130

Koebel DA, Miers PG, Nelson RA, and Steffen JM (1991) Biochemical changes in skeletal muscles of denning bears (*Ursus americanus*). Comparative Biochemistry and Physiology 100B:377–380

Kooyman GL, Cherel Y, Le Maho Y, Croxall JP, Thorson PH, Ridoux V, and Kooyman CA (1992) Diving behavior and energetics during foraging cycles in king penguins. Ecological Monographs 62:143–163

Korine C, Daniel S, van Tets IG, Yosef R, and Pinshow B (2004) Measuring fat mass in small birds by dual-energy x-ray. Physiological and Biochemical Zoology 77:522–529

Krasnov BR, Khokhlova IS, Arakelyan MS, and Degen AA (2005) Is a starving host tastier? Reproduction in fleas parasitizing food-limited rodents. Functional Ecology 19:625–631

Krockenberger A (2006) Lactation. In: Armati PJ, Dickman CR, and Hume ID (eds) Marsupials. Cambridge University Press, Cambridge, pp 108–136

Krockenberger A and Hume ID (1998) Production of the milk and nutrition of the dependent young of free-ranging koalas (*Phascolarctos cinereus*). Physiological Zoology 71:45–56

Krockenberger AK and Hume ID (2007) A flexible digestive strategy accommodates the nutritional demands of reproduction in a free-living folivore, the koala (*Phascolarctos cinereus*). Functional Ecology 21:748–756

Ksiazek A, Konarzewski M, and Lapo IB (2004) Anatomic and energetic correlates of divergent selection for basal metabolic rate in laboratory mice. Physiological and Biochemical Zoology 77:890–899

Kucera TE (1997) Fecal indicators, diet, and population parameters in mule deer. Journal of Wildlife Management 61:550–560

Kuenzel WJ, Beck MM, and Teruyama R (1999) Neural sites and pathways regulating food intake in birds: a comparative analysis to mammalian systems. Journal of Experimental Zoology 283:348–364

Kuntz R, Kubalek C, Ruf T, Tartaruch F, and Arnold W (2006) Seasonal adjustment of energy budget in a large wild mammal, the Przewalski horse (*Equus ferus przewalski*). Journal of Experimental Biology 209:4557–4565

Kurle CM and Worthy GAJ (2002) Stable nitrogen and carbon isotope ratios in multiple tissues of the northern fur seal *Callorhinus ursinus*: immplications for dietary and migratory reconstruction. Marine Ecology Progress Series 236:289–300

Kutz SJ, Hoberg EP, Polley L, and Jenkins EJ (2005) Global warming is changing the dynamics of Arctic host–parasite systems. Proceedings of the Royal Society London B 272:2571–2576

Lall SP (2002) The minerals. In: Halver JE and Hardy RW (eds) Fish nutrition, 3rd edn. Academic Press, Amsterdam, pp 260–308

Landsberg JA, O'Connor T, and Freudenberger DO (1999) The impacts of livestock grazing on biodiversity in natural ecosystems. In: Jung H-JG and Fahey GC Jr (eds) Nutritional ecology of herbivores. American Society of Animal Science, Savoy, IL, pp 752–777

Landys MM, Piersma T, Visser GH, Jukema J, and Wuker A (2000) Water balance during real and simulated long-distance migratory flight in the bar-tailed godwit. Condor 102:645–652

Lane SJ and Hassall M (1996) Nocturnal feeding by dark-bellied brent geese (*Branta bernicla bernicla*). Ibis 138:291–297

Langer P (1994) Weaning time and bypass structures in the forestomachs of Marsupialia and Eutheria. In: Chivers DJ and Langer P (eds) The digestive system in mammals: food, form and function. Cambridge University Press, Cambridge, pp 264–286

Langhans W, Rossi R, and Scharrer E (1995) Relationships between feed and water intake in ruminants. In: Engelhardt Wv, Leonhard-Marek S, Breves G, and Giesecke D (eds) Ruminat physiology: digestion, metabolism, growth and reproduction. Ferdinand Enke, Stuttgart, pp 199–216

Lautier JK, Dailey TV, and Brown RD (1988) Effect of water restriction on feed intake of white-tailed deer. Journal of Wildlife Management 52:602–606

Laverty G and Skadhauge E (1999) Physiological roles and regulation of transport activities in the avian lower intestine. Journal of Experimental Zoology 283:480–494

Lawler IR, Foley WJ, Pass GJ, and Eschler BM (1998) Administration of a $5HT_3$ receptor antagonist increases the intake of diets containing *Eucalyptus* secondary metabolites by marsupials. Journal of Comparative Physiology B 168:611–618

Lawler IR, Foley WJ, and Eschler BM (2000) Foliar concentrations of a single toxin creates habitat patchiness for a marsupial folivore. Ecology 81:1327–1338

Lawler JP and White RG (1997) Seasonal changes in metabolic rates in muskoxen following twenty-four hours of starvation. Rangifer 17:135–138

Lawler JP and White RG (2003) Temporal responses in energy expenditure and respiratory quotient following feeding in muskox: influence of season on energy costs of eating and standing and an endogenous heat increment. Canadian Journal of Zoology 81:1524–1538

Lawler JP and White RG (2006) Effect of browse on post-ingestive energy loss in an Arctic ruminant: implications for muskoxen (*Ovibos moschatus*) in relation to vegetation change in the Arctic. Canadian Journal of Zoology 84:1657–1667

Le Maho Y, van Kha HV, Koubi H, Dewasmes G, Girard J, Ferre P, and Cagnard M (1981) Body composition, energy expenditure, and plasma metabolites in long-term fasting geese. American Journal of Physiology 241:E342–E354

Leader-Williams N and Ricketts C (1981) Seasonal and sexual patterns of growth and condition of reindeer introduced into South Georgia. Oikos 38:27–39

Leafloor JO and Ankney CD (1991) Factors affecting wing molt chronology of female mallards. Canadian Journal of Zoology 69:924–928

Leighton FA, Cattet M, Norstrom R, and Trudeau S (1988) A cellular basis for high levels of vitamin A in livers of polar bears (*Ursus maritimus*): the Ito cell. Canadian Journal of Zoology 66:480–482

Leklem JE (1991) Vitamin B_6. In: Machlin LJ (ed) Handbook of vitamins, 2nd edn. Marcel Dekker, New York, pp 341–392

Lentle RG, Hume ID, Stafford KJ, Kennedy M, Springett BP, Browne R, and Haslett S (2004) Temporal aspects of feeding events in tammar (*Macropus eugenii*) and parma (*Macropus parma*) wallabies. I. Food acquisition and oral processing. Australian Journal of Zoology 52:81–95

Leopold A (1949) A Sand County almanac. Oxford University Press, New York

Lessard RB, Martell SJD, Walters CJ, Essington TE, and Kitchell JF (2005) Should ecosystem management involve active control of species abundances? Ecology and Society 10:1

Levey DJ and Cipollini ML (1996) Is most glucose absorbed passively in northern bobwhite? Comparative Biochemistry and Physiology 113:225–231

Levey DJ and del Rio CM (2001) It takes guts (and more) to eat fruit: lessons from avian nutritional ecology. Auk 118:819–831

Lifson N and McClintock R (1966) Theory of use of the turnover rates of body water for measuring energy and material balance. Journal of Theoretical Biology 12:46–74

Lindberg MS, Ward DH, Tibbitts L, and Roser J (2007) Winter movement dynamics of black brant. Journal of Wildlife Management 71:534–540

Linder MC (1991) Nutritional biochemistry and metabolism: with clinical applications, 2nd edn. Elsevier, New York

Lindsay DG (2005) Nutrition, hormetic stress and health. Nutrition Research Reviews 18:249–258

Lindström Å and Piersma T (1993) Mass changes in migrating birds: the evidence for fat and protein storage re-examined. Ibis 135:70–78

Lindström Å, Visser GH, and Daan S (1993) The energetic cost of feather synthesis is proportional to basal metabolic rate. Physiological Zoology 66:490–510

Linton SM and Greenaway P (2007) A review of feeding and nutrition of herbivorous land crabs: adaptations to low quality plant diets. Journal of Comparative Physiology B 177:269–286

Lochmiller RL, Peitz DG, Leslie DM Jr, and Engle DM (1995) Habitat-induced changes in essential amino-acid nutrition in populations of eastern cottontails. Journal of Mammalogy 76: 1164–1177

Loe LE, Bonenfant C, Mysterud A, Gaillard JM, Langvatn R, Klein F, Calenge C, Ergon T, Pettorelli N, and Stenseth NC (2005) Climate predictability and breeding phenology in red deer: timing and synchrony of rutting and calving in Norway and France. Journal of Animal Ecology 74:579–588

Loison A and Langvatn R (1998) Short- and long-term effects of winter and spring weather on growth and survival of red deer in Norway. Oecologia 116:489–500

Long RA, Martin TJ, and Barnes BM (2005) Body temperature and activity patterns in free-living arctic ground squirrels. Journal of Mammalogy 86:314–322

Lovegrove BG (1989) The cost of burrowing by the social mole rats (Bathyergidae) *Cryptomys damarensis* and *Heterocephalus glaber*: the role of soil moisture. Physiological Zoology 62:449–469

Lovvorn JR (1994) Nutrient reserves, probability of cold spells and the question of reserve regulation in wintering canvasbacks. Journal of Animal Ecology 63:11–23

Lucas PW (1994) Categorisation of food items relevant to oral processing. In: Chivers DJ and Langer P (eds) The digestive system in mammals: food, form and function. Cambridge University Press, Cambridge, pp 197–218

MacCluskie MC, Flint PL, and Sedinger JS (1997) Variation in incubation periods and egg metabolism in mallards: intrinsic mechanisms to promote hatch synchrony. Condor 99:224–228

Macfarlane WV and Howard B (1972) Comparative water and energy economy of wild and domestic mammals. Symposia Zoological Society London 31:261–296

Macfarlane WV, Morris RJH, Howard B, McDonald J, and Budtz-Olsen OE (1961) Water and electrolyte changes in tropical merino sheep exposed to dehydration during summer. Australian Journal of Agricultural Research 12:889–912

Machlin LJ (1991) Vitamin E. In: Machlin LJ (ed) Handbook of vitamins, 2nd edn. Marcel Dekker, New York, pp 99–144

Mackie RI, Aminov RI, White BA, and McSweeney CS (2000) Molecular ecology and diversity in gut microbial ecosystems. In: Cronjé PB (ed) Ruminant physiology. Digestion, metabolism, growth and reproduction. CAB International, Wallingford, Oxon, pp 61–77

Madrid JA, Boujard T, and Sanchez-Vazquez FJ (2001) Feeding rhythms. In: Houlihan D, Boujard T, and Jobling M (eds) Food intake in fish. Blackwell Science, Oxford, pp 189–215

Maloiy GMO, Taylor CR, and Clemens ET (1978) A comparison of gastrointestinal water content and osmolality in East African herbivores during hydration and dehydration. Journal of Agricultural Science Cambridge 91:249–254

Maloney SK and Dawson TJ (1998) Changes in pattern of heat loss at high ambient temperature caused by water deprivation in a large flightless bird, the emu. Physiological Zoology 71:713–719

Maltz E and Shkolnik A (1984) Lactational strategies of desert ruminants: the Bedouin goat, ibex and desert gazelle. Symposia Zoological Society London 51:198–213

Mambrini M and Guillaume J (2001) Protein nutrition. In: Guillaume J, Kaushik S, Bergot P, and Métailler R (eds) Nutrition and feeding of fish and crustaceans. Springer–Praxis, Chichester, pp 81–110

Mangione AM, Dearing MD, and Karasov WH (2000) Interpopulation differences in tolerance to creosote bush resin in desert woodrats (*Neotoma lepida*). Ecology 81:2067–2076

Mao JS, Boyce MS, Smith DW, Singer FJ, Vales DJ, More JM, and Merrill EH (2005) Habitat selection by elk before and after wolf reintroduction in Yellowstone National Park. Journal of Wildlife Management 69:1691–1707

Marini JC, Sands JM, and Van Amburgh ME (2006) Urea transporters and urea recycling in ruminants. In: Sejrsen K, Hvelpund T, and Nielsen MO (eds) Ruminant physiology. Digestion, metabolism and impact of nutrition on gene expression, immunology and stress. Wageningen Academic Publishers, Wageningen, pp 155–171

Martin LB II, Navara KJ, Bailey MT, Hutch CR, Powell ND, Sheridan JF, and Nelson RJ (2008) Food restriction compromises immune memory in deer mice (*Peromyscus maniculatus*) by reducing spleen-derived antibody-producing N cell numbers. Physiological and Biochemical Zoology 81:366–372

Martin RG, McMeniman NP, Norton BW, and Dowsett KF (1996) Utilization of endogenous and dietary urea in the large intestine of the mature horse. British Journal of Nutrition 76:373–386

Martinez del Rio C and Wolf BO (2005) Mass-balance models for animal isotopic ecology. In: Starck JM and Wang T (eds) Physiological and ecological adaptations to feeding in vertebrates. Science Publishers, Enfield, NH, pp 141–174

Mason DD, Barboza PS, and Ward DH (2006) Nutritional condition of Pacific black brant wintering at extremes of their range. Condor 108:678–690

Mason DD, Barboza PS, and Ward DH (2007) Mass dynamics of wintering Pacific black brant: body, adipose tissue, organ, and muscle masses vary with location. Condor 85:728–736

Mason VC (1969) Some observations on the distribution and origin of nitrogen in sheep faeces. Journal of Agricultural Science Cambridge 73:99–111

Mathews CK and Van Holde KE (1996) Biochemistry. Benjamin Cummings, Menlo Park, CA

Mautz WW, Kanter J, and Pekins PJ (1992) Seasonal metabolic rhythms of captive female white-tailed deer: a reexamination. Journal of Wildlife Management 56:656–661

Mayes RW and Dove H (2000) Measurement of dietary nutrient intake in free-ranging mammalian herbivores. Nutrition Research Reviews 13:107–138

McAllan BM, Roberts JR, and O'Shea T (1996) Seasonal changes in the renal morphometry of *Antechinus stuartii* (Marsupialia:Dasyuridae). Australian Journal of Zoology 44:337–354

McAllan BM, Roberts JR, and O'Shea T (1998) Seasonal changes in glomerular filtration rate in *Antechinus stuartii* (Marsupialia: Dasyuridae). Journal of Comparative Physiology B 168:41–49

McArthur C, Robbins CT, Hagerman AE, and Hanley TA (1993) Diet selection by a ruminant generalist browser in relation to plant chemistry. Canadian Journal of Zoology 71:2236–2243

McCormick SD, Cunjak RA, Dempson B, O'Dea MF, and Carey JB (1999) Temperature related loss of smolt characteristics in Atlantic salmon (*Salmo salar*) in the wild. Canadian Journal of Fisheries and Aquatic Sciences 56:1649–1658

McCue MD, Bennett AF, and Hicks JW (2005) The effect of meal composition on specific dynamic action in Burmese pythons (*Python molurus*). Physiological and Biochemical Zoology 78:182–192

McCullough DR (1985) Variables influencing food habits of white-tailed deer on the George Reserve. Journal of Mammalogy 66:682–692

McDonald JE Jr and Fuller TK (2005) Effects of spring acorn availability on black bear diet, milk composition, and cub survival. Journal of Mammalogy 86:1022–1028

McEwan EH and Koelink AFC (1973) The heat production of oiled mallards and scaup. Canadian Journal of Zoology 51:27–31

McGee H (2004) On food and cooking. The science and lore of the kitchen, 2nd edn. McMillan, New York
McGraw KJ (2006a) Mechanics of carotenoid-based coloration. In: Hill GE and McGraw KJ (eds) Bird coloration. Harvard University Press, Cambridge, MA, pp 177–242
McGraw KJ (2006b) Mechanics of melanin-based coloration. In: Hill GE and McGraw KJ (eds) Bird coloration. Harvard University Press, Cambridge, MA, pp 243–294
McKechnie AE and Wolf BO (2004) Partitioning of evaporative water loss in white-winged doves: plasticity in response to short-term thermal acclimation. Journal of Experimental Biology 207:203–210
McKenzie DJ (2005) Effects of dietary fatty acids on the physiology of environmental adaptation in fish. In: Starck JM and Wang T (eds) Physiological and ecological adaptations to feeding in vertebrates. Science Publishers, Enfield, NH, pp 363–388
McKinney T, Smith TW, and deVos JC Jr (2006) Evaluation of factors potentially influencing a desert bighorn sheep population. Wildlife Monographs 164:1–36
McLoughlin PD, Dzus E, Wynes B, and Boutin S (2003) Declines in populations of woodland caribou. Journal of Wildlife Management 67:755–761
McNab BK (2002) The physiological ecology of vertebrates. A view from energetics. Cornell University Press, Ithaca, NY
McNaughton SJ (1985) Ecology of a grazing ecosystem: the Serengeti. Ecological Monographs 53:259–294
McShea WJ, Underwood HB, and Rappole JH (1997) The science of overabundance. Deer ecology and population management. Smithsonian Institution Press, Washington, DC
McSweeney CS and Mackie RI (1997) Gastrointestinal detoxification and digestive disorders in ruminant animals. In: Mackie RI, White BA, and Isaacson RE (eds) Gastrointestinal microbiology. Chapman and Hall, New York, pp 583–634
McWhorter TJ (2005) Paracellular intestinal absorption of carbohydrates in mammals and birds. In: Starck JM and Wang T (eds) Physiological and ecological adaptations to feeding in vertebrates. Science Publishers, Enfield, NH, pp 113–140
McWhorter TJ, Martinez del Rio C, Pinshow B, and Roxburgh L (2004) Renal function in Palestine sunbirds: elimination of excess water does not constrain energy intake. Journal of Experimental Biology 207:3391–3398
McWilliams SR and Karasov WH (1998) Test of a digestion optimization model: effects of costs of feeding on digestive parameters. Physiological Zoology 71:168–178
McWilliams SR and Karasov WH (2005) Migration takes guts: digestive physiology of migratory birds and its ecological significance. In: Marra PP and Greenberg R (eds) Birds of two worlds: the ecology and evolution of migration. Johns Hopkins University Press, Baltimore, MD, pp 67–78
McWilliams SR, Guglielmo C, Pierce B, and Klaassen M (2004) Flying, fasting, feeding birds during migration: a nutritional and physiological ecology perspective. Journal of Avian Biology 35:377–393
Mead GC (1997) Bacteria in the gastrointestinal tract of birds. In: Mackie RI, White BA, and Isaacson RE (eds) Gastrointestinal microbiology. Chapman and Hall, New York, pp 216–240
Mech LD (2007) Femur-marrow fat of white-tailed deer fawns killed by wolves. Journal of Wildlife Management 71:920–923
Meldgaard M (1986) The Greenland caribou – zoogeography, taxonomy, and population dynamics. Meddelelser om Gronland Bioscience 20:1–77
Mellish JE, Iverson SJ, and Bowen WD (1999) Variation in milk production and lactation performance in grey seals and consequences for pup growth and weaning characteristics. Physiological and Biochemical Zoology 72:677–690
Mellish JE, Tuomi PA, and Horning M (2004) Assessment of ultrasound imaging as a noninvasive measure of blubber thickness in pinnipeds. Journal of Zoo and Wildlife Medicine 35:116–118
Mello JRB (2003) Calcinosis – calcinogenic plants. Toxicon 41:1–12

Meteyer CU, Rideout BA, Gilbert M, Shivaprasad HL, and Oaks JL (2005) Pathology and proposed pathophysiology of diclofenac poisoning in free-living and experimentally exposed oriental white-backed vultures (*Gyps bengalensis*). Journal of Wildlife Diseases 41:707–716

Miller JD (1997) Reproduction in sea turtles. In: Lutz PL and Musick JA (eds) The biology of sea turtles. CRC Press, Boca Raton, FL, pp 51–81

Millidine KJ, Armstrong JD, and Metcalfe NB (2006) Presence of shelter reduces maintenance metabolism of juvenile salmon. Functional Ecology 20:839–845

Milner-Gulland EJ, Bukreeva OM, Coulson T, Lushchekina AA, Kholodova MV, Bekenov AB, and Grachev IA (2003) Reproductive collapse in saiga antelope harems. Nature 422:135

Min BR, Barry TN, Attwood GT, and McNabb WC (2003) The effect of condensed tannins on the nutrition and health of ruminants fed fresh temperate forages: a review. Animal Feed Science and Technology 106:3–19

Minnich JE (1972) Excretion of urate salts by reptiles. Comparative Biochemistry and Physiology A 41:535–549

Moe B, Stolevik E, and Bech C (2005) Ducklings exhibit substantial energy-saving mechanisms as a response to short-term food shortage. Physiological and Biochemical Zoology 78: 90–104

Moen R and DelGiudice GD (1997) Simulating nitrogen metabolism and urinary urea nitrogen: creatinine ratios in ruminants. Journal of Wildlife Management 61:881–894

Moir RJ (1994) The 'carnivorous' herbivores. In: Chivers DJ and Langer P (eds) The digestive system in mammals: food, form and function. Cambridge University Press, Cambridge, pp 87–102

Mommsen TP, French CJ, and Hochachka PW (1980) Sites and patterns of protein and amino acid utilization during the spawning migration of salmon. Canadian Journal of Zoology 58:1785–1799

Moriarty F (1999) Ecotoxicology: the study of pollutants in ecosystems, 3rd edn. Academic Press, San Diego, CA

Morris JG (1999) Ineffective vitamin D synthesis in cats is reversed by an inhibitor of 7-dehydrocholesterol-delta 7- reductase. Journal of Nutrition 129:903–909

Morris JG (2002) Idiosyncratic nutrient requirements of cats appear to be diet-induced evolutionary adaptations. Nutrition Research Reviews 15:153–168

Moser U and Bendich A (1991) Vitamin C. In: Machlin LJ (ed) Handbook of vitamins, 2nd edn. Marcel Dekker, New York, pp 195–232

Mousa HM, Ali KE, and Hume ID (1983) Effects of water deprivation on urea metabolism in camels, desert sheep and desert goats fed dry desert grass. Comparative Biochemistry and Physiology A 74:715–720

Mowatt G, Slough BG, and Boutin S (1996) Lynx recruitment during a snowshoe hare population peak and decline in southwest Yukon. Journal of Wildlife Management 60:441–452

Moyle PB and Cech JJ Jr (2004) Fishes: an introduction to ichthyology. Prentice Hall, Upper-Saddle River, NJ

Mueller P and Diamond J (2001) Metabolic rate and environmental productivity: well-provisioned animals evolved to run and idle fast. Proceedings National Academy of Sciences USA 98:12550–12554

Muir PD, Sykes AR, and Barrell GK (1987) Calcium metabolism in red deer (*Cervus elaphus*) offered herbages during antlerogenesis: kinetic and stable balance studies. Journal of Agricultural Science Cambridge 109:357–364

Mundy PR (2005) The Gulf of Alaska: biology and oceanography. Alaska Sea Grant College Program, Fairbanks, AK

Munn AJ and Dawson TJ (2001) Thermoregulation in juvenile red kangaroos (*Macropus rufus*) after pouch exit: higher metabolism and evaporative water requirements. Physiological and Biochemical Zoology 74:917–927

Munn AJ and Dawson TJ (2003) How important is milk for near-weaned red kangaroos (*Macropus rufus*) fed different forages? Journal of Comparative Physiology B 173:141–148

Munn AJ and Dawson TJ (2006) Forage fibre digestion, rates of feed passage and gut fill in juvenile and adult red kangaroos *Macropus rufus* Desmarest: why body size matters. Journal of Experimental Biology 209:1535–1549
Murphy ME and King JR (1984) Sulfur amino acid nutrition during molt in the white-crowned sparrow. 2. Nitrogen and sulfur balance in birds fed graded levels of the sulflur-containing amino acids. Condor 86:324–332
Murphy ME and King JR (1990) Diurnal changes in tissue glutathione and protein pools of molting white-crowned sparrows: the influence of photoperiod and feeding schedule. Physiological Zoology 63:1118–1140
Murphy ME and Taruscio TG (1995) Sparrows increase their rates of tissue and whole-body protein synthesis during the annual molt. Comparative Biochemistry and Physiology A 111:385–396
Murray DL, Cox EW, Ballard WB, Whitlaw HA, Lenarz MS, Custer TW, Barnett T, and Fuller TK (2006) Pathogens, nutritional deficiency, and climate influences on a declining moose population. Wildlife Monographs 166:1–30
Mysterud A, Yoccoz NG, Stenseth NC, and Langvatn R (2000) Relationships between sex ratio, climate and density in red deer: the importance of spatial scale. Journal of Animal Ecology 69:959–974
Mysterud A, Langvatn R, and Stenseth NC (2004) Patterns of reproductive effort in male ungulates. Journal of Zoology London 264:209–215
Mysterud A, Meisingset EL, Langvatn R, Yoccoz NG, and Stenseth NC (2005a) Climate-dependent allocation of resources to secondary sexual traits in red deer. Oikos 111:245–252
Mysterud A, Solberg EJ, and Yoccoz NG (2005b) Ageing and reproductive effort in male moose under variable levels of intrasexual competition. Journal of Animal Ecology 74:742–754
Mörschel FM and Klein DR (1997) Effects of weather and parasitic insects on behavior and group dynamics of caribou in the Delta Herd, Alaska. Canadian Journal of Zoology 75:1659–1690
Müller MS, McWilliams SR, Podlesak DW, Donaldson JR, Bothwell HM, and Lindroth RL (2006) Tri-trophic effects of plant defenses: chickadees consume caterpillars based on host leaf chemistry. Oikos 114:507–517
Nagy KA (1994) Seasonal water, energy and food use by free-living, arid-habitat mammals. Australian Journal of Zoology 42:55–63
Nagy KA (2005) Field metabolic rate and body size. Journal of Experimental Biology 208:1621–1625
Nagy KA and Medica PA (1986) Physiological ecology of desert tortoises in southern Nevada. Herpetologica 42:73–92
Nagy KA and Montgomery GG (1980) Field metabolic rate, water flux, and food consumption in three-toed sloths (*Bradypus variegatus*). Journal of Mammalogy 61:465–472
Nagy KA, Girard IA, and Brown TK (1999) Energetics of free-ranging mammals, reptiles, and birds. Annual Review of Nutrition 19:247–277
National Research Council (1974) Nutrients and toxic substances in water for livestock and poultry. National Academy Press, Washington, DC
National Research Council (1983) Nutrient requirements of warmwater fishes and shellfishes. National Academy Press, Washington, DC
National Research Council (1996) Nutrient requirements of beef cattle. National Academy Press, Washington, DC
National Research Council (1997) Wolves, bears, and their prey in Alaska. National Academy Press, Washington, DC
National Research Council (2002) Ecological dynamics on Yellowstone's northern range. National Academy Press, Washington, DC
National Research Council (2003) Nutrient requirements of non-human primates, 2nd edn. National Academy Press, Washington, DC
National Research Council (2005) Mineral tolerance of animals, 2nd edn. National Academy Press, Washington, DC

National Research Council (2006) Nutrient requirements of dogs and cats. National Academy Press, Washington, DC

National Research Council (2007a) Nutrient requirements of horses, 6th edn. National Academy Press, Washington, DC

National Research Council (2007b) Nutrient requirements of small ruminants: sheep, goats, cervids and New World camelids. National Academy Press, Washington, DC

Naylor R, Richardson SJ, and McAllan BM (2008) Boom and bust: a review of the physiology of the marsupial genus *Antechinus*. Journal of Comparative Physiology B 178:545–562

Nevitt G, Reid K, and Trathan P (2004) Testing olfactory foraging strategies in an Antarctic seabird assemblage. Journal of Experimental Biology 207:3537–3544

Nolan JV (1993) Nitrogen kinetics. In: Forbes JM and France J (eds) International quantitative aspects of ruminant digestion and metabolism. CAB International, Wallingford, Oxon, pp 123–143

Nordoy E and Blix AS (1985) Energy sources in fasting grey seal pups evaluated with computed tomography. American Journal of Physiology 249:R471–R476

Noren DP and Mangel M (2003) Energy reserve utilization in northern elephant seal (*Mirounga anguistirostris*) pups during the postweaning fast: size does matter. Journal of Comparative Physiology B 173:443–454

Norris DO (1997) Vertebrate endocrinology, 3rd edn. Academic Press, San Diego, CA

Norris DR, Marra PP, Kyser TK, Sherry TW, and Ratcliffe LM (2004) Tropical winter habitat limits reproductive success on the temperate breeding grounds in a migratory bird. Proceedings of the Royal Society London B 271:59–64

Nussey DH, Postma E, Gienapp P, and Visser ME (2005) Selection on heritable phenotypic plasticity in a wild bird population. Science 310:304–306

Oarada M, Nikawa T, and Kurita N (2002) Effect of timing of food deprivation on host resistance to fungal infection. British Journal of Nutrition 88:151–158

Oftedal OT (1984) Milk composition, milk yield and energy output at peak lactation: a comparative review. Symposia Zoological Society London 51:33–85

Oftedal OT (1985) Pregnancy and lactation. In: Hudson RJ and White RG (eds) Bioenergetics of wild herbivores. CRC Press, Boca Raton, FL, pp 216–238

Oldham JD (1999) Genotype × nutrition interactions in herbivores. In: Jung H-JG and Fahey GC Jr (eds) Nutritional ecology of herbivores. American Society of Animal Science, Savoy, Illinois, pp 482–504

Olsen GH and Langenberg JA (1996) Veterinary techniques for rearing crane chicks. In: Ellis DH, Gee GF, and Mirande CM (eds) Cranes: their biology, husbandry, and conservation. US Department of Interior, National Biological Service, Washington, DC, and International Crane Foundation, Baraboo, WI, pp 95–104

Olsen MA, Blix AS, Utsi TH, Sormo W, and Mathiesen SD (2000) Chitinolytic activity in the minke whale forestomach. Canadian Journal of Microbiology 46:85–94

Olson JA (1991) Vitamin A. In: Machlin LJ (ed) Handbook of vitamins, 2nd edn. Marcel Dekker, New York, pp 1–57

Ormseth OA, Nicolson M, Pelleymounter MA, and Boyer BB (1996) Leptin inhibits prehibernation hyperphagia and reduces body weight in arctic ground squirrels. American Journal of Physiology 271:R1775–R1779

Orpin CG, Greenwood Y, Hall FJ, and Paterson IW (1985) The rumen microbiology of seaweed digestion in Orkney sheep. Journal of Applied Bacteriology 59:585–596

Ortiz RM, Wade CE, Costa DP, and Ortiz CL (2002) Renal responses to plasma volume expansion and hyperosmolality in fasting seal pups. American Journal of Physiology 282:R805–R817

Osawa R, Walsh TP, and Cork SJ (1993) Metabolism of tannin–protein complex by facultatively anaerobic bacteria isolated from koala feces. Biodegradation 4:91–99

Osborn RG and Ginnett TF (2001) Fecal nitrogen and 2,6-diaminopimelic acid as indices to dietary nitrogen in white-tailed deer. Wildlife Society Bulletin 29:1131–1139

Osko TJ, Hiltz MN, Hudson RJ, and Wasel SM (2004) Moose habitat preferences in response to changing availability. Journal of Wildlife Management 68:576–584

Ostrowski S, Mesochina P, and Williams JB (2006) Physiological adjustments of sand gazelles (*Gazella subgutturosa*) to a boom-or-bust economy: standard fasting metabolic rate, total evaporative water loss, and changes in the sizes of organs during food and water restriction. Physiological and Biochemical Zoology 79:810–819

Ouellet JP, Heard DC, Boutin S, and Mulders R (1997) A comparison of body condition and reproduction of caribou on two predator-free islands. Canadian Journal of Zoology 75:11–17

Owen-Smith RN (1988) Megaherbivores. The influence of very large body size on ecology. Cambridge University Press, Cambridge

Owen-Smith RN (2002) Adaptive herbivore ecology. Cambridge University Press, Cambridge

Owens FN, Secrist DS, Hill WJ, and Gill DR (1998) Acidosis in cattle: a review. Journal of Animal Science 76:275–286

Paglia DE, Kenny DE, Dierenfeld ES, and Tsu I-H (2001) Role of excessive maternal iron in the pathogenesis of congenital leukoencephalomalacia in captive black rhinoceroses (*Diceros bicornis*). American Journal of Veterinary Research 62:343–349

Parker KL (1988) Effects of heat, cold, and rain on coastal black-tailed deer. Canadian Journal of Zoology 66:2475–2483

Parker KL (2003) Advances in the nutritional ecology of cervids at different scales. Ecoscience 10:395–411

Parker KL and Gillingham MP (1990) Estimates of critical thermal environments for mule deer. Journal of Range Management 43:73–81

Parker KL and Robbins CT (1984) Thermoregulation in mule deer and elk. Canadian Journal of Zoology 62:1409–1422

Parker KL, Robbins CT, and Hanley TA (1984) Energy expenditures for locomotion by mule deer and elk. Journal of Wildlife Management 48:474–488

Parker KL, White RG, Gillingham MP, and Holleman DF (1990) Comparison of energy metabolism in relation to daily activity and milk consumption by caribou and muskox neonates. Canadian Journal of Zoology 68:106–114

Parker KL, DelGiudice GD, and Gillingham MP (1993a) Do urinary urea nitrogen and cortisol ratios of creatinine reflect body-fat reserves in black-tailed deer? Canadian Journal of Zoology 71:1841–1848

Parker KL, Gillingham MP, Hanley TA, and Robbins CT (1993b) Seasonal patterns in body mass, body composition, and water transfer rates of free-ranging and captive black-tailed deer (*Odocoileus hemionus sitkensis*) in Alaska. Canadian Journal of Zoology 71: 1397–1404

Parker KL, Gillingham MP, Hanley TA, and Robbins CT (1996) Foraging efficiency: energy expenditure versus energy gain in free-ranging black-tailed deer. Canadian Journal of Zoology 74:442–450

Parker KL, Gillingham MP, Hanley TA, and Robbins CT (1999) Energy and protein balance of free-ranging black-tailed deer in a natural forest environment. Wildlife Monographs 143:1–48

Parker KL, Barboza PS, and Stephenson TR (2005) Protein conservation of female caribou (*Rangifer tarandus*): effects of decreasing diet quality during winter. Journal of Mammalogy 86:610–622

Parmesan C (2006) Ecological and evolutionary responses to recent climate change. Annual Review of Ecology and Systematics 37:637–669

Parmesan C (2007) Influences of species, latitudes, and methodologies on estimates of phenological response to global warming. Global Change Biology 13:1860–1872

Parmesan C and Yohe G (2003) A globally coherent fingerprint of climate change impacts across natural systems. Nature 421:37–42

Parra R (1978) Comparison of foregut and hindgut fermentation in herbivores. In: Montgomery GG (ed) The ecology of arboreal folivores. Smithsonian Institution Press, Washington, DC, pp 205–230

Patterson BD, Kays RW, Kasiki SM, and Sebestyen VM (2006) Developmental effects of climate on the lion's mane (*Panthera leo*). Journal of Mammalogy 87:193–200

Pearcy SD and Murphy ME (1997) Essential amino acid metabolism in molting and non-molting sparrows in different nutritional states. Comparative Biochemistry and Physiology A 118:1157–1163

Peddie J, Dewar WA, Gilbert AB, and Waddington D (1982) The use of titanium dioxide for determining apparent digestibility in mature domestic fowls (*Gallus domesticus*). Journal of Agricultural Science Cambridge 99:233–236

Pekins PJ, Gessaman JA, and Lindzey FG (1994) Field metabolic rate of blue grouse during winter. Canadian Journal of Zoology 72:227–231

Peltier TC and Barboza PS (2003) Growth in an arctic grazer: effects of sex and dietary protein on yearling muskoxen. Journal of Mammalogy 84:915–925

Peltier TC, Barboza PS, and Blake JE (2003) Seasonal hyperphagia does not reduce digestive efficiency in an arctic grazer. Physiological and Biochemical Zoology 76:471–483

Penry DL and Jumars PA (1987) Modeling animal guts as chemical reactors. American Naturalist 129:69–96

Perga ME and Gerdeaux D (2005) 'Are fish what they eat' all year round? Oecologia 144:598–606

Perrin MR (1994) Herbivory and niche partitioning. In: Chivers DJ and Langer P (eds) The digestive system in mammals: food, form and function. Cambridge University Press, Cambridge, pp 128–149

Person BT, Herzog MP, Ruess RW, Sedinger JS, Anthony RM, and Babcock CA (2003) Feedback dynamics of grazing lawns: coupling vegetation change with animal growth. Oecologia 135:583–592

Peterson CC, Nagy KA, and Diamond J (1990) Sustained metabolic scope. Proceedings National Academy of Sciences USA 87:2324–2328

Phillips DL and Eldridge PM (2006) Estimating the timing of diet shifts using stable isotopes. Oecologia 147:195–203

Phillips DL and Gregg JW (2001) Uncertainty in source partitioning using stable isotopes. Oecologia 127:171–179

Pierce BJ and McWilliams SR (2005) Seasonal changes in composition of lipid stores in migratory birds: causes and consequences. Condor 107:271–281

Pierce BJ, McWilliams SR, O'Connor TP, Place AR, and Gugliemo CG (2005) Effect of dietary fatty acid composition on depot fat and exercise performance in a migrating songbird, the red-eyed vireo. Journal of Experimental Biology 208:1277–1285

Piersma T, Bruinzeel L, Drent R, Dersten M, and van der Meer JWP (1996) Variability in basal metabolic rate of a long-distance migrant shorebird (red knot, *Calidris canutus*) reflects shifts in organ sizes. Physiological Zoology 69:191–217

Pillans RD, Good JP, Anderson WG, Hazon N, and Franklin CE (2005) Freshwater to seawater acclimation of juvenile bull sharks (*Carcharinus leucas*): plasma osmolytes and Na^+/K^+ ATPase activity in gill, rectal gland, kidney and intestine. Journal of Comparative Physiology B 175:37–44

Pitcher TR, Buffenstein R, Keegan JD, Moodley GP, and Yahav S (1992) Dietary calcium content, calcium balance and mode of uptake in a subterranean mammal, the dmara mole rat. Journal of Nutrition 122:108–114

Place AR (1992) Comparative aspects of lipid digestion and absorption: physiological correlates of wax ester digestion. American Journal of Physiology 263:R464–R471

Place AR, Stoyan NC, Ricklefs RE, and Butler RG (1989) Physiological basis of stomach oil formation in Leach's storm-petrel (*Oceanodroma leucorhoa*). Auk 106:687–699

Podlesak DW, McWilliams SR, and Hatch KA (2005) Stable isotopes in breath, blood, feces and feathers can indicate intra-individual changes in the diet of migratory songbirds. Oecologia 142:501–510

Pond CM (1978) Morphological aspects and the ecological and mechanical consequences of fat deposition in wild vertebrates. Annual Review of Ecology and Systematics 9:519–570

Pond CM (1984) Physiological and ecological importance of energy storage in the evolution of lactation: evidence for a common pattern of anatomical organization of adipose tissue in mammals. In: Peaker M, Vernon RG, and Knight CH (eds) Physiological strategies in lacation

(Symposia of the Zoological Society of London, No. 51). Academic Press, Orlando, FL, pp 1–32

Pond CM, Mattacks CA, Gilmour I, Johnston MA, Pilinger CT, and Prestrud P (1995a) Chemical and carbon isotopic composition of fatty acids in adipose tissue as indicators of dietary history in wild arctic foxes (*Alopex lagopus*) on Svalbard. Journal of Zoology London 236:611–623

Pond CM, Mattacks CA, Gilmour I, Johnston MA, Pilinger CT, and Prestrud P (1995b) Variability in the distribution and composition of adipose tissue in wild arctic foxes (*Alopex lagopus*) on Svalbard. Journal of Zoology London 236:593–610

Poole KG, Serrouya R, and Stuart-Smith K (2007) Moose calving strategies in interior montane ecosystems. Journal of Mammalogy 88:139–150

Post E, Langvatn R, Forchhammer MC, and Stenseth NC (1999) Environmental variation shapes sexual dimorphism in red deer. Proceedings National Academy of Sciences USA 96: 4467–4471

Poulton VK, Lovvorn JR, and Takekawa JY (2002) Clam density and scaup feeding behavior in San Pablo Bay, California. Condor 104:518–527

Prange HD and Schmidt-Nielsen K (1970) The metabolic cost of swimming in ducks. Journal of Experimental Biology 53:763–777

Price MA and White RG (1985) Growth and development. In: Hudson RJ and White RG (eds) Bioenergetics of wild herbivores. CRC Press, Boca Raton, FL, pp 183–213

Primack RB (2004) A primer of conservation biology, 3rd edn. Sinauer, Sunderland, MA

Pritchard GT and Robbins CT (1990) Digestive and metabolic efficiencies of grizzly and black bears. Canadian Journal of Zoology 68:1645–1651

Prothero J (1995) Bone and fat as a function of body weight in adult mammals. Comparative Biochemistry and Physiology A 111:633–639

Provenza FD (1995) Postingestive feedback as an elementary determinant of food preference and intake in ruminants. Journal of Range Management 48:2–17

Raberg L, Sim D, and Read AF (2007) Disentangling genetic variation for resistance and tolerance to infectious diseases in animals. Science 318:812–814

Rachlow JL and Bowyer RT (1991) Interannual variation in timing and synchrony of parturition in Dall's sheep. Journal of Mammalogy 72:487–492

Raubenheimer D and Simpson SJ (1997) Integrative models of nutrient balancing: application to insects and vertebrates. Nutrition Research Reviews 10:151–179

Raubenheimer D and Simpson SJ (2007) Geometric analysis: from nutritional ecology to livestock production. Recent Advances in Animal Nutrition in Australia 16:51–63

Raveling DG (2004) Waterfowl of the world. A comparative perspective. University of Missouri, Columbia, MO

Rea LD, Rosen DAS, and Trites AW (2000) Metabolic response to fasting in 6-week-old steller sea lion pups (*Emuetopias jubatus*). Canadian Journal of Zoology 78:890–894

Rechkemmer G, Gabel G, Diernaes L, Sehested J, Moller PD, and von Engelhardt W (1995) Transport of short chain fatty acids in the forestomach and hindgut. In: Engelhardt Wv, Leonhard-Marek S, Breves G, and Giesecke D (eds) Ruminant physiology, digestion, metabolism, growth and reproduction. Enke, Stuttgart, pp 95–116

Reeder D and Kramer KM (2005) Stress in free-ranging mammals: integrating physiology, ecology, and natural history. Journal of Mammalogy 86:225–235

Regehr EV, Lunn NJ, Amstrup SC, and Stirling I (2007) Effects of earlier sea ice breakup on survival and population size of polar bears in western Hudson Bay. Journal of Wildlife Management 71:2673–2683

Regelin WL, Schwartz CC, and Franzmann AW (1987) Effects of forest succession on nutritional dynamics of moose forage. Swedish Wildlife Research Suppl 1:247–263

Reichman OJ and Smith SC (1987) Burrows and burrowing behavior by mammals. In: Genoways HH (ed) Current Mammalogy. Plenum Press, New York, pp 197–244

Reiss MJ (1989) The allometry of growth and reproduction. Cambridge University Press, Cambridge

Remington TE and Braun CE (1988) Carcass composition and energy reserves of sage grouse during winter. Condor 90:15–19

Renecker LA and Hudson RJ (1990) Behavioral and thermoregulatory responses of moose to high ambient temperatures and insect harassment in aspen-dominated forests. Alces 26:66–72

Reynolds DS and Kunz TH (2001) Standard methods for destructive body composition analysis. In: Speakman JR (ed) Body composition analysis of animals. A handbook of non-destructive methods. Cambridge University Press, Cambridge, pp 39–55

Rhind SM, Archer ZA, and Adam CL (2002) Seasonality of food intake in ruminants: recent developments in understanding. Nutrition Research Reviews 15:43–65

Richardson PRK, Mundy PJ, and Plug I (1986) Bone crushing carnivores and their significance to osteodystrophy in griffon vulture chicks. Journal of Zoology 210:23–43

Richman S and Lovvorn JR (2004) Relative foraging value to lesser scaup ducks of native and exotic clams from San Francisco Bay. Ecological Applications 14:1217–1231

Ricklefs RE and Starck JM (1998) Embryonic growth and development. In: Starck JM and Ricklefs RE (eds) Avian growth and development. Evolution within the altricial-precocial spectrum. Oxford University Press, New York, pp 31–58

Ricklefs RE, Starck JM, and Konarzewski M (1998) Internal constraints on growth in birds. In: Starck JM and Ricklefs RE (eds) Avian growth and development. Evolution within the altricial-precocial spectrum. Oxford University Press, New York, pp 266–287

Ringberg TM, White RG, Holleman DF, and Luick JR (1981) Prediction of carcass composition in reindeer (*Rangifer tarandus tarandus* L.) by use of selected indicator bones and muscles. Canadian Journal of Zoology 59:583–588

Ringelman JK, Szymczak MR, Jeske CW, and Ragotzkie KE (1992) Ulnar lipid as an indicator of depleted fat reserves in mallards. Journal of Wildlife Management 56:317–321

Risenhoover KL and Peterson RO (1986) Mineral licks as a sodium source for Isle Royale moose. Oecologia 71:121–126

Robbins CT (1993) Wildlife feeding and nutrition, 2nd edn. Academic Press, San Diego

Robbins CT, Moen AN, and Reid JT (1974) Body composition of white-tailed deer. Journal of Animal Science 38:871–876

Robbins CT, Hanley TA, Hagerman AE, Hjeljord O, Baker DL, Schwartz CC, and Mautz WW (1987) Role of tannins in defending plants against ruminants: reduction in protein availability. Ecology 68:98–107

Robbins CT, Fortin JK, Rode KD, Farley SD, Shipley LA, and Felicetti L (2007) Optimizing protein intake as a foraging strategy to maximize mass gain in an omnivore. Oikos 116:1675–1682

Roby DD, Brink KL, and Place AR (1989) Relative passage rates of lipid and aqueous digesta in the formation of stomach oils. Auk 106:303–313

Roby DD, Taylor JRE, and Place AR (1997) Significance of stomach oil for reproduction in seabirds: an inter-species cross-fostering experiment. Auk 114:725–736

Rode KD, Farley SD, Fortin J, and Robbins CT (2007) Nutritional consequences of experimentally introduced tourism in brown bears. Journal of Wildlife Management 71:929–939

Rodger RWA (2006) The fisheries of North America. Canadian Marine Publications, Halifax, NS

Roffe TJ (1993) Perinatal mortality in caribou from the Porcupine herd, Alaska. Journal of Wildlife Diseases 29:295–303

Rogers CM (1987) Predation risk and fasting capacity: do wintering birds maintain optimal body mass? Ecology 68:1051–1061

Rogers CM, Nolan V Jr, and Ketterson ED (1994) Winter fattening in the dark-eyed junco: plasticity and possible interaction with migration trade-offs. Oecologia 97:526–532

Rombach EP, Barboza PS, and Blake JE (2002a) Trace mineral reserves of muskoxen during gestation: copper, ceruloplasmin, and metallothionein. Comparative Biochemistry and Physiology C 134:157–168

Rombach EP, Barboza PS, and Blake JE (2002b) Utilization of copper during lactation and neonatal development in muskoxen. Canadian Journal of Zoology 80:1460–1469

Romero LM (2002) Seasonal changes in plasma glucocorticoid concentrations in free-living vertebrates. General and Comparative Endocrinology 128:1–24

Root TL (1988) Atlas of wintering North American birds. University of Chicago Press, Chicago, IL

Root TL, Price JT, Hall KR, Schneider SH, Rosenweig C, and Pounds JA (2003) Fingerprints of global warming on wild animals and plants. Nature 421:57–60

Roze U (1989) The North American porcupine. Smithsonian Institution Press, Washington, DC

Ruf T, Valencak T, Tataruch F, and Arnold W (2006) Running speed in mammals increases with muscle n-6 polyunsaturated fatty acid content. PLOS One 1(1):e65

Russell DE, Martell AM, and Nixon WAC (1993) Range ecology of the Porcupine caribou herd in Canada. Rangifer 13:1–168

Russell JB (2002) Rumen microbiology and its role in ruminant nutrition. JB Russell, Ithaca, NY

Russell JB and Rychlik JL (2001) Factors that alter rumen microbial ecology. Science 292:1119–1122

Russell JB and Wilson DB (1996) Why are ruminal cellulolytic bacteria unable to digest cellulose at low pH? Journal of Dairy Science 79:1503–1509

Rust MB (2002) Nutritional physiology. In: Halver JE and Hardy RW (eds) Fish nutrition, 3rd edn. Academic Press, Amsterdam, pp 368–453

Ryg M and Jacobsen E (1983) Inhibition of growth and food intake in reindeer (*Rangifer tarandus tarandus*) by immunization to thyrotropin-releasing hormone. Acta Zoological Fennica 175:75

Ryg M and Langvatn R (1982) Seasonal changes in weight gain, growth hormone, and thyroid hormones in male red deer (*Cervus elaphus atlanticus*). Canadian Journal of Zoology 60:2577–2581

Rřnnestad I and Conceição LEC (2005) Aspects of protein and amino acid digestion and utilization by marine fish larvae. In: Starck JM and Wang T (eds) Physiological and ecological adaptations to feeding in vertebrates. Science Publishers, Enfield, NH, pp 389–416

Sacks BN (2005) Reproduction and body condition of California coyotes (*Canis latrans*). Journal of Mammalogy 86:1036–1041

Said HM (2004) Recent advances in carrier-mediated intestinal absorption of water-soluble vitamins. Annual Review of Physiology 66:419–446

Saltz D and White GC (1991) Urinary cortisol and urea nitrogen responses to winter stress in mule deer. Journal of Wildlife Management 55:1–16

Sams MG, Lochmiller RL, Qualls CWJ, Leslie DM Jr, and Payton ME (1996) Physiological correlates of neonatal mortality in an overpopulated herd of white-tailed deer. Journal of Mammalogy 77:179–190

Sapolsky RM (2001) Physiological and pathophysiological implication of social stress in mammals. In: McEwen BS and Goodman HM (eds) Handbook of physiology. Oxford University Press, New York, pp 517–532

Sargent JR, Tocher DR, and Bell JG (2002) The lipids. In: Halver JE and Hardy RW (eds) Fish nutrition, 3rd edn. Academic Press, Amsterdam, pp 181–257

Savage VM, Gillooly JF, Woodruff WH, West GB, Allen AP, Enquist BJ, and Brown JH (2004) The predominance of quarter-power scaling in biology. Functional Ecology 18:257–282

Savory CJ (1999) Temporal control of feeding behaviour and its association with gastrointestinal function. Journal of Experimental Zoology 283:339–347

Schmid J and Kappeler PM (2005) Physiological adaptations to seasonality in nocturnal primates. In: Brockman DK and van Schaik CP (eds) Seasonality in primates: studies of living and extinct human and non-human primates. Cambridge University Press, Cambridge, pp 3–20

Schmidt-Nielsen B, Schmidt-Nielsen K, Houpt TR, and Jarnum SA (1957) Urea excretion in the camel. American Journal of Physiology 188:477–478

Schmidt-Nielsen K (1997) Animal physiology: adaptation and environment. Cambridge University Press, Cambridge

Schmutz JA, Hobson KA, and Morse JA (2006) Strategic decisions in preparing to breed – an isotopic assessment of protein from diet and endogenous stores: effects on egg production and incubation behavior of geese. Ardea 94:385–399

Schwartz CC (1992) Physiological and nutritional adaptations of moose to northern environments. Alces Suppl 1:139–155

Schwartz CC, Regelin WL, and Franzmann AW (1982) Male moose successfully breed as yearlings. Journal of Mammalogy 63:334–335

Schwartz CC, Hubbert ME, and Franzmann AW (1988) Energy requirements of adult moose for winter maintenance. Journal of Wildlife Management 52:26–33

Schweigert FJ, Ryder OA, Rambeck WA, and Zucker H (1990) The majority of vitamin A is transported as retinyl esters in the blood of most carnivores. Comparative Biochemistry and Physiology A 95:573–578

Schwenk K and Rubega M (2005) Diversity of vertebrate feeding systems. In: Starck JM and Wang T (eds) Physiological and ecological adaptations to feeding in vertebrates. Science Publishers, Enfield, NH, pp 1–41

Secor SM and Diamond J (1997) Determinants of the postfeeding metabolic response of Burmese pythons, *Python molurus*. Physiological Zoology 70:202–212

Sedinger JS (1986) Growth and development of Canada goose goslings. Condor 88:169–180

Setchell KDR, Gosselin SJ, Welch MB, Johnston JO, Balistreri WF, Kramer LW, Dresser BL, and Tarr MJ (1987) Dietary estrogens: a probable cause of infertility and liver disease in captive cheetahs. Gastroenterology 93:225–233

Sharbaugh SM (2001) Seasonal acclimatization to extreme climatic conditions by black-capped chickadees (*Poecile atricapilla*) in interior Alaska (64°N). Physiological and Biochemical Zoology 74:568–575

Shipley LA (2007) The influence of bite size on foraging at larger spatial and temporal scales by mammalian herbivores. Oikos 116:1964–1974

Shipley LA and Spalinger DE (1992) Mechanics of browsing in dense food patches: effects of plant and animal morphology on intake rate. Canadian Journal of Zoology 70:1743–1752

Shkolnik A, Maltz E, and Choshniak I (1980) The role of the ruminant's digestive tract as a water reservoir. Digestive Physiology and Metabolism in Ruminants. AVI Publishing, Westport, CT, pp 731–742

Shochat E, Robbins CT, Parish SM, Young PB, Stephenson TR, and Tamayo A (1997) Nutritional investigations and management of captive moose. Zoo Biology 16:479–494

Sibly RM (1981) Strategies of digestion and defecation. In: Townsend CR and Calow P (eds) Physiological ecology: an evolutionary approach to resources use. Sinauer Associates, Sunderland, MA, pp 109–139

Sinclair ARE and Krebs CJ (2003) Complex numerical responses to top-down and bottom-up processes in vertebrate populations. In: Sibley RM, Hone J, and Clutton-Brock TH (eds) Wildlife population growth rates. Cambridge University Press, Cambridge, pp 127–147

Sinclair ARE, Mduma SAR, and Arcese P (2000) What determines phenology and synchrony of ungulate breeding in Serengeti? Ecology 81:2100–2111

Skinner DC, Moodley G, and Buffenstein R (1991) Is vitamin D_3 essential for mineral metabolism in the Damara mole-rat (*Cryptomys damarensis*)? General and Comparative Endocrinology 81:500–505

Smith FA, Betancourt JL, and Brown JH (1995) Evolution of body size in the woodrat over the past 25,000 years of climate change. Science 270:2012–2014

Smith RL and Smith TM (2001) Ecology and field biology. Benjamin Cummings, San Francisco

Smith SB, McPherson K, Backer JM, Pierce BJ, Podlesak DW, and McWilliams SR (2007) Fruit quality and consumption by songbirds during autumn migration. Wilson Journal of Ornithology 119:419–428

Smith VH, Jones TP II, and Smith MS (2005) Host nutrition and infectious disease: an ecological view. Frontiers in Ecology and the Environment 3:268–274

Snyder N and Snyder H (2000) The California condor: a saga of natural history and conservation. Academic Press, San Diego

Solberg EJ and Saether BE (1994) Male traits as life-history variables: annual variation in body mass and antler size in moose (*Alces alces*). Journal of Mammalogy 75:1069–1079

Soppela P, Nieminen M, Saarela S, Keith JS, Morrison JN, MacFarlane F, and Trayhurn P (1991) Brown fat-specific mitochondrial uncoupling protein in adipose tissues of newborn reindeer. American Journal of Physiology 260:R1229–R1234

Sorensen JS, Turnbull CA, and Dearing MD (2004) A specialist herbivore (*Neotoma stphensi*) absorbs fewer plant toxins than does a generalist (*Neotoma albigula*). Physiological and Biochemical Zoology 77:139–148

Spaeth DF, Bowyer RT, Stephenson TR, Barboza PS, and Van Ballenberghe V (2002) Nutritional quality of willows for moose: effects of twig age and diameter. Alces 38:143–154

Spaeth DF, Bowyer RT, Stephenson TR, and Barboza PS (2004) Sexual segregation in moose *Alces alces*: an experimental manipulation of foraging behaviour. Wildlife Biology 10:59–72

Speakman JR (1993) How should we calculate CO_2 production in doubly labelled water studies of animals? Functional Ecology 7:746–750

Speakman JR, Visser GH, Ward S, and Krol E (2001) The isotope dilution method for the evaluation of body composition. In: Speakman JR (ed) Body composition analysis of animals. A handbook of non-destructive methods. Cambridge University Press, Cambridge, pp 56–98

Speakman JR, Stubbs RJ, and Mercer JG (2002) Does body mass play a role in the regulation of food intake? Proceedings of the Nutrition Society 61:473–487

Spicer JI and Gaston KJ (1999) Physiological diversity and its ecological implications. Blackwell Science, Oxford

Spraker TR (1993) Stress and capture myopathy in artiodactylids. In: Fowler ME (ed) Zoo and wild animal medicine: current therapy, 3rd edn. WB Saunders, Philadelphia, PA, pp 481–488

Staaland H, White RG, Luick JR, and Holleman DF (1980) Dietary influences on sodium and potassium metabolism of reindeer. Canadian Journal of Zoology 58:1728–1734

Staaland H, Bjornstad H, Pedersen O, and Hove K (1991) Radiostrontium, radiocesium and stable mineral composition of bones of domestic reindeer from Vaga, Norway. Rangifer 11:17–22

Staaland H, Garmo TH, Hove K, and Pedersen O (1995) Feed selection and radiocaesium intake by reindeer, sheep and goats grazing alpine summer habitats in southern Norway. Journal of Environmental Radioactivity 29:39–56

Stahl J, van der Graaf AJ, Drent RH, and Bakker JP (2006) Subtle interplay of competition and facilitation among small herbivores in coastal grasslands. Functional Ecology 20:908–915

Starck JM (2005) Structural flexibility of the digestive system of tetarapods – patterns and processes at the cellular and tissue level. In: Starck JM and Wang T (eds) Physiological and ecological adaptations to feeding in vertebrates. Science Publishers, Enfield, NH, pp 176–200

Starck JM and Beese K (2002) Structural flexibility of the small intestine and liver of garter snakes in response to feeding and fasting. Journal of Experimental Biology 205:1377–1388

Starck JM and Ricklefs RE (1998) Patterns of development: the altricial-precocial spectrum. In: Starck JM and Ricklefs RE (eds) Avian growth and development. Evolution within the altricial-precocial spectrum. Oxford University Press, New York, pp 3–30

Steinheim G, Mysterud A, Holand O, Bakken M, and Adnoy T (2002) The effect of initial weight of the ewe on later reproductive effort in domestic sheep (*Ovis aries*). Journal of Zoology London 258:515–520

Stephenson TR, Hundertmark KJ, Schwartz CC, and Ballenberghe VV (1998) Predicting body fat and body mass in moose with ultrasonography. Canadian Journal of Zoology 76:717–722

Stephenson TR, Bleich VC, Pierce BM, and Mulcahy GP (2002) Validation of mule deer body composition using in vivo and post-mortem indices of nutritional condition. Wildlife Society Bulletin 30:557–564

Stevens CE and Hume ID (1995) Comparative physiology of the vertebrate digestive system, 2nd edn. Cambridge University Press, Cambridge

Stevens L (1996) Avian biochemistry and molecular biology. Cambridge University Press, Cambridge

Stevenson KT and van Tets IG (2008) Dual-energy x ray absorptiometry (DXA) can accurately and nondestructively measure the body composition of small free-living rodents. Physiological and Biochemical Zoology 81:373–382

Stewart KM, Bowyer RT, Ruess RW, Dick BL, and Kie JG (2006) Herbivore optimization by North American elk: consequences for theory and management. Wildlife Monographs 167:1–24

Stirling I and McEwan EH (1975) The caloric value of whole ringed seals (*Phoca hispida*) in relation to polar bear (*Ursus maritimus*) ecology and hunting behavior. Canadian Journal of Zoology 53:1021–1027

Stott A, Davies E, Evershed RP, and Tuross N (1997) Monitoring the routing of dietary and biosynthetic lipids through compound-specific stable isotope ($\delta^{13}C$) measurements at natural abundance. Naturwissenschaften 84:82–86

Summers M, McBride BW, and Milligan LP (1988) Components of basal energy expenditure. In: Dobson A and Dobson M (eds) Aspects of digestive physiology in ruminants. Comstock Publishing Associates, Ithaca, NY, pp 257–285

Susmel P and Stefanon B (1993) Aspects of lignin degradation by rumen microorganisms. Journal of Biotechnology 30:141–148

Suttle NF (1991) The interactions between copper, molybdenum, and sulphur in ruminant nutrition. Annual Review of Nutrition 11:121–140

Swart D, Mackie RI, and Hayes JP (1993a) Fermentative digestion in the ostrich (*Struthio camelus* var. *domesticus*), a large avian species that utilizes cellulose. South African Journal of Animal Science 23:127–135

Swart D, Siebrits FK, and Hayes JP (1993b) Growth, feed intake and body composition of ostriches (*Struthio camelus*) between 10 and 30 kg live mass. South African Journal of Animal Science 23:142–150

Swihart RK, Weeks HP Jr, Easter-Pilcher AL, and DeNicola AJ (1998) Nutritional condition and fertility of white-tailed deer (*Odocoileus virginianus*) from areas with contrasting histories of hunting. Canadian Journal of Zoology 76:1932–1941

Swor RM (2002) The effect of marginal copper diet on muskox calf growth and immune function. Unpublished MS Thesis, University of Alaska, Fairbanks

Tankersley NG and Gasaway WC (1983) Mineral lick use by moose in Alaska. Canadian Journal of Zoology 61:2242–2249

Targett NM and Arnold TM (2001) Effects of secondary metabolites on digestion in marine herbivores. In: McClintock JB and Baker BJ (eds) Marine chemical ecology. CRC Press, Boca Raton, FL, pp 391–411

Tatner P and Bryant DM (1986) Flight cost of a small passerine measured using doubly labeled water: implications for energetic studies. Auk 103:169–180

Taylor CR, Heglund NC, and Maloiy GMO (1982) Energetics and mechanics of terrestrial locomotion. I. Metabolic consumption as a function of speed and body size in birds and mammals. Journal of Experimental Biology 97:1–21

Taylor I (1994) Barn owls. Predator–prey relationships and conservation. Cambridge University Press, Cambridge

Taylor RB and Steinberg PD (2005) Host use by Australasian seaweed mesograzers in relation to feeding preferences of larger grazers. Ecology 86:2955–2967

Theodorou MK and France J (1993) Rumen microorganisms and their interactions. International quantitative aspects of ruminant digestion and metabolism. CAB International, Wallingford, Oxon, pp 145–163

Thomas DW, Blondel J, Perret P, Lambrechts MM, and Speakman JR (2001) Energetic and fitness costs of mismatching resource supply and demand in seasonally breeding birds. Science 291:2598–2600

Thomas VG and George JC (1975) Plasma and depot fat fatty acids in Canada geese in relation to diet, migration, and reproduction. Physiological Zoology 48:157–167

Thomson JM, Long JL, and Horton DR (1987) Human exploitation of and introduction to the Australian fauna. In: Dyne GR and Walton DW (eds) Fauna of Australia. General articles. Australian Government Publishing Service, Canberra, pp 227–249

Thouzeau C, Massemin S, and Handrich Y (1997) Bone marrow fat mobilization in relation to lipid and protein catabolism during prolonged fasting in barn owls. Journal of Comparative Physiology B 167:17–24

Tiku PE, Gracey AY, Macartney AI, Benyon RJ, and Cossins AR (1996) Cold-induced expression of delta 9-desaturase in carp by transcriptional and posttranslational mechanisms. Science 271:815–818

Tilgar V, Ots I, and Mand R (2008) The rate of bone mineralization in birds is directly related to alkaline phosphatase activity. Physiological and Biochemical Zoology 77:530–535

Tomasi TE and Mitchell DA (1994) Seasonal shifts in thyroid function in the cotton rat (*Sigmodon hispidus*). Journal of Mammalogy 75:520–528

Torbit SC, Carpenter LH, Bartmann RM, Alldredge AW, and White GC (1988) Calibration of carcass fat indices in wintering mule deer. Journal of Wildlife Management 52:582–588

Torrell DT, Hume ID, and Weir WC (1972) Effect of level of protein and energy during flushing on lambing performance of range ewes. Journal of Animal Science 34:479–482

Tracy RL and Walsberg GE (2002) Kangaroo rats revisited: re-evaluating a classic case of desert survival. Oecologia 133:449–457

Tramontin RR, Giles RC Jr, Hong CB, and Newman LE (1983) Myodegeneration in Kentucky white-tailed deer. Journal of the American Veterinary Medical Association 183:1263–1265

Trites AW and Joy R (2005) Dietary analysis from fecal samples: how many scats are enough? Journal of Mammalogy 86:704–712

Trudell J and White RG (1981) The effect of forage structure and availability on food intake, biting rate, bite size and daily eating time of reindeer. Journal of Applied Ecology 18:63–81

Trumble SJ, Barboza PS, and Castellini MA (2003) Digestive constraints on an aquatic carnivore: effects of feeding frequency and prey composition on harbor seals. Journal of Comparative Physiology B 173:501–509

Tyler NJC (1986) The relationship between the fat content of Svalbard reindeer in autumn and their death from starvation in winter. Rangifer Special Issue No. 1:311–314

Tyler NJC, Turi JM, Sundset MA, Strom Bull K, Sara MN, Reinert E, Oskal N, Nellemann C, McCarthy JJ, Mathiesen SD, Martello ML, Magga OH, Hovelsrud GK, Hanssen-Bauer I, Eira NI, Eira IMG, and Corell RW (2007) Saami reindeer pastoralism under climate change: applying a generalized framework for vulnerability studies to a sub-arctic social-ecological system. Global Environmental Change 17:191–206

Unangst ET Jr and Wunder BA (2001) Need for species-specific models for body-composition estimates of small mammals using EM-SCAN. Journal of Mammalogy 85:527–534

Underwood EJ and Suttle NF (2001) The mineral nutrition of livestock. CAB International, Wallingford, Oxon

Urton EJM and Hobson KA (2005) Intrapopulation variation in gray wolf isotope (δ^{15}N and δ^{13C}) profiles: implications for the ecology of individuals. Oecologia 145:317–326

Vagnoni DB, Garrott RA, Cook JG, White PJ, and Clayton MK (1996) Urinary allantoin: creatinine ratios as a dietary index for elk. Journal of Wildlife Management 60:728–734

Valera F, Wagner RH, Romero-Pujante M, Gutierrez JE, and Rey PJ (2005) Dietary specialization on high protein seeds by adult and nestling serins. Condor 107:29–40

Van Der Wal R and Brooker RW (2004) Mosses mediate grazer impacts on grass abundance in arctic ecosystems. Functional Ecology 18:77–86

Van Der Wal R, Van Wijnen H, Van Wieren S, Beucher O, and Bos D (2000) On facilitation between herbivores: how brent geese profit from brown hares. Ecology 81:969–980

Van Devender TR, Averill-Murray RC, Esque TC, Holm PA, Dickinson VM, Schwalbe CR, Wirt EB, and Barrett SL (2002) Grasses, mallows, desert vine, and more. Diet of the desert tortoise in Arizona and Sonora. In: Van Devender TR (ed) The Sonoran Desert tortoise: natural history, biology, and conservation. University of Arizona Press, Tucson, AZ, pp 159–193

van Eys J (1991) Nicotinic acid. In: Machlin LJ (ed) Handbook of vitamins, 2nd edn. Marcel Dekker, New York, pp 311–340

van Oort BEH, Tyler NJC, Gerkema MP, Folkow L, and Stokkan K-A (2007) Where clocks are redundant: weak circadian mechanisms in reindeer living under polar conditions. Naturwissenschaften 94:183–194

van Schaik CP and Brockman DK (2005) Seasonality in primate ecology, reproduction, and life history: an overview. In: Brockman DK and van Schaik CP (eds) Seasonality in primates: studies

of living and extinct human and non-human primates. Cambridge University Press, Cambridge, pp 3–20
Van Soest PJ (1994) Nutritional ecology of the ruminant, 2nd edn. Cornell University Press, Ithaca, NY
Van Soest PJ, Robertson JB, and Lewis BA (1991) Methods for dietary fiber, neutral detergent fiber, and nonstarch polysaccharides in relation to animal nutrition. Journal of Dairy Science 74:3583–3597
Vanderkist BA, Williams TD, Bertram DF, Lougheed L, and Ryder JP (2000) Indirect, physiological assessment of reproductive state and breeding chronology in free-living birds: an example in the marbled murrelet (*Brachyramphus marmoratus*). Functional Ecology 14:758–765
Vest JL, Kaminski RM, Afton AD, and Vilella FJ (2006) Body mass of lesser scaup during fall and winter in the Mississippi flyway. Journal of Wildlife Management 70:1789–1795
Vispo C and Karasov WH (1997) The interaction of avian gut microbes and their host: an elusive symbiosis. In: Mackie RI and White BA (eds) Gastrointestinal microbiology. Chapman and Hall, New York, pp 116–155
Visser GH, Dekinga A, Achterkamp B, and Piersma T (2000) Ingested water equilibrates isotopically with body water pool of a shorebird with unrivalled water fluxes. American Journal of Physiology. Regulatory, Integrative and Comparative Physiology 279:R1795–R1804
Vleck CM and Bucher TL (1998) Energy metabolism, gas exchange, and ventilation. In: Starck JM and Ricklefs RE (eds) Avian growth and development. Evolution within the altricial-precocial spectrum. Oxford University Press, New York, pp 89–116
Vézina F and Williams TD (2003) Plasticity of body composition in breeding birds: what drives the metabolic costs of egg production? Physiological and Biochemical Zoology 76:716–730
Vézina F, Speakman JR, and Williams TD (2006) Individually variable energy management strategies in relation to energetic costs of egg production. Ecology 87:2247–2458
Walker B and Salt D (2006) Resilience thinking: sustaining ecosystems and people in a changing world. Island Press, Washington, DC
Walker BG, Wingfield JC, and Boersma PD (2005) Age and food deprivation affects expression of the glucocorticoid stress response in Magellanic penguin (*Sphenicus magellanicus*) chicks. Physiological and Biochemical Zoology 78:78–89
Walsberg GE (2003) How useful is energy balance as a overall index of stress in animals? Hormones and Behavior 43:16–17
Walther GR, Post E, Convey P, Menzel A, Parmesan C, Beebee TJC, Fromentin JM, Hoegh-Guldberg O, and Bairlein F (2002) Ecological responses to recent climate change. Nature 416:389–395
Wang J-M, Zhang Y-M, and Wang D-H (2006) Seasonal thermogenesis and body mass regulation in plateau pikas (*Ochotona curzoniae*). Oecologia 149:373–382
Wang LCH and Lee TF (1996) Torpor and hibernation in mammals: metabolic biochemical and physiological adaptations. In: Fregly MJ and Blatteis CM (eds) Handbook of physiology. American Physiological Society, Bethesda, MD, 507–532
Wang SW, Iverson SJ, Springer AM, and Hatch SA (2007) Fatty acid signatures of stomach oil and adipose tissue of northern fulmars (*Fulmarus glacialis*) in Alaska: implications for diet analysis of Procellariform birds. Journal of Comparative Physiology B 177:893–903
Wang T, Andersen JB, and Hicks JW (2005) Effects of digestion on the respiratory and cardiovascular physiology of amphibians and reptiles. In: Starck JM and Wang T (eds) Physiological and ecological adaptations to feeding in vertebrates. Science Publishers, Enfield, NH, pp 229–254
Wang WH (2004) Regulation of renal K transport by dietary K intake. Annual Review of Physiology 66:547–569
Wannemacher RWJ and Cooper WK (1970) Relationship between protein metabolism in muscle tissue and the concept of 'Protein Reserves'. In: Bianchi CP and Hilf R (eds) Protein metabolism and biological function. Rutgers University, New Jersey, pp 121–138
Warner ACI (1981) Rate of passage of digesta through the gut of mammals and birds. Nutrition Abstracts and Reviews Series B 51:789–820

Washburn BE, Morris DL, Millspaugh JJ, Faaborg J, and Schulz JH (2002) Using a commercially available radioimmunoassay to quantify corticosterone in avian plasma. Condor 104:558–563
Wasser SK, Hunt KE, Brown JL, Cooper K, Crockett CM, Bechert U, Milspaugh JJ, Larson S, and Monfort SL (2000) A generalized fecal glucocorticoid assay for use in a diverse array of nondomestic mammalian and avian species. General and Comparative Endocrinology 120:260–275
Watanabe H and Tokuda G (2001) Animal cellulases. Cellular and Molecular Life Sciences 58:1167–1178
Waterlow JC (1999) The mysteries of nitrogen balance. Nutrition Research Reviews 12:25–54
Weimerskirch H, Gault A, and Cherel Y (2005) Prey distribution and patchiness: factors in foraging success and efficiency of wandering albatrosses. Ecology 86:2611–2622
Weiss WP and Spears JW (2006) Vitamin and trace mineral effects on immune function of ruminants. In: Sejrsen K, Hvelpund T, and Nielsen MO (eds) Ruminant physiology. Wageningen Acadameic Publishers, Wageningen, pp 473–496
Weladji RB, Holand Ø, Steinheim G, Colman JE, Gjøstein H, and Kosmo A (2005) Sexual dimorphism and intercohort variation in reindeer calf antler length is associated with density and weather. Oecologia 145:549–555
Welch KC Jr, Bakken BH, and Martinez del Rio C (2006) Hummingbirds fuel hovering flight with newly ingested sugar. Physiological and Biochemical Zoology 79:1082–1087
Welsley DE, Knox KL, and Nagy JG (1973) Energy metabolism of pronghorn antelope. Journal of Wildlife Management 37:563–573
White GC (2000) Modeling population dynamics. In: Demarais S and Krausman PR (eds) Ecology and management of large mammals in North America. Prentice Hall, Upper Saddle River, NJ, pp 84–107
White RG (1983) Foraging patterns and their multiplier effects on productivity of northern ungulates. Oikos 40:377–384
White RG and Yousef MK (1977) Energy expenditure in reindeer walking on roads and on tundra. Canadian Journal of Zoology 56:215–223
White TCR (1993) The inadequate environment: nitrogen and the abundance of animals. Springer, Berlin
Whitford MF, McPherson MA, Forster RJ, and Teather RM (2001) Identification of bacteriocin-like inhibitors from rumen *Streptococcus* spp. and isolation and characterization of Bovicin 255. Applied and Environmental Microbiology 67:569–574
Whyte RJ and Bolen EG (1984) Impact of winter stress on mallard body composition. Condor 86:477–482
Wickstrom ML, Robbins CT, Hanley TA, Spalinger DE, and Parish SM (1984) Food intake and foraging energetics of elk and mule deer. Journal of Wildlife Management 48:1285–1301
Wiggins NL, McArthur C, and Davies NW (2006) Diet switching in a generalist mammalian folivore: fundamental to maximising intake. Oecologia 147:650–657
Wikelski M and Carbone C (2004) Environmental scaling of body size in island populations of Galapagos marine iguanas. In: Alberts AC, Carter RL, Hayes WK, and Martins EP (eds) Iguanas: biology and conservation. University of California Press, Berkeley, CA, pp 148–157
Wild SH and Byrne CD (2004) Evidence for fetal programming of obesity with a focus on putative mechanisms. Nutrition Research Reviews 17:153–162
Wilkes GE and Janssens PA (1988) The development of renal function. In: Tyndale-Biscoe CH and Janssens PA (eds) The developing marsupial. Springer, Berlin, pp 176–189
Williams JB, Bradshaw D, and Schmidt L (1995) Field metabolism and water requirements of spinifex pigeons (*Geophaps plumifera*) in Western Australia. Australian Journal of Zoology 43:1–15
Williams TM, Haun J, Davis RW, Fuiman LA, and Kohin S (2001) A killer appetite: metabolic consequences of carnivory in marine mammals. Comparative Biochemistry and Physiology A 129:785–796
Williams TM, Estes JA, Doak DA, and Springer AM (2004a) Killer appetites: assessing the role of predators in ecological communities. Ecology 85:3373–3384

Williams TM, Fuiman LA, Horning M, and Davis RW (2004b) The cost of foraging by a marine predator, the Weddell seal *Leptonychotes weddellii*: pricing by the stroke. Journal of Experimental Biology 207:973–982

Wilmshurst JF, Fryxell JM, and Hudson RJ (1995) Forage quality and patch choice by wapiti (*Cervus elaphus*). Behavioral Ecology 6:209–217

Wilson DE and Ruff S (1999) The Smithsonian book of North American mammals. Smithsonian Institution Press, Washington, DC

Wingfield JC (2005) The concept of allostasis: coping with a capricious environment. Journal of Mammalogy 86:248–254

Wingfield JC and Ramenofsky M (1999) Hormones and the behavioral ecology of stress. In: Balm PHM (ed) Stress physiology in animals. Sheffield Academic Press, Sheffield, pp 1–51

Wingfield JC, Hahn TP, Levin R, and Honey P (1992) Environmental predictability and control of gonadal cycles in birds. Journal of Experimental Zoology 261:214–231

Withers PC (1992) Comparative animal physiology. Saunders College Publishing, Fort Worth, TX

Withers PC, Cooper CE, and Larcombe AN (2006) Environmental correlates of physiological variables in marsupials. Physiological and Biochemical Zoology 79:437–453

Wittmer HU, McLellan BN, Seip DR, Young JA, Kinley TA, Watts GS, and Hamilton D (2005) Population dynamics of the endangered mountain ecotype of woodland caribou (*Rangifer tarandus caribou*) in British Columbia, Canada. Canadian Journal of Zoology 83:407–418

Woebeser GA (2006) Essentials of disease in wild animals. Blackwell, Ames, IA

Wood IS, Dyer J, Hofmann RR, and Shirazi-Beechey SP (2000) Expression of the Na^+ glucose co-transporter (SGLT1) in the intestine of domestic and wild ruminants. Pflügers Archives 441:155–162

Wooley JB and Owen RB Jr (1978) Energy costs of activity and daily energy expenditure in the black duck. Journal of Wildlife Management 42:739–745

Wright G, Peterson RO, Smith DW, and Lemke TO (2006) Selection of northern Yelowstone elk by gray wolves and hunters. Journal of Wildlife Management 70:1070–1078

Yancey PH (1988) Osmotic effectors in kidneys of xeric and mesic rodents: corticomedullary distributions and changes with water availability. Journal of Comparative Physiology B 158:369–380

Yokota SD, Benyajati S, and Dantzler WH (1985) Comparative aspects of glomerular filtration in vertebrates. Renal Physiology 8:193–221

Young V and Hume ID (2005) Nitrogen requirements and urea recycling in a bandicoot. Physiological and Biochemical Zoology 78:456–467

Zuercher GL, Roby DD, and Rexstad EA (1997) Validation of two new total body electrical conductivity (TOBEC) instruments for estimating body composition of live northern red-backed voles *Clethrionomys rutilis*. Acta Theriologica 42:387–397

Zuercher GL, Roby DD, and Rexstad EA (1999) Seasonal changes in body mass, composition, and organs of northern red-backed voles in interior Alaska. Journal of Mammalogy 80:443–459

Zug GR (1993) Herpetology. An introductory biology of amphibians and reptiles. Academic Press, San Diego, CA

Zullinger EM, Ricklefs RE, Redford KH, and Mace GM (1984) Fitting sigmoidal equations to mammalian growth curves. Journal of Mammalogy 65:607–636.

List of Common and Scientific Names of Animals and Plants

Taxa	Common name	Latin name
Alga	Blue-green alga	*Spirulina* sp.
Plant	Acorn	*Quercus* sp.
Plant	Alfalfa	*Medicago sativa*
Plant	Almond	*Prunus amygdalus*
Plant	Avocado	*Persea* sp.
Plant	Beech	*Fagus* sp.
Plant	Beet	*Beta* sp.
Plant	Blueberry	*Vaccinium* sp.
Plant	Bracken fern	*Pteridium aquilinum*
Plant	Cabbage	*Brassica oleracea*
Plant	Cassava	*Manihot esculenta*
Plant	Chili pepper	*Capsicum anuum*
Plant	Clover	*Trifolium repens*
Plant	Corn	*Zea mays*
Plant	Creosote bush	*Larrea tridentata*
Plant	False lily of the valley	*Maianthemum dilatatum*
Plant	Gemsbok cucumber	*Acanthosicyos naudinianus*
Plant	Globemallow	*Sphaeralcea ambigua*
Plant	Horsetail fern	*Equisetum arvense*
Plant	Maple	*Acer* sp.
Plant	Milk vetch	*Astragalus* sp.
Plant	Milkweed	*Asclepias currassavicas*
Plant	Nardoo fern	*Marsilea drummondii*
Plant	Oak	*Quercus* sp.
Plant	Olive	*Olea europaea*
Plant	Rice	*Oryza sativa*
Plant	Sagebrush	*Artemesia tridentata*
Plant	Salt bush	*Atriplex* sp.
Plant	Schismus grass	*Schismus barbatus*
Plant	Sea grass	*Zostera marina*
Plant	Skunk cabbage	*Lysichiton americanum*
Plant	Spreading woodfern	*Dryopteris dilatata*
Plant	Subterranean clover	*Trifolium subterraneum*
Plant	Sugar cane	*Saccharum* sp.
Plant	Sunflower	*Helianthus annuus*
Plant	Sweet clover	*Melilotus* sp.
Plant	Sweet potato	*Ipomoea batatas*

(continued)

(continued)

Taxa	Common name	Latin name
Plant	Terricolous lichen	*Cladina stellaris*
Plant	Trefoil	*Lotus pedunculatus*
Plant	Turnip	*Brassica rapa*
Plant	Wheat	*Triticum aestivum*
Plant	White spruce	*Picea glauca*
Plant	Whitebark pine	*Pinus albicaulis*
Plant	Willow	*Salix barclayi*
Fungus	Yeast	*Saccharomyces cervisiae*
Invertebrate	Amphipod	*Gammarus* sp.
Invertebrate	Blue mussel	*Mytilus edulus*
Invertebrate	Brine shrimp	*Artemia* sp.
Invertebrate	Christmas beetle	*Chrysophtharta flaveola*
Invertebrate	Edible snail	*Helix pomatia*
Invertebrate	Freshwater shell fish	*Anodonta cygnaea*
Invertebrate	Monarch butterfly	*Danaus plexippus*
Invertebrate	Subterranean termite	*Nasutitermes wolheri*
Fish	American shad	*Alosa sapidissima*
Fish	Atlantic salmon	*Salmo salar*
Fish	Barracuda	*Sphyraena baracuda*
Fish	Butterfly fish	*Chaetodon auriga*
Fish	Capelin	*Malotus villosus*
Fish	Carp	*Cyprinus carpio*
Fish	Chinook salmon	*Oncorhynchus tshawytscha*
Fish	Damsel fish	*Dascyllus albisella*
Fish	Dogfish shark	*Squalus acanthias*
Fish	Electric eel	*Electrophorus* sp.
Fish	Goldfish	*Crassus auratus*
Fish	Grouper	*Epinephelus* sp.
Fish	Herring	*Clupea pallasii*
Fish	Lamprey	*Petromyzon* sp.
Fish	Marine cod	*Gadus moruha*
Fish	Pollock	*Theragra chalcogramma*
Fish	Rainbow trout	*Oncorhynchus mykiss*
Fish	Sculpin	*Cottus* sp.
Fish	Sockeye salmon	*Oncorhynchus nerka*
Fish	Spotted tilapia	*Tilapia mariae*
Fish	Sturgeon	*Acipenser* sp.
Fish	Three-spined stickleback	*Gasterosteus aculeatus*
Fish	White shark	*Carcharodon carcharias*
Amphibian	Cane toad	*Bufo marinus*
Amphibian	Edible frog	*Pelophylax kl. esculentus*
Amphibian	Wood frog	*Rana sylvatica*
Reptile	Alligator	*Alligator mississippiensis*
Reptile	Anaconda	*Eunectes* sp.
Reptile	Burmese python	*Python molurus*
Reptile	Caiman	*Caiman latirostris*
Reptile	Chameleon lizard	*Chamaeleo* spp.

(continued)

(continued)

Taxa	Common name	Latin name
Reptile	Desert tortoise	*Xeroabtes agasszii*
Reptile	Freshwater crocodile	*Crocodylus johnstoni*
Reptile	Galapagos tortoise	*Geochelone nigra*
Reptile	Gharial crocodile	*Gavialis gangeticus*
Reptile	Green iguana	*Iguana iguana*
Reptile	Green sea turtle	*Chelonia mydas*
Reptile	Hawksbill turtle	*Eretmochelys imbricata*
Reptile	Indian cobra	*Naja naja*
Reptile	Marine iguana	*Amblyrhynchus cristatus*
Reptile	Ornate dragon	*Ctenophorus* (syn. *Amphibolurus*) *ornatus*
Reptile	Red-bellied turtle	*Pseudemys nelsonii*
Reptile	Saltwater crocodile	*Crocodylus porosus*
Reptile	Viperous snake	*Croatalus* spp.
Bird	Alpine swift	*Apus melba*
Bird	American goldfinch	*Carduelis tristis*
Bird	American kestrel	*Falco sparverius*
Bird	American robin	*Turdus migratorius*
Bird	Australian brush turkey	*Alectura lathami*
Bird	Australian wood duck	*Chenonetta jubatus*
Bird	Barn owl	*Tyto alba*
Bird	Barnacle goose	*Branta bernicla*
Bird	Black duck	*Anas rubripes*
Bird	Black vulture	*Coragyps atratus*
Bird	Black-capped chickadee	*Parus atricapillus*
Bird	Black-chinned hummingbird	*Archilochus alexandri*
Bird	Black-winged stilt	*Himantopus himantopus*
Bird	Blue grouse	*Dendragapus obscurus*
Bird	Blue jay	*Cyanocitta cristata*
Bird	Blue tit	*Cyanistes caeruleus*
Bird	Brant	*Branta bernicla*
Bird	California condor	*Gymnogyps californianus*
Bird	Canada goose	*Branta canadensis*
Bird	Canary	*Serinus canaria*
Bird	Canvasback duck	*Aythya valisineria*
Bird	Cape Griffon vulture	*Gyps cprotheres*
Bird	Chicken	*Gallus gallus*
Bird	Common eider	*Somateria mollissima*
Bird	Common merganser	*Mergus merganser*
Bird	Common raven	*Corvus corax*
Bird	Costa's hummingbird	*Calypte costa*
Bird	Emperor geese	*Chen canagica*
Bird	Emu	*Dromaius novaehollandae*
Bird	Gambel's quail	*Callipepla gambelii*
Bird	Golden eagle	*Aqulia chryasetos*
Bird	Gray jay	*Perisoreus canadensis obscurus*
Bird	Great horned owl	*Bubo virginianus*
Bird	Greater flamingo	*Phoenicopterus roseus*
Bird	Greylag goose	*Anser anser*
Bird	Gull	*Larus* spp.

(continued)

(continued)

Taxa	Common name	Latin name
Bird	Harlequin duck	*Histrionicus histrionicus*
Bird	Herring gull	*Larus argentatus*
Bird	House finch	*Carpodacus mexicanus*
Bird	Japanese quail	*Coturnix japonica*
Bird	King parrot	*Alisterus scapularis*
Bird	King penguin	*Aptenodytes patagonicus*
Bird	Kiwi	*Apteryx* sp.
Bird	Lesser scaup	*Aythya affinis*
Bird	Magellanic penguin	*Spheniscus magellanicus*
Bird	Mallard duck	*Anas platyrhynchos*
Bird	Mew gull	*Larus canus*
Bird	Mute swan	*Cygnus olor*
Bird	Northern fulmar	*Fulmarus glacialis*
Bird	Northern shoveler	*Anas clypeata*
Bird	Olive sunbird	*Cyanomitra olivacea*
Bird	Oriental white-backed vulture	*Gyps bengalensis*
Bird	Ostrich	*Struthio dromedarius*
Bird	Pacific black brant	*Branta bernicla nigricans*
Bird	Peregrine falcon	*Falco peregrinus*
Bird	Pigeon	*Columba livia*
Bird	Ptarmigan	*Lagopus* sp.
Bird	Rainbow lorikeet	*Trichglossus haematodus*
Bird	Red grouse	*Lagopus lagopus scotica*
Bird	Red knot	*Calidris canutus*
Bird	Red-eyed vireo	*Vireo olivaceous*
Bird	Rhea	*Rhea pennata*
Bird	Ring dove	*Streptopelia risoria*
Bird	Sage grouse	*Centrocercus urophasianus*
Bird	Scarlet tanager	*Piranga olivacea*
Bird	Snow goose	*Chen caerulescens*
Bird	Spectacled eider	*Somateria fischeri*
Bird	Spruce grouse	*Dendragapus canadensis*
Bird	Tree swallow	*Tachycineta bicolor*
Bird	Turkey	*Meleagris gallopavo*
Bird	Turkey vulture	*Cathartes aura*
Bird	White-crowned sparrow	*Zonotrichia leucophrys*
Bird	Whooping crane	*Grus americana*
Bird	Wood stork	*Mycteria americana*
Bird	Zebra finch	*Taeniopygia guttata*
Mammal	African buffalo	*Synercus caffer*
Mammal	African wild dog	*Lycaon pictus*
Mammal	American badger	*Taxidea taxus*
Mammal	American mink	*Mustella vison*
Mammal	Antarctic fur seal	*Arctocephalus tropicalis*
Mammal	Arabian oryx	*Oryx leucoryx*
Mammal	Arctic ground squirrel	*Spermophilus paryii*
Mammal	Arctic hare	*Lepus arcticus*
Mammal	Baboon	*Papio hamydryas*
Mammal	Baikal seal	*Pusa sibrica*

(continued)

(continued)

Taxa	Common name	Latin name
Mammal	Beluga whale	*Delphinapterus leucas*
Mammal	Bighorn sheep	*Ovis canadensis*
Mammal	Bilby	*Macrotis lagotis*
Mammal	Black bear	*Ursus americanus*
Mammal	Black rhino	*Diceros bicornis*
Mammal	Black-footed ferret	*Mustells nigripes*
Mammal	Black-tailed deer	*Odocoileus hemionus*
Mammal	Black-talied jack rabbit	*Lepus californicus*
Mammal	Boar	*Sus scrofa*
Mammal	Bowhead whale	*Balaena mysticetus*
Mammal	Brown bear	*Ursus arctos*
Mammal	Brown rat	*Rattus norvegicus*
Mammal	Brushy tailed woodrat	*Neotoma cinerea*
Mammal	Camel	*Camelus dromedarius*
Mammal	Caribou	*Rangifer tarandus*
Mammal	Cattle	*Bos taurus*
Mammal	Cheetah	*Acinonyx jubatus*
Mammal	Common brushtail possum	*Trichosurus vulpecula*
Mammal	Common marmoset	*Callithrix jachus*
Mammal	Common ringtail possum	*Pseudochirus peregrinus*
Mammal	Cotton tail rabbit	*Sylvilagus floridanus*
Mammal	Coyote	*Canis latrans*
Mammal	Dall's sheep	*Ovis dalli*
Mammal	Damara mole rat	*Cryptomys damarensis*
Mammal	Deer mouse	*Peromyscus maniculatus*
Mammal	Dingo	*Canis lupus dingo*
Mammal	Domestic cat	*Felis domesticus*
Mammal	Donkey	*Equus asinus*
Mammal	Dormouse	*Glis glis*
Mammal	Dugong	*Dugong dugong*
Mammal	Eastern gray kangaroo	*Macropus giganteus*
Mammal	Echidna	*Tachyglossus aculeatus*
Mammal	Eland	*Taurotragus oryx*
Mammal	Elephant	*Loxodonta africana*
Mammal	Elk	*Cervus elaphus*
Mammal	Ethiopian wolf	*Canis simensis*
Mammal	European hare	*Lepus europaeus*
Mammal	European rabbit	*Oryctolagus cuniculus*
Mammal	Fallow deer	*Dama dama*
Mammal	Field vole	*Microtus agrestis*
Mammal	Flying squirrel	*Glaucomys* spp.
Mammal	Giant panda	*Ailuropoda melanoleuca*
Mammal	Giraffe	*Giraffa cemelopardalis*
Mammal	Goat	*Capra hircus*
Mammal	Gray wolf	*Canis lupus*
Mammal	Greater glider	*Petauroides volans*
Mammal	Gray seal	*Halichoerus grypus*
Mammal	Grizzly bear	*Ursus arctos*
Mammal	Hairy-nosed wombat	*Lasiorhinus latifrons*
Mammal	Hamydryas baboon	*Papio hamydryas*

(continued)

(continued)

Taxa	Common name	Latin name
Mammal	Harbor seal	*Phoca vitulina*
Mammal	Hartebeest	*Alcephalus buscephalus*
Mammal	Hippo	*Hippopotamus amphbius*
Mammal	Hoary marmot	*Marmota caligata*
Mammal	Honey possum	*Tarsipes rostratus*
Mammal	Hooded seal	*Cystophora cristata*
Mammal	Hopping mouse	*Notomys alexis*
Mammal	Horse	*Equus cabalus*
Mammal	Impala	*Aepyceros melampus*
Mammal	Indian mongoose	*Herpestes javanicus*
Mammal	Indian rhino	*Rhinoceros unicornis*
Mammal	Japanese serow	*Capricornis crispus*
Mammal	Javelina	*Pecari tajacu*
Mammal	Killer whale	*Orcinus orca*
Mammal	Kit fox	*Vulpes macrotis*
Mammal	Koala	*Phascolarctos cinereus*
Mammal	Lemming	*Lemmus* spp.
Mammal	Lion	*Panthera leo*
Mammal	Long-tailed macaque	*Macaca fasicularis*
Mammal	Lynx	*Lynx canadensis*
Mammal	Maned wolf	*Chrysocyon brachyurus*
Mammal	Marmot	*Marnota* sp.
Mammal	Meerkat	*Suricatta suricatta*
Mammal	Merriam's kangaroo rat	*Dipodomys merriami*
Mammal	Montane vole	*Microtus montanus*
Mammal	Moose	*Alces alces*
Mammal	Mormon cricket	*Anabrus simplex*
Mammal	Mountain goat	*Oreamnos americanus*
Mammal	Mountain hare	*Lepus timidus*
Mammal	Mountain lion	*Puma concolor*
Mammal	Mule deer	*Odocoileus hemionus*
Mammal	Mule deer	*Odocoileus hemionus hemionus*
Mammal	Muskox	*Ovibos moschatus*
Mammal	Northern elephant seal	*Mirounga angustirostris*
Mammal	Northern red-backed vole	*Clethrionomys rutilus*
Mammal	Pig	*Sus srcofa*
Mammal	Pika	*Ochotona* sp.
Mammal	Plains bison	*Bison bison*
Mammal	Plains zebra	*Equus burchelli*
Mammal	Plateau pika	*Ochotona curzoniae*
Mammal	platypus	*Ornithorhynchus anatinus*
Mammal	Polar bear	*Ursus maritimus*
Mammal	Prairie vole	*Microtus orchogaster*
Mammal	Prezwalski's horse	*Equus ferus*
Mammal	Pronghorn antelope	*Antilocarpa americana*
Mammal	Pygmy rabbit	*Brachylagus idahoensis*
Mammal	Quokka	*Setonix brachyurus*
Mammal	Raccoon	*Procyon lotor*
Mammal	Red deer	*Cervus elaphus*
Mammal	Red fox	*Vulpes vulpes*

(continued)

(continued)

Taxa	Common name	Latin name
Mammal	Red kangaroo	*Macropus rufus*
Mammal	Red squirrel	*Tamiasciurus hudsonicus*
Mammal	Rhesus monkey	*Macaca mulatta*
Mammal	Roe deer	*Capreolus capreolus*
Mammal	Rufous rat kangaroo	*Aepyprymnus rufescens*
Mammal	Saiga antelope	*Saiga tartarica tartarica*
Mammal	sand gazelle	*Gazella subgutturosa*
Mammal	Sea otter	*Enhydra lutris*
Mammal	Sheep	*Ovis aries*
Mammal	Sitka black-tailed deer	*Odocoileus hemionus sitkensis*
Mammal	Snowshoe hare	*Lepus americanus*
Mammal	South American sea lion	*Otaria byronia*
Mammal	Spotted hyena	*Crocutta crocutta*
Mammal	Springbok antelope	*Antidorcas marsupialis*
Mammal	Steller sea lion	*Eumetopias jubatus*
Mammal	Sumatran rhino	*Dicerorhinus sumatrensis*
Mammal	Swamp wallaby	*Wallabia bicolor*
Mammal	Tammar wallaby	*Macropus eugenii*
Mammal	Tasmanian devil	*Sarcophilus harrisii*
Mammal	Thirteen lined ground squirrel	*Spermophilus tridecemlineatus*
Mammal	Thompson's gazelle	*Gazella thompsonii*
Mammal	Tiger	*Panthera tigris*
Mammal	Tree squirrel	*Sciurus* spp.
Mammal	Tule elk	*Cervus elaphus nannodes*
Mammal	Virginia opossum	*Didelphis virginianus*
Mammal	Wallaroo	*Macropus robustus*
Mammal	Wapiti	*Cervus. elaphus canadensis*
Mammal	Water buffalo	*Bubalis bubalis*
Mammal	Water vole	*Arvicola scherman*
Mammal	Wedell seal	*Leptonychotes weddellii*
Mammal	White rhino	*Ceratotherium simum*
Mammal	White-tailed deer	*Odocoileus virginianus*
Mammal	Wild goat	*Capra aegagrus*
Mammal	Wildebeest	*Conochaetes* sp.
Mammal	Woodland caribou	*Rangifer tarandus caribou*
Mammal	Woodrat	*Neotoma* sp.
Mammal	Yellow-footed rock-wallaby	*Petrogale xanthopus*

Index

G

N

Printing: Krips bv, Meppel, The Netherlands
Binding: Stürtz, Würzburg, Germany